Making Public Enterprises Work

Making Public Enterprises Work
From Despair to Promise: A Turn Around Account

William T. Muhairwe

Published by IWA Publishing
Alliance House
12 Caxton Street
London SW1H 0QS, UK
Telephone: +44 (0)20 7654 5500
Fax: +44 (0)20 7654 5555
Email: publications@iwap.co.uk
Web: www.iwapublishing.com

First published 2009
© 2009 IWA Publishing

Printed by Lightning Source.
Typeset in India by Alden Prepress Services Private Limited.

Apart from any fair dealing for the purposes of research or private study, or criticism or review, as permitted under the UK Copyright, Designs and Patents Act (1998), no part of this publication may be reproduced, stored or transmitted in any form or by any means, without the prior permission in writing of the publisher, or, in the case of photographic reproduction, in accordance with the terms of licences issued by the Copyright Licensing Agency in the UK, or in accordance with the terms of licenses issued by the appropriate reproduction rights organization outside the UK. Enquiries concerning reproduction outside the terms stated here should be sent to IWA Publishing at the address printed above.

The publisher makes no representation, express or implied, with regard to the accuracy of the information contained in this book and cannot accept any legal responsibility or liability for errors or omissions that may be made.

Disclaimer
The information provided and the opinions given in this publication are not necessarily those of IWA and should not be acted upon without independent consideration and professional advice. IWA and the Author will not accept responsibility for any loss or damage suffered by any person acting or refraining from acting upon any material contained in this publication.

British Library Cataloguing in Publication Data
A CIP catalogue record for this book is available from the British Library

Library of Congress Cataloging-in-Publication Data
A catalog record for this book is available from the Library of Congress

ISBN 10: 1843393247
ISBN 13: 9781843393245
ISBN: 9789970027422 (Fountain Publishers)

Co-Published by
Fountain Publishers
Kampala - Uganda
www.fountainpublishers.co.ug

Dedication

*To colleagues trying to make public enterprises fruitful,
my late 'brother' Laro Wod Ofwono – the chairman of Uganda Public Employees Union
who always worked for harmony between the management and workers,
and my dear parents – Cecelia, Theresa and Nazarius for good up-bringing*

Contents

About the Author ix
Preface xi
Introduction xv

Part One	**TOUGH JOB: COULD IT BE DONE?** 1
1	Serving Against all Odds 3
Part Two	**DOWN TO WORK: THE TURNAROUND** 17
2	Setting the Priorities 19
	The 100 Days Programme 19
3	Focusing on the Customer and Financial Viability 47
	Service and Revenue Enhancement Programme 47
4	A Pat on the Back 71
	First Government Performance Contract 71
5	Swinging the Pendulum from the Centre to the Grassroots 81
	Area Performance Contracts 81
6	The Group Incentive Mechanism 107
	Stretch-out Programme 107
7	The Individual Incentive Mechanism 137
	One-Minute Management Programme 137
8	Reshaping the Partnership with the Government 167
	Second Government Performance Contract 167
9	An Alternative Approach to Privatisation 175
	Internally Delegated Area Management Contracts 175
10	A New Approach to Performance Monitoring 201
	The Checkers System 201
Part Three	**SO FAR SO GOOD** 221
11	Performance Balance Sheet 223
Part Four	**NOT ALONE: PARTNERS' ROLE** 243
12	Winning the Much Needed Goodwill 245
	The Donors 245
13	In the Shadow of Privatisation 275
	The Government 275
14	Public Partnership Reconstructed 303
	The People's Voice 303
15	Behind the Scenes 331
	The Invisible Partners 331
16	Friends in Need 361
	Suppliers and Service Providers 361

© 2009 William T. Muhairwe. *Making Public Enterprises Work*, by William T. Muhairwe.
ISBN: 9781843393245. Published by IWA Publishing, London, UK.

Part Five	**YES WE DID**	379
17	**Can Public Enterprises Perform?**	381
	Final Thoughts	381

Acknowledgements ... 397
Acronyms ... 401
Bibliography ... 405

About the Author

Dr William T. Muhairwe is a Management Specialist trained in Economic and Business Management. Since 1998 to date, he has been the Managing Director (MD) of National Water and Sewerage Corporation (NWSC). While at NWSC, he has instituted high performance turnaround initiatives that have transformed the organisation into a profit-making government parastatal. Dr Muhairwe has spearheaded the formation of an External Services arm in NWSC that presents an excellent model on how a utility can help its peers to benchmark best practice solutions, at affordable costs. Apart from Uganda, some of the beneficiary countries include Kenya, Tanzania, Rwanda, Zambia, Nigeria, India, Mozambique and Ethiopia. Formerly, Dr Muhairwe worked for the Uganda Investment Authority as its Deputy and eventually Ag Executive Director for three years, and was responsible for attracting inward private investments in Uganda. Prior to that he was the MD of the East African Steel Corporation, a Joint Venture Company between the government of Uganda and a private company (The Madhvani Group). Dr Muhairwe has key competences in strategic planning, performance turnaround policy formulation, private sector development and stakeholder management.

© 2009 William T. Muhairwe. *Making Public Enterprises Work*, by William T. Muhairwe. ISBN: 9781843393245. Published by IWA Publishing, London, UK.

Winner of East Africa's Most Respected Public Enterprise, 2007

♦ Nation Media Group   PRICEWATERHOUSECOOPERS

"This is the most sincere and pragmatic approach to dealing with a public enterprise that is about to drop the ball. It is worth every minute of your read."

Dr. Gitahi Githinji, Nation Media Group

Preface

This book is about the people and processes that transformed a moribund state-owned enterprise into one of Africa's most successful publicly owned organisations. The author does not purport to provide a recipe for success, but he describes one organisation's response to the roadblocks that have slowed reform in other developing countries. He surveys the strategies adopted by a state-owned water and sewerage company in Africa – hardly the kind of business expected to serve as a model for decision makers in low-income countries. Nevertheless, the volume provides important insights into how organisational reform can promote high performance.

In the case of National Water and Sewerage Corporation (NWSC), Uganda, there was no 'grand design', just a commitment to improving performance – so that those in political power, current customers, citizens without service and those controlling donor funds would give managers the latitude to change the organisation's culture. Without the support (and trust) of these key stakeholders, the transformation process could not have been successful. The NWSC 'story' provides inspiration to those attempting similar tasks in other countries and contexts. The steps required to strengthen public organisations that have become fragmented, unfocused and unproductive are described in the context of a 'sick' organisation. The steps (with appropriate modifications) can be effective in other situations. Four basic ingredients for organisational transformation are identified here: thoughtful leadership, careful measurement, open communication channels and well-designed implementation strategies.

The book has several audiences: managers, informed citizens, academics and policy makers. As a comprehensive case study, it provides the context and detail that allows managers and top executives to see what it takes to succeed in a world that often creates barriers to good performance. Opinion-leaders will better understand the difficulties of turning around poorly performing organisations, and will find inspiration and hope in the way people can rise to meet the challenges they face. Academics will find that the book provides rich examples of how organisational strategy and structure are mutually reinforcing. The details illustrate the ways organisations change over time – given the right leadership and institutional support. Finally, policy makers and infrastructure specialists will see how crucial implementation is to the ultimate success of reforms. Information, incentives and institutional governance determine whether an organisation delivers on its promises. Initially, some readers may skim the institutional specifics, but I suspect they will return to the details to understand the types of strategies that were effective in this particular setting.

© 2009 William T. Muhairwe. *Making Public Enterprises Work*, by William T. Muhairwe. ISBN: 9781843393245. Published by IWA Publishing, London, UK.

I witnessed first-hand NWSC's emphasis on capacity building: how an organisation trains the next generation of leaders. Several years ago, I attended a NWSC dinner in Kampala, held to celebrate Makerere University's conferring of a PhD to one of its top managers. After Dr Muhairwe had said some words expressing the organisation's justifiable pride in the accomplishment, he asked five other managers to stand up. He proceeded to describe the accomplishments of each individual in turn, and indicated his confidence that each would be able to take further steps in their professional development – completing an MBA or starting a PhD programme. The language was positive and supportive, and by publicly setting goals for members of his management team, he was also underscoring how he valued the acquisition of knowledge that would enhance organisational performance. Then he named five more individuals, and went through the same exercise – emphasising how the additional skills would make them and the organisation more effective. I was stunned and impressed by this event. In my 35 years in academia, I had never seen a dean or department chair publicly show such familiarity with faculty accomplishments, nor had I seen explicit goals set for me and my colleagues. I suspect that few business leaders would be able to go through such an exercise. For some organisations, perhaps it is unnecessary to underscore the expectation of excellence. However, I believe that we would do well to emulate leaders like Dr Muhairwe who both celebrate and challenge individuals to continue on paths that promote personal growth. His words of encouragement set targets for individuals and underscored the organisation's commitment to continuous learning – a key element in the new corporate culture.

So with that event still vivid in my own memory, I can testify to the way NWSC supports professional staff and people in the field. In subsequent days during that visit to Uganda, I had the opportunity to travel to some NWSC sites. Dr Muhairwe wanted me to meet some workers, so he stopped by a billing and collections office. Receptionists were certainly aware that their CEO was in the building, but one could still see authenticity in the bustle of people at work, dealing with customers. He took me to an office where leaks and other problems were tracked. He asked the young woman and young man at the computer station to describe their procedures to me. Then he turned away, as though a final exam was just starting, and I was to be the external evaluator. He listened to their overview of the process; then he turned around to face them again, asking them to explain the detailed map that showed problems, such as reported leaks. I was very impressed with the way these two very articulate staff members understood how their jobs fit into the overall organisational objectives. When they had completed their summary of the activities triggered by various types of reports, Dr Muhairwe thanked the two of them and smiled at me – revealing the pleasure he took in observing his staff pass the test and the pride he took in their accomplishments.

Another day, during a field trip to a water distribution system operations site, I met a Jinja Area manager in her office. The room was spartan – desk, computer, bookshelves and files. She proudly pointed to several trophies her district had won in team competitions with other systems. In earlier conversations with NWSC professionals, I had been skeptical of using such incentives to improve system performance, but I am now convinced that non-monetary rewards have their place as inducements for exceeding benchmark targets. She took a group of us to the water plant, and professionals showed us the steps necessary for delivering clean water to homes. The facilities were immaculate and the individuals were clearly proud of 'their' system.

If we went back to 1998, such encounters could not have happened. In 1998, NWSC was a failing organisation – threatened with financial non-sustainability, facing severe operational problems, and lacking credibility with citizens. This book documents the initiatives that transformed the organisation. In addition, the author describes his own journey – the colleagues and books that influenced dialogues within NWSC and affected the business strategies adopted at various stages of the reform process.

Anyone who hears advice from consultants and reflects on personal experience knows that the way to change an organisation is to engage its leaders in a set of tasks. These tasks involve at least eight elements:

1. *Identify trends* so that past performance is understood;
2. *Establish baselines* documenting current performance;
3. *Select measurable goals* as challenging (but achievable) targets;
4. *Design internal incentives* to reward teams for meeting those objectives;
5. *Establish lines of communication* to promote information-sharing internally;
6. *Develop and implement strategies* for dealing with external threats;
7. *Ensure accountability* by assigning responsibilities to leaders and teams and
8. *Review results* within a reasonable timeframe to evaluate process implementation – which takes us back to identifying trends again.

These elements appear again and again in the NWSC story. This process was not top-down – it was inclusive and bottom-up. At NWSC, the focus was on key performance indicators. The issue was to establish acceptable ranges for key indicators and identify reasonable targets. Problems were diagnosed and remedies suggested by teams of specialists. Those responsible for developing and implementing strategies had to determine the nature of the organisation's 'health' problem. Identifying and implementing strategies promoting the patient's recovery and (ultimately) revitalisation were the fundamental issue facing NWSC managers and staff at the start of the reform process in 1998.

Recall the physician's injunction to 'do no harm.' Managers, like physicians, can make mistakes, causing injury. They can misdiagnose a problem; they can let their egos get in the way of assigning appropriate treatments; they can abuse the doctor-patient relationship or they can become overworked – leading to inattention to crucial details. Fortunately, the NWSC case is one where the initial diagnoses and subsequent treatments have proven to be highly effective. The success story warrants wide attention – not because an individual rescued an enterprise, but because a critical mass of professionals came together in common purpose: to turn around an unhealthy organisation.

Sanford V. Berg
Distinguished Service Professor
Director of Water Studies, Public Utility Research Center
University of Florida

Introduction

In writing this book, I intended to examine the operations of National Water and Sewerage Corporation (NWSC) from its establishment at the beginning of the 1970s, throughout the country's turbulent history, to its eventual turnaround into the respected service provider it is today.

I have attempted to tell the story of the women and men who worked with me to ensure that the Corporation is restored to health, enabling it fulfil its mandate. I explain how we managed to achieve what appeared, and still seems, impossible to managers in many parastatals across the developing world. We defied the widely held view that public enterprises in the developing world cannot work, the view that such organisations should be privatised to make them work. The book explains how eminent economists and planners of the global economy eventually accepted the prescriptions we proposed and administered various doses to treat and heal the Corporation.

The book is arranged in five parts. Part One is entitled Tough Job: Could it be Done? In this part, I explain how the idea of a national water and sewerage body, with a social mission objective to provide clean and safe water to the population, was conceived and set up.

I examine the impact that Idi Amin's 'economic war' had on the Corporation after the expulsion of an estimated 80,000 Ugandan Asians, who were a big part of the Corporation's customer base. I look at how the Amin regime impacted negatively on the country's ethos – the culture of the high and mighty consuming services, such as water, without paying for it. I explain how this culture persisted long after Amin had left the scene and how this culture led to the near collapse of NWSC and other parastatal organisations.

I further explain how various management teams tried their best to keep the company afloat, under very difficult conditions, and how I eventually assumed the reins of power, though not without drama. Hardly had I been in office for two days than I was locked out by court bailiffs, acting on behalf of the company's numerous creditors.

In Part Two of the book, Down to Work: The Turnaround, I discuss how, together with my managers, we mooted many ideas to turn the Corporation's fortunes around. I explain that no formula is sacrosanct, that people or managers should not be stuck in the conventional ways of doing things; that ideas ought to be sought from an array of sources, places and people; that besides textbooks on management and knowledge acquired from training institutions, a manager should remain up to date through reading and networking locally, regionally and internationally.

© 2009 William T. Muhairwe. *Making Public Enterprises Work*, by William T. Muhairwe. ISBN: 9781843393245. Published by IWA Publishing, London, UK.

I demonstrate that after collecting the ideas, one should brainstorm with colleagues and come up with a blend of the best interventions that will propel an organisation to change its work ethic and lead to profitability. I highlight the importance of team building and teamwork. Furthermore, I underscore the importance of involving members of staff, from top to bottom, in the discussion of the challenges facing an organisation and how this approach helps to understand the problems clearly and to devise effective corrective measures. In this section I explain how, together with my management team, we managed to rally all NWSC 'troops' to achieve what appeared to be impossible.

In Part Three, So far so Good, I provide a logical break from the long list of actions and strategies implemented to improve the performance of the Corporation. My intention here is to provide a snapshot overview of the net effects of the various sub-programmes implemented and how they contribute to the overall improved performance.

In Part Four, Not Alone: Partners' role, I recognise the importance of various development partners, civil servants, politicians and the business community, who helped catapult us to the top. One of the major lessons we learnt is that, however intelligent a manager might be, he or she can never succeed without cooperating with all the stakeholders. In cases of differences of opinion, one should use the power of persuasion over and over again, till the other parties appreciate one's opinion as manager and realise that you are headed in the right direction.

Lastly, Part Five, Yes We Did, answers the question I posed to introduce Part One. I note that, although privatisation may frighten organisations into changing their work ethic, it is certainly not the only way of promoting efficiency. In this part, I also demonstrate that ideas which lead to development are not set in stone.

Part One
TOUGH JOB: COULD IT BE DONE?

Chapter One
Serving Against all Odds

Uganda's National Water and Sewerage Corporation (NWSC) was conceived by the first Apollo Milton Obote government (1962–71) in collaboration with the African Development Bank, the World Bank and the World Health Organisation. After the overthrow of Obote's government in January 1971, Idi Amin's military regime (1971–79) established the Corporation under Decree No. 34 of February 1972. The NWSC became operational in February 1973 when the Minister of Minerals and Water Resources signed the Statutory Instrument No. 14, which paved the way for the first national water utility in Uganda. Originally a merger of the water boards of Kampala, Jinja and Entebbe, the Corporation was envisaged to extend its services gradually to other urban areas in the country. As of 2008, NWSC was serving 22 major urban areas across the country.

To run the new Corporation, the government appointed a Board of Directors comprising permanent secretaries (or their representatives) from the Ministries of Local Administration, Mineral and Water Resources, Health, Finance and Works and Communications, a representative from Kampala City Council and two representatives each from the industry and the general public, as well as the chief executive. The composition of the Board meant that, from its inception, the NWSC was to operate as a government department rather than a commercial public utility out to make money. The government appointed Mr Christopher Kasozi-Kaya, who had been the Head of the Engineering Unit at the Kampala District Water Board, as the first Managing Director (MD), and Mr Francisco Pinychwa Openyto as the Deputy MD and Chief Engineer.

The NWSC was established for several reasons. To start with, the government believed that a unified water and sewerage services utility for Kampala, Jinja and Entebbe Areas, which would gradually serve all urban areas in the country, would benefit from economies of scale and deliver more efficient services to the public. The Kampala, Jinja and Entebbe water boards were considered to be too small in terms of operations and customer base to benefit from economies of scale. For example, in Kampala, the population served was less than 12,900 people out of the 500,000 residents at the time, while in Jinja, only 6,500 people were served out of a total population of 50,000. Secondly, the fragmented water boards did not have the capacity to attract, train and retain sufficiently skilled manpower to maintain, sustain and expand the delivery of water and sewerage services. For example, according to the 1970 Annual Report of the Kampala

© 2009 William T. Muhairwe. *Making Public Enterprises Work*, by William T. Muhairwe. ISBN: 9781843393245. Published by IWA Publishing, London, UK.

District Water Board, the authority had only three engineers, including the manager.

Thirdly, the waterworks infrastructure and distribution networks, some of which had been installed in the 1920s and 1930s, were outdated and could not cope with the growing demand for water, especially taking into account the rapid urbanisation process. Moreover, the water boards did not have the capacity to raise loans to expand the water and sewerage services to urban dwellers in Kampala, Entebbe and Jinja who did not have safe piped water and sanitation facilities.

Indeed, the 1960s and 1970s were prime times for corporationalisation. Developing countries, including Uganda, believed that only governments, through public enterprises, were capable of mobilising local and foreign resources for large-scale and capital-intensive national projects. During those years, the private ownership of utilities such as water, electricity and telecommunication networks was inconceivable and out of the question for most policy makers and planners. In any case, in the newly independent Africa, foreign private investment, even where it was available, was not welcome, because the consensus in nationalist circles was that political independence without economic emancipation, which meant Africanisation and nationalisation of the economy, was meaningless. It was in this same spirit that NWSC was established.

In principle, the creation of the NWSC was a good and farsighted idea. Water service delivery is unique, challenging and expensive, requiring concerted efforts in planning, management, resource mobilisation and investment. Experiences elsewhere in Africa have revealed the daunting challenges of managing fragmented water and sewerage systems with acute shortages of capital, skilled human resources and technology. Indeed, there is an ongoing debate in Africa regarding the case for and against centralised water utilities regardless of the nature of ownership. Many water experts argue that a national water utility such as NWSC is better placed to deliver services than the municipal or even regional water boards. Others believe that 'small is beautiful' because it is flexible and responsive to local needs and aspirations, and that is why countries such as Kenya and Tanzania have established regional and municipal water boards, respectively. But for all the good intentions of establishing a national water utility, the NWSC experience shows that the enlargement of scale *per se* does not necessarily guarantee efficient and effective service delivery. Usually, the road to disaster is more often than not paved with good intentions, as the turbulent history of NWSC from 1972 to 1998 reveals.

This chapter explores the historical perspective of the management of NWSC during the politically turbulent times of Uganda during the 1970s and early 1980s. The chapter underscores the importance of perseverance, careful dealing and stakeholder mapping amidst immense political log rolling. Some reflections resulting from these experiences are presented at the end of the chapter.

In the eye of the storm

The early years of the Corporation (1970s and early 1980s) were marked by political instability and economic decline in the country. The Corporation was born in the eye of the storm, in that its establishment coincided with the launch of Amin's disastrous 'economic war' involving the expulsion of Asians from Uganda and the confiscation of their property. With over 80,000 urbanised Asians evicted, the infant Corporation immediately lost a big chunk of water consumers in Kampala, Entebbe and Jinja, which significantly affected its revenue base. To make matters worse, most of the beneficiaries of Asian properties assumed that whatever they took over, water and sewerage services was free of charge; therefore only a few of them registered with the Corporation as *bona fide* consumers and even then not all bothered to pay their water bills. Some of them were politically too 'powerful' to pay anyway! This inevitably culminated in financially crippling bad debts for the NWSC.

The general economic and political atmosphere during the 1970s was not conducive to the growth and consolidation of the NWSC, letalone any possibility that it would achieve its corporate objectives as set out in the 1972 decree. The economy and, in particular, the Uganda shilling were in free fall. The formal economic sector gave way to *magendo* (black market economy). Uganda lost its creditworthiness around the world and all imports, including those of NWSC, had to be paid for in advance and in cash. The revenue base of NWSC shrank and its collection declined sharply. As a result, the Corporation did not have enough money to pay its workers, buy inputs to produce water and maintain its infrastructure as well as sustain the delivery of water and sewerage services in accordance with its statutory mandate and consumers' expectations. Foreign aid, which had been anticipated when the Corporation was in the making, was not forthcoming owing to the breakdown of relations between the Amin regime and the donors. For its part, the government did not inject any seed capital for the development of the Corporation. The expected benefits of a national water utility became an illusion, which goes to show that public enterprises or utilities do not operate in isolation. In other words, their success (or failure) depends very much on the social, political and economic environments. In Uganda's case, the 1970s and early 1980s were characterised by instability and economic hardships.

The Amin regime created an atmosphere of fear, anxiety and uncertainty among NWSC employees, which meant that the Corporation could not build a coherent, productive and sustainable corporate culture. Morale among the workers sank so low that the 'I-don't-care attitude' became the order of the day. Professionalism gave way to cynicism and patronage. To cope with mounting economic hardships, employees resorted to all sorts of coping mechanisms, including moonlighting, petty trade and 'chasing lines' (looking for alternative ways of making ends meet) in order to survive. The virtues of punctuality, integrity, hard work and commitment, which are essential to the cultivation of a productive corporate culture, were abandoned. Coming late and absenteeism, dishonesty and corruption, abuse of office, malingering, intrigue, rumour mongering and double-dealing set in. This deprived NWSC of an enabling environment and starved it of the capital required to build it into a viable public water utility.

Interference from the high and mighty

As if economic hardships and fear, and uncertainty among workers were not bad enough, NWSC became a playground for the high and mighty in government circles. At its establishment, NWSC was ideally supposed to operate as an autonomous public utility with commercial as well as social objectives. The Corporation was supposed to focus on the gradual development and expansion of water and sewerage services to other urban centres across the country by, among other ways, ploughing back its profits. However, the Corporation was not allowed to operate as an autonomous entity. As already mentioned, the Board of Directors was composed of senior government functionaries – permanent secretaries or their representatives – whose capacity to operate independently was out of the question. Even the industry and general public representatives were not independent because they too were political appointees. In fact, corporate independence in Amin's Uganda was just an illusion. Thus, the Corporation made a false start in its institutional development; this crippled its capacity to deliver services effectively in accordance with its statutory obligations.

From inception, NWSC managers did not enjoy the freedom to run the Corporation without constantly having to look over their shoulders. For example, Mr Kasozi-Kaya, the first MD, was dismissed within two years of his appointment. Similarly, although Mr Openyto, who took over as MD, outlived the Amin regime, he had to endure constant interference from senior army and security officials. Whenever there was no water, they would storm the offices or even residences of the Corporation's managers, including the MD, and order them to fix whatever had caused the interruption of water supply. Quite often, the managers and staff were accused of sabotage, for which they could summarily lose their jobs or even lives. For example, in 1976, the Public Safety Unit arrested the entire transport and stores staff of the Corporation, alleging that one of the drivers had been involved in robbery. This kind of treatment demonstrates how dangerous it was to live and work in Amin's Uganda. Mr Openyto recalls that he was arrested 15 times between 1975 and 1979; an average of three times a year. How he managed to outlive the Amin regime is still a mystery.

Nothing illustrates the extent and intensity of government interference in the management and operations of NWSC better than the example of a senior army officer, who was also the Governor of Eastern Province in Jinja in the 1970s. When he was told that water consumers in Jinja were not paying their water bills on time, he arrogated himself the power to collect revenue on behalf of the Corporation. He ordered the consumers to pay all their arrears immediately without hesitation or face dire consequences. One morning, the army officer, accompanied by soldiers, stormed the homes of Jinja residents to collect revenue on behalf of NWSC, whipping consumers. His unsolicited interference was, of course, not only contrary to the law, but also harmful to the relationship between the Corporation and its customers. Consumers began to see the Corporation as part and parcel of Amin's regime. Government interference was not limited to the Jinja Area; it was widespread in Entebbe and Kampala too.

In the 1970s, the Corporation became a hub of patronage, nepotism and favouritism. Appointments were hardly based on merit, experience and qualification. Promotion was not based on performance. Staff with 'powerful' connections within and outside the Corporation could ignore their bosses' directives, including those of the MD, with impunity. Discipline and the chain of command broke down to the detriment of professional management. By 1979, water and sewerage services in Kampala, Entebbe and Jinja were in a worse state than before NWSC was established. Given the state of service delivery then, the consumers must have regretted the dissolution of the original water boards.

From bad to worse

The fall of the Amin regime in April 1979 created short-lived euphoria among Ugandans, including NWSC managers and staff. There was understandable cause for optimism that the situation would improve. Indeed, many donors, such as the Islamic Development Bank and the World Bank, were prepared to assist the NWSC to rehabilitate its infrastructure. Unfortunately, this promising start was soon shattered by a new wave of political violence and uncertainty. The Yusuf Lule government, which took over from Amin, lasted a mere 68 days. Godfrey Binaisa government, which replaced Yusuf Lule government in June 1979, limped on for 10 months. He was removed from office by the Military Commission under Paul Muwanga in May 1980. The December 1980 elections, which brought Obote back to power, did not resolve the political crisis; it simply paved the way for the 1981–85 civil war.

Just like other public institutions, NWSC was badly affected by the wave of violence and lawlessness that ensued during the Obote II regime. The Corporation's staff continued to live in a state of fear and to endure growing economic hardships. Partisanship augmented patronage in the recruitment and promotion of the NWSC staff. Uganda Peoples' Congress (UPC) party cadres were posted to all public corporations, including NWSC, to look out for 'subversive' elements. In this environment, it was difficult, if not impossible, for the Corporation's services to improve. If anything, matters could only get worse. In many areas, some consumers went without water for months and even years. Even in prime residential areas, the water taps were often dry. Customers had to store water in jerry cans, drums and even bathtubs. Other people resorted to fetching water from wells and rainwater harvesting. Those who could afford it, especially government officials and diplomats, bought standby tankers to deliver water to their homes. Even Mr Yoweri Kaguta Museveni, the current president of Uganda, during the commissioning of Gaba III water treatment plant, recalled how he used to fetch water for his family in 1979.

Ministers, senior army officers and other government functionaries continued to demand prompt and continuous services at whatever cost, as if the Corporation could do so by magic. The Corporation's managers were put under pressure and ordered to do the impossible. A good example is the case of a senior minister who once directed the Corporation to deliver water to his residence in Entebbe (the Entebbe waterworks had shut down due to lack of spare parts) without fail. The big man could not stand the 'nonsense' that the standby water tanker used for such purposes was out of order. While talking to the MD on the phone, he asked: 'Do you know who you are talking to?' Without waiting for an answer, he quickly quipped: 'I fought the war and won it. I will not tolerate incompetent officials sabotaging government efforts to rebuild the country. Do you hear? If you can't do it right, then you'd better pack your bags and face the border.'

Later in the evening, Mr Openyto was summoned to the Uganda Club, then an exclusive senior army and government officers' club, to explain the shortage of water in Entebbe. When Mr Openyto explained that the pumps at Entebbe waterworks had broken down and that the Corporation did not have the money to buy spare parts, the political boss just said: 'Simple. Go to Bank of Uganda tomorrow and pick the money.' Mr Openyto could not believe his ears. Did the big man really mean what he said? 'But Sir, how do I just walk into the Bank of Uganda and ask for money like that?' Mr Openyto asked with genuine incredulity. By then, irritated and impatient, the senior cabinet minister shouted at the NWSC boss: 'I said, go to Bank of Uganda, do you hear me? Just go to Bank of Uganda and get the money, okay? Nobody doubts my word. Just say I sent you.' True to the big man's word, when the NWSC chief executive went to the central bank, he was not only given the money but also a return air ticket to Nairobi to buy the required spare parts.

The above example is quite revealing about the attitudes of the high and mighty towards public corporations. Constant political pressure and interference compromise the principles of accountability and transparency. The fact that the Bank of Uganda could be ordered to release funds to an individual without proper documentation, disregarding the standing public finance instructions, would appear to suggest that public funds were not immune to misappropriation. Indeed, handling public finances in this way may well explain the persistent widespread rumours in the developing world about the diversion of public funds to private accounts at home and abroad. In addition, the insensitive handling of corporate executives like Mr Openyto has tended to erode the confidence of the executives to make long-term decisions for the good of enterprises under their care without fear or favour. It is a pity that many public companies in developing countries are still subject to such high-handed and arbitrary political inference. As a matter of fact, managers operating under such conditions are left with very few options; *one either quits, dances to the tune* or *gets fired*. Indeed, in 1982 Mr Openyto got to know over the radio that he had lost his job without due recognition for his long service to the NWSC. He was replaced by one of his juniors, then a young engineer, Hilary Onek.

Serving against all odds

In spite of all the trials and tribulations of the 1970s and early 1980s, the fact that the Corporation survived and continued to render some services under extreme hardships was a remarkable achievement. Some customers continued to get water; however, the supply was inadequate and erratic. To their credit, some Corporation's staff improvised ingenious coping mechanisms to keep it afloat and to prevent its complete shutdown. For example, faced with the lack of imported spare parts, the Corporation's staff made gears from Elgon Olive timber to keep the waterworks running! The Ministry of Finance intervened from time to time to reduce the indebtedness of government departments by paying directly from source, and this somehow saved the Corporation from bankruptcy. To the pleasant surprise of the Corporation, the Ministry of Defence paid its water bills promptly, and these payments turned out to be a lifesaver on many occasions.

Between 1981 and 1986, there were some commendable efforts to rehabilitate the water and sewerage infrastructure. The appointment of Mr Onek as the chief executive in 1982 gave the Corporation a new sense of dynamism, purpose and direction. He was a young and energetic professional engineer who did his best to revamp the Corporation under extremely difficult circumstances. Indeed, he still holds a record for being the longest serving NWSC chief executive (1982–1998) and one of the few, if not the only one, to have served from one regime to another until his retirement in 1998. His active participation in the implementation of the World Bank-funded First Water Supply Project, which was approved in 1981, laid the foundation for 1986 post-conflict rehabilitation and provided leadership continuity for the subsequent NWSC–the World Bank collaboration during the late 1980s and the 1990s. Resourceful and energetic, he steered the Corporation through the turbulence of the early 1980s and ensured its survival during the trying era of structural adjustment programmes of the 1990s. Indeed, in those dark days of instability and war, Mr Onek did a commendable job.

During the early 1980s, NWSC received funding from the World Bank, the Islamic Development Bank, and the African Development Bank. In 1981, the World Bank approved the Water Supply Engineering Project to rehabilitate the water infrastructure in Kampala, Entebbe, Jinja, Tororo, Mbale, Masaka and Mbarara. Later, in 1984, the World Bank funded the Water Supply and Sanitation Rehabilitation Project, better known as the Seven Towns Water Supply and Sanitation Rehabilitation Project, as a follow-up to the work that had begun in 1981. Although the implementation of this project was delayed by the intensification of the National Resistance Army (NRA) guerrilla war and the temporary withdrawal of the World Bank officials from the country, it was eventually completed in 1987.

Efforts to rehabilitate the Corporation's infrastructure continued after the National Resistance Movement (NRM) had come to power in January 1986. In the ensuing years, the World Bank and other donors, including Austria, Italy, and Germany and the European Community, provided financial and technical assistance towards recovery of the Corporation. Mr Onek was the brain and the driving force behind a number of bold initiatives introduced in the late 1980s and during the 1990s, which were aimed at turning the Corporation around and transforming it into a viable public utility. In 1989, the Corporation launched a block-mapping exercise intended to collect accurate information on its customers. In order to generate more revenue and expand its services, the Corporation decided to raise its tariff in 1994.

In 1995, the National Resistance Council (NRC), Uganda's parliament of the day, passed the NWSC Statute, granting the Corporation institutional autonomy to provide water and sewerage services in urban areas under its jurisdiction on 'sound and viable commercial basis'. With this law in place, there was a radical shift in the composition of the Board of Directors, from one dominated by government officials, as was the case under the 1972 decree, to a broadly representative Board consisting of professional people. This law also mandated the Corporation to develop water and sewerage systems in order to provide efficient and cost-effective services to its customers in a *commercially* and *financially sustainable manner*. The new legal framework paved the way for the introduction of more systematic and structured corporate planning. The First Corporate Plan (July 1998–June 2001) spelt out the vision, mission, strategic goals, annual targets and tactical activities, as well as the operational responsibilities and time frames for implementation. The vision of the Corporation at that time was NWSC – The Pride of the Uganda Water Sector. It aimed at giving the Corporation direction into the 21st century and beyond as a viable public utility.

During the 1990s, NWSC's donor-funded projects focused on physical infrastructure rehabilitation and development, technical assistance, capacity building and human resource development. The Uganda Second Water Supply Project (1990–99) was by far the most important of all the donor-funded projects. It was funded by the World Bank, Austria, the European Community and the government. The Second Water Supply Project was designed to

- improve water and sewerage services in Kampala, Entebbe, Jinja, Masaka, Mbale, Mbarara and Tororo;
- establish delivery mechanisms for subsequent large-scale on-site sanitation programmes;
- develop NWSC into a financially viable utility;
- strengthen the Corporation in all its statutory functions and
- prepare NWSC to take over water supply and sanitation operations in other towns.

During the implementation phase of the Second Water Supply Project, water supply and sewerage systems in the seven towns were overhauled and expanded. For example, Gaba II Water Works was constructed and commissioned in 1993. Similarly, about 170 km of distribution mains and 17 km of sewerage network were installed in Kampala. Apart from the extension of the water and sewerage systems, the project upgraded the NWSC management systems and operations. Accordingly, the project provided staff training and development, computerised billing and established a new financial accounting system. A training centre, a block-mapping section, a laboratory and a central workshop were also established in Kampala. The total cost of this project rose from USD 118 million to USD 128 million, culminating in a controversial debate as to who was responsible for the cost overruns.

In addition, there were supplementary projects, such as the Kampala Water Network Rationalisation Project, which was completed in 1997. Under this project, 40 km of water mains were added to the existing network, 60 km of water mains and associated service connections were replaced and over 10,000 customer meters were installed. In Masaka, a new waterworks system was constructed, including the laying of 36 km of distribution mains and the completion of waste stabilisation ponds. In the same year, a similar work was done in Mbarara: 35 km of distribution pipes were added and new waterworks were constructed at Ruharo. Complimentary programmes in other areas were initiated to strengthen the computer-based commercial accounting and billing system, to improve revenue collection and to boost planning and capital development. Efforts to enhance customer care were intensified by gathering and compiling more accurate information on consumers' physical locations. Furthermore, NWSC stepped up its human resource and transport facilities by recruiting more engineers and buying more vehicles.

The most important management initiative during the 1990s was the Corporation's venture into private sector participation (PSP) for billing and revenue collection in the Kampala Area through the Kampala Revenue Improvement Programme (KRIP). Although revenue collection in Kampala increased from UGX 300 million in 1993 to UGX 1.2 billion per month in 1997, the Corporation's management was keen on enhancing the performance and productivity, and on reducing non-revenue water (NRW) to below 65%. In addition, there was a dire need for curbing illegal connections, improving revenue-collection efficiency and expanding the customer base. The quest for increased financial viability through a combination of revenue-enhancement and cost-reduction measures called for more effective billing methods and corporate loyalty by sensitising its customers.

Accordingly, in 1997 a German company, H.P. Gauff Ingenieure GmbH & Co (JBG), was commissioned to develop and put in place financial systems and to commercialise the operations in the Kampala Area. In November 1997, this company was contracted for three years to handle billing and revenue collection and to recover arrears in the Kampala Area at a monthly management fee of approximately 20% of the collected revenue. With private sector participation, the Corporation was expected to generate more revenue, reduce costs and improve service delivery significantly. By shedding billing, revenue collection and water distribution functions to a private contractor, the Corporation, henceforth, hoped to concentrate on its core business of water production and provision of sewerage services. The intention was, if private sector participation worked according to plan in Kampala, to replicate the management approach in other Areas of operation as a transition to full-fledged privatisation.

The Corporation's reform initiatives laid the foundation for the future recovery of the NWSC. Mr Onek's single-minded concentration on the rehabilitation and development of the physical infrastructure (the hardware issues) was a major achievement that created excess water production capacity of 42%. By 1998, the NWSC was in a position to produce more water than could be consumed. The challenge was no longer water production capacity but rather increasing access, quality of service and customer care. This called for a shift of emphasis from infrastructure (hardware issues) to commercial (software issues) aspects of the water and sewerage services such as marketing and customer care. For all its noble intentions, the KRIP private sector participation experiment did not live up to the expectations partly because it was conceived, prepared and implemented in a hurry without creating adequate ownership among the Kampala Area employees.

Despite the concerted efforts of all the stakeholders to revitalise NWSC and transform it into a viable public utility between 1986 and 1998, the results generally fell short of expectations. True enough, there were notable improvements in water production and the development of infrastructure but the overall performance of NWSC remained below expectations. Operational and cost inefficiencies were still remarkably high. There was no marked improvement in NRW performance. Also, NWSC annual turnover had stagnated at UGX 22 billion. The customer base was growing at a very slow rate, and the Corporation was not being run as a commercial enterprise with a coherent marketing strategy to become financially sustainable. By 1998, a lot of money, amounting to almost UGX 30 billion, was held up in arrears, of which the government share was nearly UGX 12 billion, or 40% of total arrears.

While NWSC was not making much progress in revenue collection, its indebtedness to local and international creditors was mounting. Local creditors included Uganda Electricity Board and Uganda Revenue Authority. NWSC owed almost UGX 2.8 billion and just over UGX 2.6 billion to the two creditors, respectively. It also owed nearly UGX 4.9 billion to other local creditors and suppliers. On top of that, the Corporation owed international creditors over UGX 52 billion at an interest rate computed as the sum of the Treasury bill rate plus 3% per year. The amount of money that was required to service these international loans was about UGX 8 billion per year. This was more than 35% of the gross annual revenue of approximately UGX 22 billion. All in all, according to the World Bank report of 1998, NWSC was in 'serious financial trouble'. Clearly, it was imperative for the Corporation to sort out its debt obligations.

In general, NWSC's performance in 1998 was lacklustre. All the stakeholders, especially government and the donor community, were concerned that the Corporation was not making much progress. As the World Bank reported in 1998, despite the investment of well over USD 100 million over 10 years, 'these investments have not been matched with the necessary commercial and financial management capacity that can ensure the delivery of sustainable services in the medium to long-term'. For its part, the government was getting tired of throwing good money after bad money. The consumers were not getting value for all the money that had been injected into the Corporation. As a result, the government, like the World Bank and other donors, had come to the conclusion that the only remaining option was to overhaul the Corporation's management in preparation for privatisation.

Accordingly, in July 1998, the government appointed a new NWSC Board of Directors to spearhead NWSC reform. The new Board was given freedom to get on with the job without interference; it was mandated to appoint a new chief executive to steer the Corporation to recovery. In November 1998, the Board appointed me as the new chief executive. All stakeholders held their breath to see whether, so to speak, the new brooms would sweep the Corporation clean. Could the new Board and management do it? Would the new chief executive – an economist rather than a professional engineer as had been the tradition – turn the Corporation around when he had no previous experience in managing water business? Could the new chief executive provide the much needed vision and leadership to no-nonsense, down-to-earth, professional engineers? The rest of this book attempts to answer these and many other questions relating to what had to be done to turn the Corporation around.

Reflections

After examining the turbulent history of NWSC, the following reflections provide food for thought, especially for managers running public enterprises in the developing world:

- Under all circumstances, change is the only constant. Policies are no exception to this. Managers should anticipate debate and then react to change but not just jump onto the bandwagon.
- Political interference and patronage of any sort are a cancer to public enterprises. They harm the relationship between the enterprise and the customers, and undermine the ability of management to take long-term decisions.
- Enterprises need new blood at some point in time to steer and inject new ideas. The challenge is knowing the point at which the new ideas are required and how to infuse them in the organisations.
- Efficient service delivery does not depend only on massive investment but also requires the right human resource, operations management and strategy.
- Managers need to be bold and take decisions without fear or favour as long as they are well thought through with a clear course of action.
- Understanding the operating environment is vital for any manager of any business enterprise. In the case of public enterprises, managers need to, specifically, understand that some sections of policy makers may have detrimental ambitions, which may not necessarily be in the interest of the general public. Constructive engagement of such stakeholders may pave the way for sustained public management, and here the use of emotional intelligence is the key.

Tough job! The Board that was charged with the responsibility of turning the Corporation around, Seated from left, Dr William Muhairwe, Mr Samuel Okec, Ms Hawa Nsubuga, Dr Aryamanya Mugisha Henry. Standing from left, Mr David Kakuba, Dr Abdullai Shire, Mr Lawrence Bategeka, Eng. Fransisco Openyto and Mr Patrick Kahangire (November 1998).

At the drawing board: Dr William Muhairwe in his office, prioritising the interventions for turning around the performance of the Corporation (November 1998).

Part Two
DOWN TO WORK: THE TURNAROUND

Chapter Two

Setting the Priorities

The 100 Days Programme

Water consumers owe [the] National Water and Sewerage Corporation over UGX 30 billion worth of unpaid bills. The Corporation's failure to raise enough money is due to [the] high volume of non-revenue water and the non-profitability of providing water to eight upcountry towns. Only three out of eleven towns where the Corporation supplies water are breaking even...

The New Vision, 10 February 1999

In 1998, when I became the chief executive of NWSC, the state of the Corporation was comparable to that of a critically ill patient, demanding painful but unavoidable radical surgery, strong medication and intensive nursing care to bring it back to health. To continue with the medical analogy, although the surgical ward was properly stocked with life-saving machines and professional nurses and doctors, the patient's life, after almost a decade of costly and painstaking treatment, was still in danger. Accordingly, all the stakeholders, particularly the government and donors, were waiting earnestly to see what the new Board of Directors and management would do to address the Corporation's sickness. Were the 'new doctors and nurses' prepared to carry out radical surgery, however painful and risky, to heal NWSC and nurse it back to full health? Did they have the right ideas and the will to transform the Corporation into a viable utility?

Given the multifaceted challenges that the Corporation was facing on the one hand, and the stakeholders' expectations on the other hand, the new managers were placed in an unenviable position, which we can figuratively compare to being between a rock and hard place. The new Board and management had no choice but to rise to the challenge, take measured risks and set priorities intended to turn the Corporation around as a matter of utmost urgency. That is why we initiated two-pronged strategies and approaches, as well as new methods of work. Firstly, we introduced incremental radical reforms to address the organisational culture and human resources management. Secondly, we initiated a series of successive short-term and action-oriented programmes that were consistent with internationally accepted change management principles and practices.

© 2009 William T. Muhairwe. *Making Public Enterprises Work*, by William T. Muhairwe. ISBN: 9781843393245. Published by IWA Publishing, London, UK.

In this chapter, I present a number of approaches and actions implemented within the framework of *setting the priorities* first. To accomplish this task, I show how NWSC was able to make the first performance landmark through the 'first high-impact' intervention, popularly known as the 100 days Programme, in the wake of my promise to turn NWSC around in 100 days during my first address to headquarters staff at the Corporation's training centre in Kampala.

Shocked into a promise

Before and after being appointed as the chief executive, I had heard many unflattering and even shocking stories about the goings-on in the Corporation. But even then, I had not anticipated what I saw on the first day I reported for duty. As I entered the dark and ugly reception area, my spirits and expectations went crashing; I immediately sensed that all was not well. The shabbily dressed receptionists looked indifferent and insensitive. They did not seem to care who was coming in or going out of the building. None of them seemed to notice my presence, let alone to enquire about my business and to offer assistance. The reception desk did not look like a front office of a serious organisation that could give a positive impression about the image of the Corporation. The three or four dusty chairs in the reception area were in tatters. The reception area looked very much like a marketplace with buyers and sellers exchanging greetings and pleasantries, and bargaining loudly over all sorts of merchandise.

This shocking first impression prompted me to hurry upstairs to my office on the first floor. What I encountered in the corridor was even more shocking, it was full of people who appeared to have nothing useful to do. They were buying and selling eats, and taking tea long before the tea break! From this crowd, it was hard to distinguish staff from visitors or the Corporation's customers. As I passed through the crowd, most of the people I saw looked totally unperturbed by what was going on around them. Some of the people I saw in the corridor looked as if they had spent the previous night on a drinking spree and did not appear to be in a sane state to perform their duties. Indeed, the crowd I encountered in the corridor was so appalling that I instinctively yearned for the security and privacy of the four walls of my office, to which I hurried without looking back.

Unfortunately, the state of my office was not anymore reassuring! Though spacious, it was dusty and disorganised presumably because it had not been occupied for the past six months following the retirement of my predecessor. But even then one would ordinarily have expected the MD's office to be clean, especially when a new chief executive was reporting for duty for the first time. The state of the office was a negative mark against those in charge of office management and cleanliness in the Corporation. The furniture was old and shabby and some pieces were broken. The walls were clearly in need of a coat of paint. The windowpanes too were broken, letting in noise and dust from Jinja road. Eventually, I discovered that compared to the state of other offices in the Corporation, both at the headquarters and in the field, that of mine was very good indeed! I also discovered that though the refurbishment of the offices had been budgeted for, the fact that none of the offices had received a facelift was as puzzling as the sorry state of the offices themselves. These shocking impressions were a rude awakening to me. I realised immediately that, if first impressions were anything to go by, the situation in NWSC was much worse than I had imagined originally. Accordingly, these impressions shocked me into promising on my first day on duty to turn the Corporation around in 100 days even before obtaining the facts and realities on the ground and gaining a deeper understanding of the malaise – both at headquarters and upcountry.

Oh! Why did I say it?

As part of my initiation into the office, arrangements had been made for me to meet and address the headquarters staff at the training centre in Kampala. During this meeting, I declared that I would resign if I did not turn the Corporation around in 100 days. This rather hasty and impulsive off-the-cuff remark – which I subsequently almost regretted, and which was certainly influenced strongly by my first impressions on that first day – sent shock waves inside and outside the Corporation. My immediate listeners could not believe their ears. How could the new Managing Director dare to promise to do the impossible! This is something that had eluded the Corporation for so long. What made him think that he could move mountains overnight? What miraculous solutions did he have up his sleeve for the longstanding problems of the Corporation? Or was he just out of his mind?

Some of my listeners pronounced my 100 days promise doomed to failure because it was 'the biggest joke of the year' with regard to both the turbulent past of the Corporation and the scope and intensity of its current challenges. Others asserted that I was 'daydreaming' or that I had run wild with my imagination before coming to grips with the facts.

Some customers brushed away my promise as a mere marketing gimmick. One customer, Mr Isabirye Musoke, for instance, in *The Monitor* of Saturday 6 March 1999, wrote: '*The special offer for 100 days and the system which has been prevailing in the Corporation are basically the same... There is no offer whatsoever. It is just an advertising gimmick by the new MD. It is a hoax.*'

Within the Corporation, my promise was dismissed with contempt by some junior and senior staff. This was overtly expressed in the staff recollections of their 100 days experience. For example, Mr Lawrence Onega, a Masaka laboratory technician, did not believe that it was possible to change the state of the Corporation, believing that this was a ploy to sell the Corporation. '... *this was imagination and day dreaming by the new MD. Changing the face of the Corporation in 100 days? No! It is impossible. They are getting ways of selling the Corporation...*'

One of the senior managers, Mr Charles Ekure, was shocked by the audacity with which I made such a sweeping promise before I found my feet. He wondered whether my promise stemmed from the excitement of taking on a new job or the desire to impress the Board of Directors. All in all, the consensus among the managers, who had worked in the Corporation for years and were aware of the facts on the ground, was that it was an impossible task. They were convinced that as I settled into the job, I would come to terms with the stark realities in the Corporation and abandon my fanciful dream!

The reaction to my off-the-cuff declaration was much the same outside the Corporation. While addressing staff, I had overlooked the presence of journalists, whose paramount interest, if not obsession, is to look out for sensational stories. The following day, my promise was splashed across the front pages of the daily newspapers. For example, *The Monitor* of 19 November 1998 reported that in my first address I had promised not only to turn the Corporation around in 100 days but also to sweep it clean, especially of liars and thieves. It also reported that I had promised to ensure the flow of water to every home under our jurisdiction and to forthwith stop the unfair and inconvenient practice of disconnecting customers whose bills were overdue on Fridays at 5.00 p.m. My address also hit the headlines of radio and television networks.

My wife, Vicky, as well as many friends who read the papers or listened to the electronic media news broadcasts, phoned me to find out whether what had been reported was true. If so, had I lost my senses? How could I make such an outrageous statement? They warned me that if the excitement of a new job had sent me to the clouds of fantasy, reality would soon bring me down to earth. Even my wife found it hard to accept that I had made such a hasty promise before I could come to grips with the reality. I tried to convince her that it was all a slip of the tongue which the press had exaggerated out of proportion. I am not sure whether or not she accepted my lame explanation. Fortunately, she did not press the issue. If she continued to harbour misgivings, she kept them to herself.

Public scepticism about, if not total disbelief in, what I had said sent my mind racing with anxiety and self-doubt. Should I deny ever making the statement and pass on the blame to 'those funny journalists who misquoted me?' But if I denied the statement, would those who were at the meeting ever trust me? I decided that it would be cowardly and irresponsible to deny what I had said when the whole world had been listening. In any case, deep down, I was convinced that something radical was needed to be done and done quickly to change the situation in the NWSC. I resolved that my promise was a matter of honour on which my name and career would stand or fall. Although turning the Corporation around in 100 days was bound to be an uphill task, it could be done and it had to be done. For this reason, I began to see the 100 days promise not as my Waterloo but as a useful yardstick against which all our actions to rescue the Corporation would be focused and against which our performance could henceforth be measured. In retrospect, I am glad I made that off-the-cuff promise. It created a new sense of urgency and momentum, which helped to focus our minds and energies on the challenges that the NWSC was facing in a programmed and prioritised manner.

More shocking encounters

Before I settled down to strategise how to turn the 100 days promise into reality, more shocks hit me with unrelenting mercilessness. To start with, when I reported to my office on my second day, I found that the Uganda Revenue Authority (URA) had overnight – and without prior warning – closed our headquarters for non-payment of Value Added Tax (VAT); our debt was in excess of UGX 2.6 billion. Soon after the URA had locked our offices, the Uganda Electricity Board (UEB) threatened to disconnect power to our waterworks if we did not immediately settle our accumulated debt which was UGX 2.8 billion in worth. As if all these shocks were not menacing enough, auctioneers, on behalf of the creditors whose debts were long overdue, impounded my official vehicle. It took a lot of time and negotiations, and government intervention, to ease the pressure put by our creditors. These shocks diverted us temporarily from getting down to business, but at the same time they reinforced my conviction and commitment to the necessity of seeing through my 100 days promise.

As I began to settle down to business in the subsequent days and weeks, it dawned on me that the work culture and the mindset of the NWSC employees, from the senior managers to the lowest ranks, were bound to be the greatest obstacles to turning the Corporation around. While commendable efforts had been made in the 1990s to improve the physical infrastructure, from waterworks and distribution mains to tariff reforms and the overhaul of the legal framework, the corporate culture remained trapped in decades of an 'I-don't-care' mentality. Coming late, shabbiness, drunkenness, rumour mongering and gossiping, ethnic-based cliques, godfathering, insubordination and absenteeism, just to give some examples, were the order of the day. While people were officially required to report for work at 8.00 a.m., many of them reported after ten o'clock and left for the day after lunch. Even those who stayed at work throughout the official working hours tended to mark time rather than engage in productive work.

Some unscrupulous workers had ingenious ways of pretending to be at work when they were actually out of the office pursuing private business. Some claimed that they were attending official meetings. The standard answer for such absences was that 'so and so is not at his/her desk or that he/she has stepped out'. In extreme cases, some employees had the audacity to take up two (or even more) jobs – one at NWSC and another one (or more) elsewhere. As I settled down in my job, I discovered that such workers used to report for duty at NWSC up to about 11.00 a.m. or lunchtime and then take off for work elsewhere in the afternoon. One such member of staff confessed to me that he had been doing so for two years! This employee had even managed to get two scholarships – one from NWSC and the second from his other employer. This duplicity, which was confirmed by his colleague at his other place of work, was discovered when we introduced morning signing-in and evening signing-out registers, as well as movement books for working hours. Although the man lost both the jobs for his duplicity, he managed to get away with the money that the NWSC had paid him for two years for no work done. Such was the price of the Corporation's work culture!

This debilitating corporate culture tended to undermine productivity and performance as well as organisational harmony and cohesion. It also tended to cripple personnel management by creating a sense of fear, timidity and favouritism among senior officers regarding the control or discipline of subordinate staff. Conversely, it created a sense of arrogance and insubordination among junior staff who had godfathers higher up the senior management ladder. Supervisors tended to turn a blind eye to wrongdoing for fear of reprimanding subordinates who enjoyed protection in the corridors of management. The consequence was a breakdown in the chain of command. For example, I was informed that in the Mbarara Area, an unruly driver had been recommended for transfer to headquarters because his boss feared the possible consequences of taking disciplinary action against him. In Jinja Area, one senior manager preferred to be transferred to a lower position at the headquarters due to constant interference from above in favour of an unruly subordinate. In Gulu, the Area manager, instead of grappling with the challenging affairs of the Corporation, including personnel management, had reportedly turned to Christ as his personal saviour, devoting most of his working time in sending Christian messages to his brethren. Obviously, such a work culture and environment required radical transformation.

When the shocks and challenges began to sink in and to shape my thoughts, I decided to tour some of the upcountry NWSC Areas to acquaint myself with the state of the business there. I soon confirmed that the vices that I had detected in the corporate culture at the Head Office were even more crudely pronounced and entrenched in the operational Areas. In Tororo – the first Area I visited – the condition of the office was so bad that the meeting had to be held in the premises of a primary school.

When I arrived at the venue, there was nobody; I was all by myself for about an hour. I became anxious and started pacing up and down like a schoolmaster whose pupils have failed to show up. Fortunately, after this long wait, the workers, who were shabbily dressed and some of whom were visibly drunk, started arriving one by one as if the presence of the new chief executive officer did not matter. Looking at the people, I could not imagine that these were actually the members of the staff I had come to meet. The staff too looked surprised and confused. They seemed not to be sure whether they were in the right place, and most of them seemed not to have any idea who I was. For a moment I wondered whether I was not smart enough to fit their expectations of their new Managing Director. The apparent suspense was eventually broken by the arrival of Dr Abdallahi Shire, a Board member from Tororo, who called the meeting to order and introduced me to the staff. During the meeting, most of the staff were clearly absentminded, they did not pay attention to what I had to say. What shocked me more was the rush for the drinks. The drinks, especially the beers, took centre stage, with staff completely oblivious of the issues I had raised. By the time I left Tororo, I had no illusions about the magnitude of the reforms and restructuring challenges ahead of us.

After the incident in Tororo, I did not expect pleasant surprises from our second batch of familiarisation tours to Masaka and Mbarara Areas, this time in the company of a few Board members, including the Board Chairman. In Masaka, we found that the waterworks were badly maintained, the buildings were dirty and the grounds were bushy. I was so incensed that I threatened to sack the Superintendent of Works there and then. When I cooled down, I instructed the Superintendent, who was then shivering for fear of losing his job, that the waterworks surroundings must henceforth not only be properly and regularly maintained but also that the lawn grass should never be higher than 4 cm. Later, the same instructions were repeated to the Area manager who was not present at the time we were inspecting the waterworks. At a subsequent working lunch, we addressed the entire Masaka Area team and cautioned them to pull up their socks and to realise that the future of their jobs in the era of privatisation was dependent on turning the Corporation around through enhanced performance, productivity and, above all, making money, not only for the government, but for themselves.

After lunch, we headed for Mbarara. As it had been in Masaka, the Mbarara waterworks premises were bushy and the buildings had not seen a coat of paint for a long time. While meeting the members of the staff, we witnessed an ugly incident which was quite revealing about the state of the water business and the Corporation as a whole. During the meeting, one member of the staff who had earlier limped into the room collapsed before us. We were shocked. The Board Chairman ordered that the poor man be taken to a health facility, and in his speech stressed that it was the duty of the Corporation to provide medical care to every worker, especially the sick and people with disabilities. Little did we know that the poor fellow actually collapsed not because he was sick but drunk! We had mistaken his unsteady gait for a limp. He had taken one glass too many the previous night and had even continued drinking in the course of the morning, including working hours. What we had seen in the course of our tour was just a hint of the sorry state in which the Corporation was. We needed to take action immediately. Our answer was to concretise the 100 days promise into a programme to arrest and reverse the downward trend in NWSC, restore morale and improve financial performance.

Marketing the 100 days Programme

Having decided that it was imperative to push forward with the 100 days promise, the next crucial step was to sell the idea to the Board and to NWSC members of the staff, from top to bottom, and to concretise the idea into a realistic and feasible programme within the specified time frame. When I met the Board Chairman after my famous off-the-cuff declaration, I expected him to reprimand me for making hasty promises for he too had received a flood of telephone calls expressing doubts about the Board's judgement in appointing me as the chief executive. To my pleasant surprise, the good old man was supportive. Of course, he reminded me of the gravity and full implications of my hasty remarks. He stressed that I had made a one-sided contract with all the NWSC stakeholders, including the workers, for the breach of which I would be solely responsible. Accordingly, he advised me to pull up my socks, mobilise my senior managers and get on with what I had promised to do, and he reassured me of his full backing. He also undertook to persuade other Board members to endorse the idea of the 100 days Programme. These were his first but not the last encouraging words. This reassuring vote of confidence was music to my ears. I knew I had nothing to worry about from the Board.

The next challenge was to bring the Corporation's senior managers, Area managers and other members of the staff on board. I was not sure how all these people would react to the proposed ambitious programme. Would they resist or embrace it once it was explained to them? What did we have to do to win over the doubting Thomases? How could we ensure that the members of the staff 'owned' the Programme? Who would spearhead the preparation and implementation of the Programme? What would KRIP's role be? These and

Setting the Priorities | 27

many other questions had to be answered correctly, if the 100 days promise was to be kept.

I knew that Kampala operations were partly under the management of JBG, but I was aware that ultimately I was responsible and accountable for the overall managerial performance of the whole Corporation. The incorporation of the objectives of KRIP into our design of the 100 days plan was agreed upon with my senior managers so as to have one corporate programme. We wanted Kampala, which constituted 65–70% of the Corporation's business, to benefit from the Programme, but we knew that we could not enforce it without interfering with the operations of the private operator.

When it came to the actual specifics of the Programme, of all the people in the NWSC, none was more critical about the takeoff and success of the Programme than Dr Wana Etyem, the Director of Technical Services and my *de facto* deputy. He was perhaps the most influential man in the Corporation. He had been the acting MD for six months or so before I took office. He was a water engineer with 10 years' experience in the Corporation. He knew the business inside out. He knew all the managers at the headquarters and in the Areas. In early December 1998, I had a brainstorming session with him about the idea of the 100 days Programme. I informed him that the Board Chairman was supportive of the idea and had cleared the management to get on with the job, but that the chairman had also warned that my credibility, reputation and my job, as well as the future of the Corporation, would be on the line if we did not turn NWSC around in 100 days as promised. Since there was no turning back on what we had set out to do, we had to take the bull by the horns, which required full cooperation and participation of all the members of the staff of the Corporation.

I must confess that I was not sure that he would support the Programme. After all, he too had had an eye on the MD's job and had been the acting chief executive for the past six months. Naturally, he must have been disappointed that he did not get to keep the job. Fortunately, whatever his private feelings, Dr Etyem turned out to be the right officer to kick-start the Programme.

The question was no longer whether there should be a 100 days Programme but rather what its contents, objectives and priorities, and expected outcomes should be and how soon it would be implemented. For these reasons, I told him that I would count on him as my deputy to spearhead the preparation and implementation, and the sooner he started working on it, the better.

I then requested him to address a number of questions that were pertinent to the preparation and launch of the 100 days Programme. It was imperative to know what problems required our immediate intervention. What key priority areas needed to be addressed by the Programme? What inputs, in terms of human and financial resources, were required to prepare and implement the Programme? What were the likely reactions of managers and members of the staff to the Programme? What challenges and constraints were likely to be encountered in preparing and implementing the Programme?

In response, Dr Etyem highlighted a diversity of problems that the Corporation was facing. He talked about plant capacity under-utilisation, non-revenue water, low billing and collection inefficiencies, arrears, water leaks and bursts and so on. Given our different professional backgrounds, Dr Etyem and I had divergent ideas and approaches to what should be the core priority areas of the 100 days Programme. As a water engineer, he argued that the Programme should focus on water production and distribution as well as on the reduction of NRW. While I agreed that these areas deserved attention, as an economist and business manager, I was more interested in the money side of the water business. For this reason, I insisted that revenue generation, cost reduction and, above all, customer care should be at the heart of this emergency programme.

After our brainstorming session, we zeroed in on five core areas, namely water production, water distribution, revenue collection, cost reduction and customer care. I then charged Dr Etyem with the responsibility to form and chair a task force to look at each of the identified areas in great detail, to identify the priorities and to prepare the Programme. I requested him to play a leading role in marketing the Programme to his colleagues at the headquarters and in the field, emphasising the importance of teamwork, hard work, consensus, consultation and working within tight deadlines. On my part, I undertook to do everything possible to facilitate the work of his task force, and to be available for consultation whenever I was needed. I was delighted when he agreed to take up the challenge and to start work immediately.

I then turned much of my attention to the transformation of the work culture of the Corporation as an indispensable prerequisite for the implementation and success of the Programme. In order to transform the NWSC's work culture, I had to lead by example. With regard to punctuality, I made sure I was in the office by 7.30 a.m. and I did not leave for home until 6.00 or 7.00 p.m., when most of the other workers had left. During the working hours, I never left the office except when I was out on official duties; otherwise I usually had a working lunch of fruit, tea or coffee and snacks. As a matter of policy, my office was open at all times for all, including senior managers, ordinary workers and customers. My telephone numbers, both at the office and at home, were made known to all those who wished to reach me for official business, day and night. All this was intended to encourage the spirit of openness, consultation, discussion and free flow of information, as well as to cultivate a culture of confidence and mutual trust across the Corporation. This personal working style

Setting the Priorities | 29

has continued unabated since 1998 and I dare say it has worked very well for me and the Corporation.

Once I had set out my working method, I insisted that others follow my example. As has been mentioned already, we introduced morning signing-in, evening signing-out and office hours' movement registers to be signed by the entire Corporation's staff, including myself, to ensure punctuality in the morning and continuous work until the official closing hour (5.00 p.m.). Through these reporting and working hours control mechanisms, we managed to catch habitual absentees. We reprimanded latecomers and those who absconded from duty. Those who did not mend their ways after several reprimands were shown the exit. Apart from strict timekeeping, we insisted that staff be tidy and smart at all times and in due course we introduced T-shirt uniforms for every Friday and identity cards for all the members of the staff, from top to bottom. We also refurbished the offices, and banned the selling and buying of food in corridors. To combat gossiping and rumour mongering, we insisted that all workers not only were to be seen to be working at all times but also had to actually work to complete the tasks assigned to them by their supervisors.

Most corporate bosses tend to manage *in situ* from their offices, where they receive reports and hand-down directives. My preferred management method has always been to be a man on his feet and to manage by walking around. Therefore, from the outset of my work at the NWSC, I chose to visit each and every office, to inspect stores and waterworks and to know workers of the Corporation. By using this approach, I was able to observe who was doing what work and what progress was being made. This way I was able to detect whatever was going wrong, and could easily take appropriate corrective action before it was too late. I was also able to talk to the staff and resolve their problems promptly, on the spot, without waiting for briefings and reports from my managers. Within a few months the whole working atmosphere, from ambience to working rhythm, had changed for the better. Thus, by February 1999, we had gone a long way towards transforming the corporate cultural landscape in preparation for the takeoff and implementation of the 100 days Programme.

On the drawing board

The preparation of the 100 days Programme began on 4 December 1998. Initially, the preparation of the Programme was supposed to last two weeks, but we soon realised that the sheer magnitude of the work, as well as its scope and complexity, required much more time than we had originally anticipated. The programme was not ready for launching until 8 February 1999. The task force, which had the overall responsibility for the preparation of the Programme, was broken into five committees, one for each of the priority areas. Chaired by the head of the department responsible for the particular priority area, each committee consisted of specialists who worked on the Programme in addition to their routine duties. From the outset, we adopted a participatory approach. Accordingly, all Area managers and trade union officials were invited to participate in the deliberations at committee stage. All the members of the five committees were also members of the task force. All in all, the task force had 33 members.

From December 1998, the task force committees worked day and night in cooperation with 'field' managers and staff to identify operational problems, constraints and bottlenecks in each of the key priority areas. They had to pinpoint the weaknesses and to recommend ways and means of rectifying them. Since not all the problems could be addressed at once, it was the duty of the committees to prioritise the problems in the order in which they were to be addressed. Committee chairmen met every two days to review progress, share information and harmonise work efforts. The task force met in plenary sessions every fortnight to receive and discuss the progress reports from the committees, as well as to suggest improvements and set new targets for the committees for the next two weeks. Every second task force plenary session also served as a senior management meeting, which I attended personally.

I closely followed the work of the task force and its committees, and received regular oral briefings and written progress reports from Dr Etyem, as well as from individual committee chairmen. As a matter of practice, I listened to the briefings, read reports and gave feedback to the task force and its committees, especially in cost reduction and customer care – my favourite priority areas. By and large, I operated behind the scenes without intruding in the work of the task force and its committees. I was anxious to show Dr Etyem and his team that they enjoyed my total confidence and support in what they were doing. Therefore, throughout the preparation of the 100 days Programme, my role was limited to guidance, consultation, encouragement and facilitation rather than to omnipresent supervision and direction. This 'invisible hand' approach worked very well and eventually yielded the desired results.

The work done by the task force in preparation of the Programme was very impressive. Dr Etyem proved to be a single-minded and dedicated task force team leader. He energised and pushed the task force and its committees to the limit. He used his vast knowledge and experience to focus the minds and deliberations of his team. Under his leadership, other members of the task force and its committees plunged into the preparation of the Programme with enthusiasm that went beyond my original expectations. They worked day and night to meet tight deadlines. From the outset, they behaved as if they had been waiting for an opportunity to take on the challenge of preparing the Programme.

By the end of its preparation in February 1999, I had no doubt that the NWSC's human resources potential was as good as any. What was needed was to tap this potential to transform the Corporation.

By the beginning of February 1999, all was set for implementation. The programme was based on the 1997–2000 corporate plan. The draft of the programme identified the problems in each of the five priority areas, as spelt out in the Programme objectives, and made projections in terms of costs, targets and expected outcomes. The task force also proposed an operational plan to implement the Programme within three months from 8 February to 31 May 1999. The Programme was supposed *'to serve as a beacon of performance measure that must be maintained to ensure the viability of NWSC irrespective of the looming privatisation'*. It was also expected *'to create a basis for sustainable corporate management'*. The quantitative targets of the Programme are summarised in Table 2.1.

Table 2.1: Targets of the 100 days Programme

Priority area	Baseline: February 1999	100 days Target: May 1999
NRW (%)	55	45
Staff Productivity (staff/1,000 connections)	36	32
Collection Ratio (%)	60	96
Metering Efficiency (%)	78.1	85
Response time to leaks and bursts (hours)	>48	<24
Capacity Utilisation (%)	58	68
Operating Surplus (UGX million/month)	(348)	300
Number of towns covering operating costs except depreciation (out of 11)	3	5

As mentioned in the introductory chapter, the Corporation had adequate installed capacity to produce enough water in all the Areas. In fact, the biggest problem was the under-utilisation of the existing capacity. Capacity utilisation for the water treatment plants in Areas, measured as the percentage of the actual water produced from the plant compared to the maximum that the plant can produce, ranged from 85% in Entebbe to as low as 11% in Lira and slightly below 50% in five other Areas (Tororo, Masaka, Jinja, Mbale and Fort Portal). During the period, the target was to improve the overall capacity utilisation of the plants to about 68%.

With regard to water distribution, the main problem was the high water losses commonly known as Non-revenue Water (NRW) – the proportion of water delivered into the distribution system that is not billed. Priorly, the Corporation was losing about 55% of the water produced, and Area-to-Area variations ranged from 23% in Gulu and Fort Portal to as high as 57% in Mbale and Kampala. The high NRW was largely due to the rampant illegal connections, poor response to leaks and bursts, poor maintenance practices and inadequate metering of customer accounts resulting in inaccurate water consumption measurement. The overall aim of the Programme was therefore to improve operating efficiency in all Areas by reducing the high level of NRW. As part of this effort, the Programme was aimed at improving response time to leaks and bursts and the metering efficiency (percentage of metered customer accounts to the total number of accounts).

It also aimed at improving the financial status of the Corporation by increasing the bill collection ratio (a percentage of bills collected compared to the total bills given to customers/clients in a particular month). The low collection ratio meant that at least 40% of NWSC bills were tied up in arrears totalling to almost UGX 30 billion. The combined effect of the high NRW and the low collection ratio meant that the Corporation was recouping cash from only 27–28% of the total water produced; the remaining 72–73% was either lost as NRW or accumulated in arrears.

Closely related to revenue collection enhancement was the issue of cost reduction. All organisations must live within their means. In the long run, you cannot survive if you spend more than you earn. And yet, in 1998, operational costs were prohibitively high, resulting in a monthly deficit of UGX 348 million. This was compounded by a heavy external debt. The Corporation was in danger of bankruptcy; efforts to enhance revenue collection had to go hand in hand with cost-reduction measures. Accordingly, the Programme was aimed at reducing electricity, security, transport and employee costs. Employee costs were to be reduced through a restructured medical scheme, a voluntary retirement scheme and a car loan scheme for entitled officers. We anticipated that these cost-reduction measures would save the Corporation a total of UGX 998 million per year. The target was to reverse the loss situation to a surplus of about UGX 300 million per month. In this way, the number of towns breaking even was projected to increase from three to five. Furthermore, staff productivity measured as a ratio of staff per 1,000 water connections was anticipated to improve mainly through increased number of customers and reduction in the staff numbers.

Furthermore the 100 days Programme recognised the importance of customer care in turning the Corporation around. The programme underscored the problems that had led to poor NWSC–customer relations. These included late responses to complaints, over-billing and estimated bills, customers' failure to pay bills on time and, above all, NWSC staff's indifference, insensitivity and even rudeness to customers. The objective was to improve customer relations through enhanced customer care, sensitisation, prompt service delivery and mutual respect to create customer satisfaction and compliance.

The design process, therefore, had to look at financial aspects critically. Since the Programme was initiated in the middle of the 1998/99 financial year, it had to be implemented within the budget limits and within the context of planned activities of that year. In addition to this, one of the key priority areas was to reduce costs, and for this reason there were no supplementary votes for its various components. Under the Programme, we could relocate funds from one budgetary item without changing the total sum of the capital expenditure budget. Accordingly, a total of UGX 564 million was earmarked for the implementation. In contrast, the financial benefits to be gained from the Programme through increased billings and cost containment were projected to be more than UGX 5 billion.

All in all, the draft of the 100 days Programme developed by Dr Etyem and his task force was thorough and comprehensive. I was excited when I received the draft and went through it with youthful enthusiasm. It provided useful benchmarks and targets against which to work. It provided a 'solid platform' on which to launch the NWSC's recovery. When I submitted the Programme to the Board of Directors for consideration, it was quickly approved with minor amendments. Indeed, the Programme, which had started as one man's reaction to the unflattering condition of the Corporation, had now been adopted as an official NWSC initiative. I was proud of it. We were all proud of it. A hurdle had been cleared.

From dream to reality

Once the Board endorsed the 100 days Programme, we hastened to translate our dream into reality. We decided that the Etyem task force and its committees, which knew it inside out, would oversee its implementation within the agreed 100 days time frame. However, the actual implementation automatically shifted from headquarters to the operational Area offices, which became the operational battlefronts where it was destined to succeed or fail. The role of the task force and its committees was reduced to that of supervisors, monitors and evaluators. With the shift of emphasis from the headquarters to the Areas, the methods of work and reporting mechanisms also changed to conform to the realities on the ground.

During the implementation stage, Area managers worked out their own performance indicators in each of the five priority areas in consultation with the task force and its committees. The Area managers devised activity sheets to record, on a regular basis, what was going on in their respective places. They submitted fortnightly reports to the heads of the committees and also made monthly briefings to the task force plenary session-cum-senior management meeting. On their part, the chairmen of the committees designed the monitoring and evaluation sheets to record the performance of individual areas in each of the five priority areas. The task force and the committees evaluated the Area activities on a monthly basis in an open and fair manner. Board members, local dignitaries and the public at large were, to varying degrees, involved in the Area activities of the Programme throughout the three months of its implementation.

To kick-start, promote and sustain the momentum of the Programme, we publicised the activities relating to its implementation in each of the priority areas, putting emphasis on customer care. To start with, a press conference was held to launch the Programme and to spell out its objectives and expectations. This conference was attended by all the Board members, senior management, members of the task force and its committees and Area managers. In addition, we introduced weekly radio programmes, and radio and television commercials, as well as regular press releases, as part of the Programme publicity and public awareness campaign. For example, in one of the media releases, on 18 February 1999, we informed our customers and the general public that the Corporation had launched it to enhance water service delivery by being more responsive to customer complaints and to improve revenue collection. In the same release, we offered to reconnect disconnected customers without claiming a reconnection fee provided they agreed to pay outstanding bills in negotiated instalments. We granted an amnesty of 100 days to illegal water consumers to regularise their connections and consumption without the fear of fines.

Unfortunately, the implementation of the Programme was slightly affected in its early stages of implementation by the retirement of its lead champion, Dr Etyem, from the Corporation after years of dedicated service. He is remembered as one of the most dedicated and outstanding officers to have served the Corporation. The reigns were passed on to Mr Charles Odonga, who also assumed the role of Technical Director of the Corporation, to see through the implementation of the Programme.

Trophies and bull roasting

One of the strategies we introduced during the Programme to enhance performance and deliver the expected results was competition between NWSC Areas. We all know that competition is the engine of human endeavour. People love to compete for all sorts of reasons. For example, they compete for money, for fame and prestige, for power and glory, for victory, for success in various spheres of life and for fun. Indeed, life without competition would be dull, boring and monotonous. That is why people love to be part of, or to support, a winning team whether in politics, business or sports. And yet competition, though an inherent part of human nature, has never been fully exploited in the world of business in Uganda. It presented a golden opportunity to practice real competition among the Areas in a transparent manner to enhance the performance of the Corporation.

The Board and management agreed to introduce monthly competitions in which the winner in each of the five priority areas, as well as the overall winner, would be declared at the end of the month. The question that remained was how to recognise and reward the winners. Would we simply say: Thank you, well done, God bless your Area, and leave it at that? True enough, words of recognition and appreciation can be encouraging in the world of business, and we should always thank those who do well and excel in their work. But achievers also expect something more tangible over and above a verbal recognition. We could give the winning Area a golden handshake; however, for a Corporation that was trying to cut corners to make ends meet, such spendthrift incentives could be misunderstood and provoke public outcry and negative publicity.

By chance, when I met Mr Jack Wavamunno, an old friend and businessman in town, I picked his brains. We considered the pros and cons of various options of rewarding the winners of competitions until he came up with the practical and brilliant idea of a trophy. 'Why don't you treat your Areas like teams in a football league?' he asked. He added: 'Look at the fight that takes place between Express and Villa football clubs[1] over a trophy! If trophies are fought over in sport, why can't it happen in the water business?' As he talked, I realised he had hit the nail on the head. A trophy would be the ideal way of recognising winners in the Programme's monthly competitions. It would be wonderful and fun to see the Area managers fighting tooth and nail to improve their water service delivery performance to win a trophy. The only likely hitch was whether my engineers would accept such a seemingly crazy idea. All the same, I decided to try out this novel experiment.

As usual, I sounded out my management team about the idea. Trophies would be given to the winners in each category, and to the overall winner and runner-up – seven trophies in all. After a heated but lively debate, the management bought my idea. Presenting the trophies would be the climax of the monthly

---

[1] The two most famous clubs with a historic rivalry in Ugandan Football.

competitions and we hoped that this would boost the morale of the staff of the winning Areas. In addition, modest cash bonuses (UGX 300,000) for the winner in each priority Area, UGX 1 million for the overall runner-up, and UGX 1.5 million for the overall winner, were agreed upon to complement the trophies. Although I was personally uneasy about the idea of cash bonuses, I accepted the fact that UGX 4 million for all winners was not really out of proportion, provided our performance improved.

Trophies and cash handover ceremonies were hosted by the winning Area, and they turned out to be very popular occasions. Board members and senior management from NWSC headquarters would be present at the occasions. Area dignitaries, such as the area MP, the mayor and the town clerk as well as a cross-section of Area customers would be invited to attend the occasions. As part of the ceremonies, the host Area would organise 'bull roasting' and other merrymaking activities during which the Board members and senior management would mingle freely with local dignitaries, customers and workers in an atmosphere of relaxation and informality. Initially, the Board Chairman was somewhat uncomfortable with bull roasting for fear of provoking public criticism at a time when we were committed to reducing costs. However, we convinced him that the end-of-month competition ceremonies were good public relations whose gains would overshadow any criticism.

For the first time in the history of the Corporation, the end-of-month competition ceremonies also brought together the NWSC fraternity and enabled them to mix work with pleasure. NWSC workers had never had competitions relating to their work. Maybe they used to have end-of-year bull roasting, but these events had nothing to do with one Area outshining other Areas in performance. For NWSC Area staff to mix and dance with their chairman and MD in recognition of their performance was a source of inspiration of revolutionary magnitude. It set a good example to other Areas. Every Area looked forward to the following month's results. All the workers pulled up their socks in order to not let their Area down. As I had expected, the competitive spirit and atmosphere worked wonders for the Corporation.

Winners and losers

During the 100 days Programme from February to May 1999, there were three end-of-month Area competitions and one overall competition. The task force and its committees monitored and evaluated the performance of all NWSC Areas in each of the priority areas and chose the winners and runners-up. The results of the evaluation process, in which some Board members participated to ensure fairness and impartiality, was kept top secretalmost up to the end-of-month competition functions. Like in all competitions, there were winners and losers. For the winners there was cause for celebration and the determination to keep winning. For the losers, there was always a 'next time' when fortune would possibly smile their way.

Entebbe emerged the winner of the first month's competition. Mr Johnson Amayo, the Area manager, was the host of the ceremonies at the Area office. He invited area dignitaries, including the mayor and other civic and political leaders, as well as a cross-section of our customers to grace the occasion. He hired the best bull roaster and disco in town.

On the 'D-day', Mr Charles Odonga, the chairman of the overall evaluation committee and the manager in charge of operations in NWSC, took charge of the ceremonies. As soon as dignitaries, staff and other guests took their seats, he announced the winners in each of the five key priority areas. The winning Area managers came forward to the high table, shook hands with the Board Chairman, the MD, the Mayor and other area dignitaries, received their trophies and cheques and made little speeches about how they had made it. The thrilled workers became the cheering crowd.

Then it was the turn of Mr Amayo, the manager of Entebbe Area, the overall winner. Accompanied by his entire management team, he marched to the high table with pride and a telling smile to receive the Area trophy and cheque. In his speech, he assured his audience that the trophy that Entebbe had won was there to stay. He urged the members of the staff not to relax; there was no room for complacency. But other Areas warned him and Entebbe 'to wait and see' what would happen the following month when the trophy would be snatched from them, whether he liked it or not. One of the Area managers was so impressed by the glamour of this first competitive event that he resolved that his Area would become 'number one'. What Entebbe Area had achieved, other Areas could too. For me, this was a promising start to the 100 days Programme and, of course, for the future of the NWSC. The spirit of competition and of Area pride was beginning to take root and to yield the desired performance results.

The Entebbe trophy handover ceremony was certainly a great and memorable day. The Area office workers were naturally very happy. They had won. It was their day. They had never experienced an occasion like this. The Area customers and local dignitaries were very happy that the occasion had brought them face to face with the Board and management. The Board members too were happy with the occasion. Whatever misgivings they might have had about the idea of competitions and bull roasting were put to rest. The competitions were here to stay and they augured well for the Corporation. For my part, I was happy and proud that what had begun as a novel experiment had made a giant step forward. No wonder we all enjoyed the subsequent bull roasting and dancing, which went well into the late hours of the night.

Of course, not everyone was happy with the Entebbe ceremony. Some thought that this was a self-serving waste of time and money. Public criticism was reflected in the newspapers the following day. 'National Water on bull roasting spree', screamed one headline. The story was accompanied by a picture of me enjoying a generous stick of roast meat. To our critics, the Entebbe merrymaking function was a sign of irresponsibility. Here was a corporation whose performance was poor and whose finances were shaky at best, and yet the management seemed to be bent on extravagantly spending money on trivial activities. The scores of vehicles parked at the Entebbe Area office and reports of celebrating with abandon seemed to say it all. How on earth could we justify such extravagance? Nobody seemed to be interested in going beyond the surface to find out the intended long-term benefits of such competitions.

Unlike some of my colleagues, who were alarmed by the negative reports, I was not worried by this unjustified criticism of our programme activities. On the contrary, such criticism was welcome. It was one way of publicising what we were doing. It demonstrated that the public was interested in our work, which was much better than being indifferent. The public had the right to know what we were up to. This show of public, albeit critical, interest could be turned into an opportunity to strengthen our customer care strategy. We too could use the electronic and print media to answer our critics, sensitise the customers and highlight our achievements.

Lira Area was the winner of the second month's competition. Therefore, all the roads led to Lira for the trophy handover ceremony. Mr Charles Ekure, the Lira Area Manager, was the man of the moment and the centre of attention. During the handover ceremony, Mr Odonga, the master of ceremonies, called on Mr Amayo, the Entebbe Area Manager, to surrender the trophy, which was then handed over to Mr Ekure. Before parting with the trophy, Mr Amayo, a jolly and interesting fellow, advised Mr Ekure to keep it safely in temporary custody until the following month when it would return to Entebbe, where it belonged. On his part, Mr Ekure vowed that Lira would never let go of the trophy ? it would never cross Karuma Falls[2] again. The spirit of competition was gaining momentum.

We had done it again. Staff morale in the Lira Area was buoyant. Customers were happy with our performance because service delivery and customer care were steadily improving since the commencement of the 100 days Programme. The Lira ceremony was crowned with another bull roasting function during which the invited guests mingled and relaxed with the workers and customers.

---

[2] Karuma Falls is one of the most exciting white falls in Africa, located 265 km from Kampala along the Kampala–Gulu highway.

Setting the Priorities | 39

During the third round of evaluation, Masaka Area emerged winner and the Area hosted the trophy handover ceremony. Mr Andrew Sekayizi, the Masaka Area Manager, an outspoken and outgoing young man, was the star of the occasion. For him, the Masaka event was a dream come true for it gave him a golden opportunity to display his excellent organisational skills and talent. The Board Chairman was so impressed by the monthly competition that he decided to introduce the Chairman's Shield at his own cost, earmarked for the most innovative and creative Area manager.

The grand finale was held in Jinja. Mr Joel Wandera, the Jinja Area Manager, was the host of the occasion. The overall pre-evaluation process for the entire programme was conducted at NWSC headquarters. All the previous evaluation reports were taken into account. The last three days of the final evaluation process took place in Jinja where the Area managers together with senior management and Board members reviewed progress, shared experiences and lessons, and mapped out strategies for the future. The Board also used the meeting in Jinja to discuss the 1999/2000 budget – the first since they took office – taking into account the experiences and recommendations of the 100 days Programme.

In terms of organisation, Mr Wandera broke the record that had been set the previous month by Mr Sekayizi in Masaka. Mr Wandera bought suits for his Area's front office staff and decorated the reception area with flowers. He invited the local dignitaries in Jinja, including the District Council Chairman, the Resident District Commissioner and local MPs, as well as the most important customers in Jinja Area. Mr Akika Othieno, the minister of state in the Ministry of Water, Lands and Environment, was the chief guest. World Bank officials, dignitaries from Germany and a representative from the African Development Bank (ADB) were also present. As on previous 100 days monthly competitions, all the NWSC Board members, senior managers and Area managers were there too. Mr Odonga was once again the master of ceremonies and the presenter of prizes to the winning Areas.

The invited guests and the NWSC staff listened with anticipation as Mr Odonga read out the winners in each of the five key priority areas. While Masaka and Entebbe won the trophies for water production and revenue generation, respectively, Jinja scooped the trophies for water distribution, cost reduction and customer care to emerge as the overall winner, followed by Masaka. Mr Wandera stole the show. With his favourite music playing in the background, and accompanied by his winning team, he proudly marched back and forth to collect the trophies and prize money amidst cheers and clapping from the audience. It was certainly a joyous moment not only for Mr Wandera and the Jinja Area, but also for the entire NWSC family.

Apart from the trophies for the winners of each of the key priority areas and the overall winner of the 100 days Programme, the Chairman's Shield for the best

Area manager was also up for grabs. Since Jinja Area had won in three of the five key priority areas to emerge as the overall winner, we all expected Mr Wandera to be announced as the most creative and innovative manager. To the credit of Mr Odonga and his evaluation team, they kept the results a top secret. The climax of the trophy awarding ceremony was perfectly set as we all waited for the announcement of the best Area manager. Surely, it had to be Mr Wandera!

Of course, the world of competition is always full of surprises. Mr Wandera and, indeed, the rest of us were surprised when Mr Odonga announced Mr Sekayizi as the best Area manager. In addition to the shield and the prize money, Mr Sekayizi won himself a trip to visit a sister water utility in Accra, Ghana, fully sponsored by the Africa Development Bank representative who attended the occasion. Mr Sekayizi was as excited and jubilant as he was surprised. For me, the specifics of the winner of the Chairman's Shield and the trip to Accra were not as important as the growing spirit of competition and improvement in performance. In addition, NWSC was getting known across the continent.

Winning and losing are two sides of the competition coin of whatever nature. The 100 days Programme competitions between the NWSC Areas were no exception. While Areas like Entebbe, Lira, Masaka and, above all, Jinja won trophies and cash prizes, the others were not so lucky. It is also natural for the losers to be disappointed and to be a little jealous, if not envious, of the winners. But losing, in our case, was not the end of the world. There was always a next time. Indeed, the experience of the 100 days Programme shows that an Area which lost one month could come out to win in the next. Even those Areas that did not win trophies or cash prizes during the competition did not give up. They continued to fight the good fight to improve their performance. In other words, instead of a 'win-lose' situation, the competitions became a 'win-win' situation, not only for the Areas but also for the entire NWSC family.

Reaping the fruits

The 100 days Programme was largely a successful story in all the five key priority areas. By the end of May 1999, we were already reaping the fruits of the Programme. The key performance indicators of our achievements are presented in Table 2.2. The results show the performance of the Areas where the Programme was implemented without any restrictions and that of Kampala, where there were some limitations because it was operating under a private operator.

Table 2.2 shows that the 100 days Programme resulted in significant efficiency gains. Specifically, the Corporation registered improvements in both operational and financial performance. Even where the 100 days Programme targets were not 100% achieved, the Corporation still registered significant improvements compared to the performance prior to the introduction of the Programme.

Setting the Priorities

Table 2.2: Performance during 100 days Programme (February–May 1999)

Indicator	Before 100 days (February 1999)			After 100 days (May 1999)			Target
	Kampala	Others	Overall	Kampala	Others	Overall	
NRW (%)	57	48	55	51	42	49	45
Staff Productivity (staff/ 1,000 connections)	26	47	36	20	37	27	32
Collection Ratio (%)	58	65	60	93	99	96.2	96
Metering Efficiency (%)	75	83	78.1	77	85	80.7	85
Response time to leaks and bursts (hours)	>48	>48	>48	36–48	24–36	36–48	<24
Capacity Utilisation (%)	68	49	58	70	53	59	68
Operating Surplus (million UGX/month)		(348)			350		300
Number of towns covering operating costs except depreciation (out of 11)		3			5		5

The ultimate output indicators presented in Table 2.2 could not have been achieved without a significant improvement in activity-related indicators. Notably, total arrears were reduced by 3%, suppressed accounts (inactive) were reduced by a little over 10%, over 400 illegal connections were regularised and integrated into the billing and revenue collection systems, and estimated bills were reduced by 52% by the end of the Programme. Furthermore, as a result of the cost-reduction strategies, we were able to reduce our average monthly expenses by UGX 431 million.

Apart from the output indicators given above, the 100 days Programme achieved a number of process indicators. For example, we installed bulk flow meters in Masaka, Jinja, Tororo and Mbale to ensure more accurate and reliable measurement of flows. The Masaka wetland treatment system was refurbished to meet the National Environment Management Authority (NEMA) effluent standards, and as a short-term measure, we constructed a sewage pit in Kasese pending the establishment of a modern sewerage system. All these gains were achieved at an affordable cost. The actual cost of the water production improvement activities turned out to be UGX 58.6 million, which was far below the original estimated cost of UGX 172 million. All accounts in Gulu, Masaka and Fort Portal were metered during the 100 days Programme. Apart from Kampala and Mbale, there was considerable improvement in water distribution in all NWSC Areas, which contributed towards reduced NRW.

The 100 days Programme also made a promising start in improving customer care. Customers were involved in NWSC end-of-month competition ceremonies. By holding regular Area customer consultative meetings and briefings, customers were sensitised on what the Corporation was up to. Our staff became more sensitive and responsive to customers. They promptly and politely responded to customer queries and complaints. By the end of the Programme, customers had begun to appreciate that a wind of change was blowing through the Corporation. Invariably this wind of change brought the desired customer satisfaction and gratification. It also brought about a change of mind and attitude in the way things were being done in the new NWSC. For instance, the *Sunday Monitor* of 11 July 1999 praised the NWSC for a job well done and commended the Corporation's management for the improvement in water and sewerage services. Another customer, whose letter was published in the *Sunday Monitor* of 8 August 1999, commended the Corporation with the following words:

> NWSC employees were used to being handled with kid gloves. Things are now changing. The heat is on. Everybody has to deliver or Mr William Muhairwe will crack the whip. Look at the estimated bills we have been getting yet we have functional meters. Meter readers

were too lazy to move around. Look at the dormant accounts that have been activated. Look at the leakages – both sewage and treated water – that have been minimised. Open your eyes and see the NWSC offices (KRIP) on 6th street, step inside and see many achievements in this short time. Muhairwe, crack the whip if this is what delivering entails.

This was an encouraging compliment. Customers were responding positively to our customer care strategy. There was cause for optimism that turning the NWSC around was within our reach. Indeed, with the progressive improvement in the work environment and ambience, as well as in staff cleanliness and punctuality, customers were beginning to feel welcome and to develop confidence in the Corporation's service delivery.

The 100 days Programme caused a revolution in the thinking, attitudes, work culture and actions of NWSC staff in all Areas. The programme lifted the spirits of the workers, who began to appreciate that change was possible and beneficial. Competitions created an upbeat atmosphere in the Corporation. Every staff member and every Area wanted to win and to get things done, and everyone looked forward to a bright future. Workers began to appreciate that the future of their Corporation was in their hands. The shadow of privatisation over the Corporation appeared to be receding. The secret of success was hard work, teamwork, competition, fair play and transparency. No wonder most members of the staff in the Corporation, including senior managers, were enthusiastic about the achievements of the 100 days Programme, which they variously described as 'a great awakening', 'an eye opener' and 'a great source of inspiration' that had enhanced good housekeeping, inculcated the right, positive attitudes and laid a solid foundation for subsequent change management initiatives. For example, Mr Kaamu, the Mbarara Area manager and one of the original sceptics of the success of the 100 days Programme, was astonished by its success:

> This is an eye opener! Changing a parastatal is like moving a tractor. It's hard to move the first inch; it is very difficult to move it at all. But after it starts moving, however slightly, it becomes easier.

Of course, the 100 days Programme was not an unqualified success. Invariably, the Programme encountered many operational challenges and constraints. There were many hurdles to clear along the way. For one thing, some of the Areas were too slow to internalise the essence and urgency of the Programme. In Mbarara, for example, only a small percentage of the workers initially understood what the Programme was about. As a result, it took more than one month for the Programme to kick off in Mbarara. Such a delay was bound to affect the Area's performance. More importantly, Kampala Area, which accounted for 60–70% of NWSC's operations, was not adequately covered by the 100 days Programme because KRIP's management was hesitant to get involved in the end-of-month performance competitions. Therefore, Kampala's performance achievements in all the priority areas around which the 100 days Programme revolved were lower than the performance achievements of most other Areas. For example, the overall evaluation of the 100 days Programme placed Kampala in the seventh position out of 12 Areas. The reason Kampala Area did not do as well as other Areas will be explained in some detail in Chapter 4, but for the moment it suffices to say that, given its centrality in NWSC operations, its lacklustre performance dampened the overall achievements of the Programme.

Secondly, some of the required inputs were not mobilised fast enough to keep pace with the implementation of the Programme. Since time was of utmost urgency, any delay in the procurement of materials, such as laboratory chemicals, equipment and tools, impacted negatively on the Programme. A number of suppliers were unable to deliver the required materials on time. Some of the Programme targets and expected outputs turned out to be too ambitious to be realised within the 100 days. Government did not pay its water bills, including arrears, regularly letalone promptly. Uganda Electricity Board could not guarantee reliable and continuous supply of power to the waterworks and treatment plants. URA's insistence that the Corporation pays VAT on bills rather than on collections was a major problem that was yet to be resolved. Some of the cost-reduction measures needed more time and money to bear fruit. Although the 100 days turned out to be too short to address the problems that had bedevilled the Corporation for decades, it nevertheless demonstrated that the work could be done.

Good ending, but what next?

The 100 days Programme ended well. It was a success, thanks to the tireless efforts and competitive spirit that had been unleashed in all the Areas.

However, the success of the 100 days Programme presented a new challenge. Indeed, as we celebrated the climax of the Programme in Jinja, the Board chairman pointedly asked me, 'What next?' I had already been asking myself this and other important questions. Had we gained a sustainable and irreversible momentum during the Programme? What needed to be done to consolidate and build on the achievements of the Programme? How could we use the experience and lessons drawn to address other challenges that the Corporation was facing? What set of priorities required immediate attention? Addressing these questions culminated in the launching of the follow-up Service and Revenue Enhancement Programme (SEREP), which is the subject of the next chapter. However, some of the important lessons that can be drawn from the 100 days Programme are summarised in the box given below.

Lessons learnt

- As a leader or manager, set the example of culture change in your organisation and always dance to your tune so that others can follow.
- Be bold and do not hesitate to act on impulse. First impressions can be useful guides to action as detailed briefings and careful study of an organisational environment can create excuses for inaction and equivocation.
- Stick to your actions with single-minded determination and commitment. Do not be discouraged by detractors.
- Listen, consult and introduce participation across the board because organisational change requires not only strong and firm leadership but also ownership of programmes.
- Identify lead champions to drive change and let them get on with the job while giving them all the institutional support and confidence they need and deserve.
- In effecting organisational change, go for short, sharp, high-impact programmes with clearly set priorities, inputs and outcomes.
- Do not allow capacity and resource constraints to stand in the way of necessary action. Exploit the existing opportunities and strengths, and maximise the use of available resources to achieve the desired results.
- Celebrate every achievement, however small, by involving high profile persons and all stakeholders to create maximum impact possible for change.
- Always remember that the secret of success is hard work, teamwork, competition, fair play, accountability and transparency.

Chapter Three

Focusing on the Customer and Financial Viability

Service and Revenue Enhancement Programme

One of the insidious problems for managers and their organisations is the failure phenomenon called the "Success Trap". The term has been used in studying organisations which show great promise and then hit a plateau or decline. A single underlying pattern – a great success followed by a big fall. Often managers as well as organisations grow complacent and short-sighted as a result of success. Thus, blinded by past and current success, individual managers and indeed organisations often fail to recognise new business realities in time to adapt, leaving others to wonder where the brilliance went as the once-promising organisation stalls or fails.

Gary R. Casselman & Timothy C. Daughtry
How Leaders Can Avoid the Success Trap

During the evaluation workshop of the 100 days Programme in Jinja in June 1999, the participants did not only review the Programme's achievements but also identified its shortfalls, gaps, and constraints which required immediate attention in order to sustain the change management momentum. By the end of the workshop, there was a general consensus that management should initiate a follow-up programme to consolidate the achievements and to ensure that the NWSC's turnaround became sustainable and irreversible.

Although the 100 days Programme had been successful, for us there was no room for complacency. There was no time to lose. A lot of work remained to be done to achieve the ultimate objective of transforming the Corporation into a commercially viable public utility on a sustainable basis. Of the 11 Areas operated by the Corporation, only five – Kampala, Jinja, Entebbe, Mbale and Mbarara – were breaking even. Others, Tororo, Lira, Gulu, Fort Portal and Kasese, could barely cover their personnel-related costs.

The culture of change and setting priorities on the basis of short-term and focused programmes had to be internalised and integrated into the regular working life of the Corporation. All this meant that the shortfalls and constraints that had hindered the Programme had to be prioritised and addressed in the follow-up programme. The question then was: What kind of programme was this going to be?

© 2009 William T. Muhairwe. *Making Public Enterprises Work*, by William T. Muhairwe. ISBN: 9781843393245. Published by IWA Publishing, London, UK.

In this chapter, I present how the change management process in NWSC incorporated increased focus on the customer and financial viability. In this regard, I describe in greater detail the conception and eventual implementation of a customer-oriented programme code-named Service and Revenue Enhancement Programme (SEREP).

Searching for the next step

As pointed out earlier, the Board Chairman wanted to know what management was planning next. Was the management going to implement the recommendations of the Jinja evaluation workshop? If so, when and how would this happen? He stressed the importance of moving forward fast to sustain the momentum we had generated and to avoid backsliding into the bad old days. He was preaching to the converted. I was already thinking along the same lines. I knew that it is fatal to be complacent and to take things for granted in the management. The very thought of slipping back to the old days was frightening. I assured him that the management would soon come up with another programme to consolidate and build on the gains of the just concluded programme and address shortcomings that had been identified during the Jinja evaluation workshop.

Fortunately, to our advantage, we did not have to start from scratch. Senior managers at the Head Office and Area managers were already used to working as teams within tight timeframes; they had mastered the techniques of programme planning and implementation; and they knew how to work under pressure and to meet tight deadlines. A sound foundation upon which the next programme had to be launched already existed. What was required was to work out its contents and priorities. How different would the new programme be from the 100 days Programme? What was the new programme going to focus on? What financial, material and human resources were required to implement it? How long would it take to implement? What challenges were the management and the task force likely to encounter during the course of its implementation?

To address these questions, I convened a meeting of senior managers to brainstorm on our next course of action. Having considered several options, we decided to call the new programme Service and Revenue Enhancement Programme (SEREP). As the name indicates, the new programme was designed to put emphasis on customer care and corporate financial viability. The key priority areas of the 100 days Programme were maintained. The only difference was that in every priority area, SEREP would focus on customer satisfaction and transformation of the Corporation into a financially viable entity. In other words, the customer was to be the focal point around which all our resources, operations and management efforts would rotate under the often quoted catchphrase 'the customer is always right and the reason we exist'.

A 39-member task force, divided into five committees in line with the key priority areas, was then set up to work out detailed implementation strategies and activities of a new programme based on SMART strategic planning principles. The Director of Technical Services chaired the task force, while relevant heads of departments chaired the committees. As in the case of the 100 days Programme, a participatory approach was adopted, whereby Area managers and trade union officials would be fully involved in the planning and implementation processes throughout the Programme period. All SEREP activities were worked out and integrated within the framework of the 1999/2000 budget and the 1999–2002 Corporate Plan. One of the key findings from the evaluation of the 100 days Programme was that its implementation period was too short to accomplish all the planned programme activities. SEREP was initially planned to be implemented within six months, from 1 August 1999 to 31 January 2000, but it was subsequently extended by another six months to end by July of the same year because it became obvious that more time was needed if its key objectives were to be achieved. The lead champions, as well as the planning and implementation strategy and procedures, were the same as those of the 100 days Programme. The objectives of SEREP are listed in Box 3.1.

Box 3.1: SEREP objectives at a glance

In brief, SEREP objectives were to

- ensure reliable, sustainable and increased water production in all Areas;
- minimise sewage spillage and maintain international effluent standards and practices;
- increase service coverage through water mains extensions and construction of water kiosks;
- ensure prompt response to bursts and leaks;
- render efficient and effective customer service to attain and sustain customer satisfaction and willingness to pay;
- enhance revenue collection through efficient and accurate billing, as well as expand the customer base through the reactivation of suppressed accounts, regularisation of illegal connections and installation of new connections and
- to enhance cost efficiency and achieve cost savings.

In addition, the culture of change and the new corporate spirit of competition, teamwork, hard work, commitment and loyalty that had already been kick-started under the 100 days Programme had to be nurtured, entrenched and sustained as an integral and indispensable part of ordinary working life across the Corporation in order to improve and sustain productivity, service delivery and customer care. In particular, SEREP was to address the constraints, shortfalls and weaknesses of the 100 days Programme, such as high electricity costs, and unfriendly and unreliable suppliers that impacted negatively on the Corporation's performance and service delivery as well as the challenges, which had not been covered by the 100 days Programme due to time and resource constraints.

In line with the SEREP objectives, the task force and its five committees worked out the targets and expected outputs that had to be accomplished within six months. At the end of the six months, a performance review was carried out, and new strategies were formulated for the next phase of SEREP. In addition to the strategies, the task force and committees also drew up the cost estimates for each of the five operational areas. The key performance targets for SEREP are presented in Table 3.1.

Table 3.1: SEREP performance targets

Perfzormance Indicator	100 Days Performance	SEREP Target
NRW (%)	49	32
Staff Productivity (staff/1,000 connections)	27	18.5
Collection Ratio (%)	96.2	100
Metering Efficiency (%)	80.7	100
Response time to leaks and bursts (hours)	36–48	<24
Capacity Utilisation (%)	59	72
Operating Surplus (UGX million/month)	350	630
Number of towns covering operating costs except depreciation (out of 11)	5	6

The overall objective of SEREP was to enhance service delivery and sustain the performance momentum that had been registered during the 100 days Programme. This had to be done in a manner that cultivated customer confidence and satisfaction. As demonstrated in Table 3.1, we set ourselves ambitious targets that were well above what we had achieved during the 100 days Programme.

SEREP implementation strategies

Redirecting our efforts to customer satisfaction was henceforth to become the underlying principle of all our activities and operations. Accordingly, we had to work out several implementation strategies, which are summarised in Box 3.2.

Box 3.2: SEREP implementation strategies

- Putting the customer first because 'the customer is the reason we exist'.
- Establishing customer care units at the Head Office and in all Areas.
- Empowering and enhancing the capacity and skills of employees to serve customers effectively and efficiently.
- Enhancing customer confidence, satisfaction and willingness to pay.
- Identifying customer needs and addressing them accordingly.
- Talking and listening to customers and following up their suggestions, comments and complaints.
- Improving the quality of water and sewerage service delivery through increased accessibility and reliability.
- Enhancing revenue generation by increasing billing and collection efficiencies and reducing operational costs.

Putting the customer first: It is a generally accepted fact that a good business enterprise must strive to ensure that the quality of its products, packaging, mode and speed of delivery, its price structure and public relations are geared towards the satisfaction of the needs and expectations of its customers. Workers, especially in a poor country like Uganda, must realise that customers, not the investors and their managers, are the ultimate employers in any business organisation. If the consumers do not buy your products for whatever reason, the business will go bust and the workers will lose their jobs. That is why workers and managers alike must appreciate that 'the customer is king and the reason we exist'. Without a satisfied customer who is willing to pay for what we produce, whether it is water or something else, no business can survive, letalone flourish.

However, before the era of divestiture and privatisation, employees in public organisations used to take their jobs for granted. Once their appointments were confirmed, employees assumed that their jobs were permanent and virtually guaranteed until retirement, whether they performed well or not. They had no cause to pay much attention to customer care and satisfaction, maybe because public utilities like the NWSC were monopolies guaranteed by the law. Instead of being at the customers' beck and call, workers tended to see themselves as the masters, whose service delivery was a favour rather than an obligation. The members of the staff were notoriously rude and shouted down their customers. In the case of NWSC, when customers reported to settle their bills, complain or report faults, they were often ignored even if the front desk officers were doing nothing at all. Before the SEREP initiative, the members of the staff could even refuse to accept part-payment of water bills. A good example is the case of a cashier in Mbarara who, having ignored a customer for almost 30 minutes (without any explanation), demanded to know why he was not settling his bill in full. She directed the poor customer to go back home and return with all the money the following day!

All this had to change. Workers and managers had to deliver to the satisfaction of customers or face dire consequences, including dismissal. The customer was to become the centre of attention, as it should have been. This change of mood and shift of emphasis to customers invariably compelled public utilities like the NWSC to behave and operate as business enterprises. For these reasons, it was imperative for NWSC to put the customer first by delivering affordable water and sewerage services efficiently to customers' expectations and satisfaction.

Accordingly, under SEREP, our customer care strategy was to make 'the customer (is) the reason we exist' a practical operative guide rather than a high-sounding slogan to impress customers. For this reason, this catchphrase was displayed prominently in all our offices. At NWSC, this catchphrase meant exactly what it said. It was going to be the yardstick of our performance as an organisation, as Areas and as individual workers. All NWSC workers and their managers were required to internalise and practise it as part of their daily working routine. They were also required to master the techniques and procedures of handling customer queries and complaints in a sensitive, responsive and courteous manner and to address whatever issues that were raised promptly and efficiently.

In addition to the orientation of the attitudes and behaviour of the employees towards customers, NWSC customer care units were established in all Areas to meet the demands of the new customer-centred working environment. Customer front offices and counters were refurbished to give them a new, welcoming ambience, and workers with pleasant personalities and friendly dispositions were employed to manage the counters. Dressed in smart new uniforms, front office workers were strictly instructed to handle customers with utmost care and respect. The NWSC field staff in all Areas, including plumbers and meter readers, received training in the best customer-handling practices. They were also briefed on the nature and common types of complaints, which revolved around the accuracy of bills and meter readings, as well as the modalities of payment. Resolving queries and complaints expeditiously became one of the core components in our customer relations management.

In our training programmes, we emphasised that NWSC staff should always ask themselves the following questions. How should you respond to customers when they come for help or service? How do you disconnect customers gently without offending their pride or sensibilities? How would you like to be treated or served if you were in the customer's shoes? How do you cultivate the trust and confidence of the customer? What do you think are the likely costs to the Corporation and individual workers of ignoring, neglecting or being rude to customers? By addressing these questions, we inculcated customer-friendly attitudes and behaviour in our staff. We warned that ignoring or neglecting customers would not be tolerated.

Those workers whose temperaments were not suitable for serving customers were re-trained or deployed in other less customer-sensitive units, or where contact with customers was minimal. Workers who were deliberately and repeatedly rude to customers were reprimanded and directed to mend their ways or face the consequences. Workers whose attitudes and conduct did not change in the face of the new realities were asked to retire.

Building and enhancing customer confidence: In addition to putting the customer first, building and enhancing customer confidence and awareness were additional pillars around which all other SEREP activities revolved. As part of the Programme, NWSC management introduced and implemented a new customer care policy to empower customers to fight for their rights and to demand the provision of effective and efficient services. The new customer care policy was incorporated into a new water charter, which clearly spelt out the rights and obligations of customers, on the one hand, and those of the Corporation on the other hand. This pro-customer charter set the standards against which the success of the SEREP would be measured.

To start with, we initiated a concerted publicity campaign through newspapers, radio and television to sensitise customers about their rights and responsibilities. Posters detailing customer rights and how to pursue those rights were displayed in every NWSC office. Customers were also given information leaflets. The central message of our publicity campaign was that customers were entitled to prompt, efficient and affordable water and sewerage services. It was the customers' right and privilege to demand accurate information regarding, for example, how to access water and sewerage services, the cost of water production and distribution, the rationale for the water tariff and the basis on which it was computed, as well as the modalities of paying accurate and regular water bills. Customers had the right to point out the shortcomings in the delivery of our services and to demand immediate rectification.

On their part, customers had the duty to report leakages and bursts in the distribution system. They were expected to pay their bills immediately to enable the NWSC to improve its services, and to report any misbehaviour or misconduct by any field staff. They were obliged not only to refuse to give bribes to dishonest staff but to also report any such demands to the Corporation authorities promptly. Customers were also duty bound to refrain from collaborating with wrongdoers or to condone wrongdoing. Our publicity campaigns explained painstakingly that by stealing water the culprits were passing on the burden to legitimate consumers and that, therefore, it was imperative to break 'the culture of silence' by reporting the black sheep in the community.

To expand our customer base, it was important not only to attract new customers but also to woo back those customers who, for one reason or another, had failed to pay their bills. One way of doing so was to invite customers with suppressed accounts to come and negotiate new, flexible terms of payment, so that they could rejoin the system without feeling the pinch and could pay their outstanding bills in negotiated instalments. In addition, management reduced the reconnection fee from UGX 38,000 to UGX 10,000 in order to encourage willingness to pay, especially for those customers whose bills were much lower than the original reconnection fee.

Another initiative was to reduce or, better still, eliminate the number of illegal connections. For this purpose, NWSC management granted amnesty to all illegal consumers who voluntarily reported themselves. The main objectives were to eliminate wrongdoing without scaring potential customers away and to reduce non-revenue water and, by so doing, to generate more revenue. ADAPT, a public relations company, was commissioned to implement this amnesty initiative. It worked very well indeed and, by the end of the SEREP in July 2000, many illegal customers had officially been regularised into the Corporation's accounts.

Some innovative managers even devised ingenious ways of encouraging customers to settle their bills as well as of implementing the amnesty initiative. For example, Mr Joshua Kibirige, the Mbale Area Manager, combined personal contacts with customers and publicity to speed up the payment of bills and to regularise illegal connections. He sent out a written communication to all NWSC customers in Mbale thanking them for their support but, at the same time, reminding them of their contractual duty to pay their bills regularly to enable NWSC to continue rendering and improving water and sewerage services in the Mbale Area. He also went around Mbale town with a loudspeaker mounted on a pick up, appealing to people with suppressed accounts to report to his office to sort out their payment problems and reactivate their accounts. He encouraged illegal consumers to take advantage of the amnesty and regularise their connections. He continued to appeal to illegal consumers to come forward, even after the expiry of the official amnesty period. For his efforts, Mr Kibirige not only won the Chairman's Shield but was also given an accelerated promotion to Commercial Manager and posted to Kampala to assist the international private operator then managing the Kampala Area. This demonstrated that the Corporation had a pool of highly motivated and innovative staff and also that the management was committed to rewarding exemplary performers, both at headquarters and in the operational Areas.

In order to consolidate customer confidence in NWSC service delivery and to boost the image of the Corporation, it was imperative to know what the customers actually felt and wanted. Accordingly, several measures were adopted to address this important requirement. Two methods were instituted to determine customers' views, criticisms and suggestions regarding NWSC service delivery and how it could be improved. NWSC field staff in all Areas started visiting customers regularly, and the Corporation set up customer complaints desks and boxes to gather information. More importantly, for the first time in the history of the Corporation, management initiated comprehensive customer service surveys in all Areas to establish customer perceptions of our services and operations. The information was analysed and the findings were fed into the ongoing implementation of the SEREP. The shortcomings that were identified were rectified.

Striving to satisfy the customer: Focusing on the customer invariably demands creating and sustaining customer interest and satisfaction in the goods received or services rendered. In practice, this means that a producer of goods or provider of services must offer quality products or services in sufficient quantities or to acceptable specifications, meeting the tastes and expectations of consumers. The price of the products or services on offer must be affordable and competitive. Consumers must feel that they are getting value for money. The products or services must always be available as long as there is demand for them. The producer or service provider must also be prepared to meet any shifts in demand or in customer tastes. The goods or services must be packaged and delivered to the customer's specifications on time and in good order. A good business enterprise should also develop mechanisms to monitor the distribution of its products or the provision of its services continuously; to detect any changes in the market and to adapt to change accordingly, and to get feedback about the goods or services rendered.

In order to satisfy our customers, we strived to introduce the best service delivery practices in all our operations. Since the Corporation already had adequate capacity to meet effective consumer demand in all its Areas, the main challenge was not water production as such but how to optimally utilise the existing capacity by connecting more customers and, therefore, increasing water sales. This meant expanding the distribution network, wooing customers with suppressed accounts back into service and significantly reducing or eliminating illegal connections.

While the NWSC's capacity to meet the current demand was adequate, the issue of water quality and reliability were other factors that required attention. Among the key activities under SEREP were systematic maintenance of water and sewerage installations, the timely purchase of adequate spare parts and chemical inputs, the day-to-day monitoring of operations, expeditious repairs of leaks, bursts, overflows and sewage spillage, and compliance with the World Health Organisation (WHO) as well as with the Uganda National Bureau of Standards (UNBS) requirements regarding effluent from sewage treatment plants. We also tightened our quality control measures to meet water supply quality standards and to ensure more accurate measurement of our water production and reliability in all Areas.

Improving water distribution to satisfy the customer was more challenging than water production. Under the SEREP, NWSC Area managers did their best to connect more customers by extending the distribution mains and installing as many water kiosks as possible. We had to reduce NRW by repairing leaks and bursts as quickly as they were reported. In addition, we had to service and maintain the distribution system regularly to ensure continuous and reliable flow of water to our customers, and drastically reduce illegal water consumers in the system. A block-mapping exercise, which had begun during the 100 days Programme, had to be completed so that an accurate record of customers on the distribution network was readily available to NWSC workers.

Making financial ends meet: A business enterprise that aspires to deliver high-quality services to its customers must at least be able to break even financially. Utilities like the NWSC require considerable resources to operate, maintain, develop and sustain services. So how did the NWSC management intend to finance the SEREP? Since the Corporation was already heavily indebted to external creditors, we were determined to avoid piling up more debts. For this reason, we chose to finance the Programme from our own internal resources and within the framework of the 1999/2000 Budget. We had to make internal savings and budgetary reallocations in order to meet the SEREP financial requirements. This called for the intensification of revenue-enhancement efforts coupled with a whole range of cost-reduction measures.

With regard to revenue-enhancement efforts, SEREP emphasised improving billing and revenue-collection efficiencies, reducing arrears and attracting more customers into the system. Besides, all Areas were required to recover debts during the SEREP period and the bad debts, like those of the Departed Asians Custodian Board, were written off. At least 50% of the suppressed accounts were to be brought back into the system. Amnesty was granted to all illegal users and customer-friendly payment regimes were introduced. By implementing these measures, we hoped to increase our revenue by 15% and to achieve a minimum monthly collection of UGX 2.2 billion. We also hoped to persuade the URA to begin demanding VAT on revenue collections rather than on bills. Our aim was to plough back the surpluses from our revenue into the improvement and extension of water and sewerage services.

Furthermore, we took a deliberate initiative to reduce NRW by granting amnesty to all illegal consumers. It started in August 1999 and was later renewed twice, in September and in December. We looked at the amnesty as both a carrot and a stick approach to our customers. In granting it, we targeted not only illegal customers but also those who had been disconnected for failure to settle their outstanding bills. The carrot was that illegal consumers who took advantage of the amnesty would be forgiven and asked to pay the usual reconnection fee to regularise their accounts. The stick was that any illegal consumer who was apprehended after the end of the amnesty period would pay a fine of UGX 400,000, an estimated water bill for water consumed over the past 24 months, and also be liable for criminal prosecution. Therefore, through successive advertisements in the local electronic media from August to December 1999, we appealed to illegal consumers in all Areas to take advantage of the amnesty before it was too late.

Not all NWSC stakeholders appreciated the rationale behind the amnesty to illegal water consumers. On the face of it, the amnesty appeared to condone the malpractices of lawbreakers. Many people demanded to know why NWSC was 'favouring' illegal consumers at the expense of law-abiding consumers who faithfully paid their bills and kept the Corporation's operations afloat. Why was NWSC giving indiscriminate blanket forgiveness to wrongdoers? Wouldn't this amnesty set a bad precedent, which would encourage other people to default? Shouldn't illegal consumers be required to pay for their sins as a lesson to other would-be offenders that crime does not pay? Indeed, after reading one of the amnesty adverts in the newspapers, Mrs Miria Matembe, who was then minister for Ethics and Integrity, was outraged and demanded to know who had given the NWSC power to 'forgive the water thieves?' To her and many others, the culprits deserved to be brought to justice.

While I understood the sentiments of the critics of the amnesty for illegal water consumers, I did not believe that 'an eye for an eye' was the answer. In a poor country like Uganda, where a culture of non-compliance with the law is the norm and where civic virtue hardly exists, it was important to handle the issue of illegal consumers with the utmost care and sensitivity. Without effective detection mechanisms and public willingness to expose them, illegal consumers would have continued to operate underground and NWSC would have continued to lose water and revenue. My argument was that if illegal consumers voluntarily came forward and confessed their sins, so to speak, past misdeeds should be forgiven, if not forgotten, paving the way for a fresh chapter in their dealings with the Corporation. This realistic approach was likely to be more beneficial in terms of reducing NRW, bringing more customers on board and increasing revenue, than sticking on legal technicalities and formalities, which were difficult, if not impossible, to enforce.

Under SEREP, we introduced a wide range of cost-reduction measures. A car loan scheme to enable senior managers buy personal vehicles was introduced and a special allowance was established to cover the running and maintenance costs of those vehicles. The repair of the Corporation's operational vehicles was decentralised and outsourced. The new transport policy enabled the Corporation to determine its transport costs accurately in advance and to budget accordingly, leaving no room for hidden costs or manipulation.

Furthermore, the medical scheme was overhauled, limiting it to individual employees and their immediate families (spouse and a maximum of six children below 18 years of age).

In addition, through a voluntary retirement scheme during SEREP, the number of employees was reduced from about 1,430 to 1,260 in the first phase and then further down to 1,199 in the second phase. NWSC management phased out security guards as Corporation employees and instead outsourced security services. This gradual process of staff downsizing, as well as the concurrent restructuring of the Corporation, reduced not only the wage bill but also other staff-related expenses, such as medical and telephone, transport and staff welfare provisions. Of course, in the short run, the voluntary retirement exercise entailed terminal benefits. This partly explains why, compared to other activities, the cost of implementing SEREP I was very high.

SEREP rewards, gains and constraints

The SEREP was another big step forward in the Corporation's effort to sustainable service delivery. During the SEREP, the Corporation registered considerable gains but also encountered a number of bottlenecks, some of which were not anticipated. The culture of celebrating every achievement and rewarding outstanding performers, introduced during the 100 days Programme, was continued.

Rewards: During the SEREP period, monthly performance competitions between the Areas were sustained along the lines of the 100 days Programme. The management continued to reward the best performing Areas with trophies and cash prizes. The trophy and cash prize handover occasions continued to provide a convivial and relaxed atmosphere for board members, senior management, Area staff and customers to fraternise. The management used these occasions not only to recognise exemplary performance and to urge all Areas to aim higher but also to reiterate the importance of focusing on the customer and revenue collection enhancement on the one hand, and a general improvement of the NWSC corporate image and public relations on the other hand. The joy and pride with which NWSC staff and managers greeted achievements under SEREP were best captured by Mr Harrison Mutikanga, then Fort Portal Area manager, in his recollections:

> I remember very well the first mid-term evaluation workshop that took place in Fort Portal in November, and Fort Portal had emerged the best overall performing Area. I had indeed lived up to my expectations and proved those who were against my promotion wrong. From this day, management had confidence in my staff and me, and this motivated us to work even harder. We put Fort Portal on the map... from a chaotic and problematic Area to the best performing Area... The trophies in my office and the many workshops we had in Fort Portal at that time were a clear testimony that we were performing well. The public, local leaders and politicians in the Area were all proud of us and became 'raving fans' ... This kept us on our feet to work harder...

What Mr Mutikanga said of Fort Portal was true of other NWSC Areas across the country as well!

Gains: In general, SEREP achieved its prime goal of consolidating the gains of the 100 days Programme and sustaining the momentum of revitalising the NWSC through concerted and progressive improvements in corporate work methods, habits and attitudes. The main quantitative performance achievements of SEREP as summarised in Table 3.2 clearly show that it maintained and in some cases exceeded the achievements of the 100 days Programme. First of all, SEREP surpassed its original target of increasing the number of break-even Areas to six. By the end of SEREP, in addition to Kampala, Jinja, Entebbe, Mbale and Mbarara, both Masaka and Fort Portal were all breaking even, raising the total number of break-even Areas to seven. Besides, by the end

Table 3.2: SEREP I and SEREP II quantitative performance

Indicator	100 days Performance			SEREP Performance		
	Kampala	Others	Overall	Kampala	Others	Overall
NRW (%)	51	42	49	46	37	43
Staff Productivity (staff/1,000 connections)	20	37	27	15	25	18
Proportion of Inactive Accounts to Total Accounts (%)	35	25	32	30	21	26
Collection Ratio (%)	93	99	96.2	94	81	89.3
Metering Efficiency (%)	77	85	80.7	85	84	85
Capacity Utilisation (%)	70	53	59	74	54	65
Operating Surplus (UGX million/month)		350			250	
Number of towns covering operating costs except depreciation (out of 11)		5			7	

of the Programme, all Areas were able to meet their personnel-related costs. This was a commendable achievement and an indication that the Corporation was moving in the right direction in our quest for financial viability and sustainability.

Secondly, we recorded substantial improvements in capacity utilisation of our water treatment plants, the quality of water delivered to our customers, reduction of the frequency of sewage spillages, effluent quality of sewage treatment works and the procurement of equipment for water and sewage treatment plants as well as for our laboratories.

During SEREP, we ensured that the distribution system was maintained regularly to guarantee continuous flow of water to our customers. Response time to technical failures like bursts and leaks in the distribution network was maintained below 24 hours on average in all Areas. The overall NRW performance was reduced significantly and the customer base increased by 13.5%.

We also extended the water distribution network by 59 km, installed 97 water kiosks, especially for the benefit of poor communities, and added over 4,200 new connections. Through a combination of increased water connections and the implementation of the voluntary retirement scheme, substantial gains were made in staff productivity.

On the other hand, the gains in revenue improvement during SEREP were not as good as we had hoped. The collection ratio during SEREP was lower than what we achieved at the end of the 100 days Programme. This was partly due to low staff morale because of the contemporaneous retrenchment exercise, which created uncertainty about job retention and hence dampened efforts to collect revenue. Similarly, relaxed payment from government ministries also hampered the overall revenue collection performance. During this period, government institutions accounted for about 30% of total collections and the corresponding collection ratio from government institutions averaged only 79% compared to 95% from other customer categories.

The cost-reduction measures that we implemented during SEREP resulted in reduction of our monthly expenses from UGX 1.858 billion during the 100 days Programme to UGX 1.744 billion during SEREP. As a result of these initiatives, we continued to make progress with respect to the operating surplus during the SEREP as shown in Table 3.2.

The biggest gains of SEREP were more qualitative than quantitative in that by the end of the Programme customer confidence and satisfaction in NWSC had been enhanced beyond the most optimistic expectations or predictions. None of the architects of the Programme could have anticipated a fundamental shift in customer attitudes and perceptions in favour of the Corporation within one year. The results of a comprehensive customer service survey, carried out at the end of

SEREP in October 2000, were very impressive. These results revealed that the vast majority of our customers were satisfied with the quality, reliability of our water and sewerage services, the concerted management efforts to improve the corporate image, the frequency and accuracy of meter reading and billing and, above all, the quality of customer services.

The customer service survey was useful in two ways. Not only did it show the success of our strategy of focusing on the customer, but it also revealed some weaknesses that required attention. For example, more than 50% of our customers complained that they did not get sufficient interruption notice. Only Kasese and Mbale had satisfactory ratings of over 60%. Similarly, 45% of all respondents complained that the Corporation's officials did not respond to complaints expeditiously. According to the customer service survey, water services in the Gulu Area were still poor probably due to the frequent interruption of power supply as well as insecurity – factors that were, needless to say, beyond the Corporation's control.

The customer service survey also taught us that we had to continuously improve our communication to customers to get reliable feedback and to address those complaints as promptly as possible. Customer surveys have since become part and parcel of our operations. The management uses customer perceptions to inform its policies, decisions and activities in order to incrementally and consistently improve our services and respond to customer demands.

Constraints: It would be misleading to give the impression that SEREP was a smooth, faultless operation which went according to plan. Like all programmes, progress during the implementation was bumpy, owing to a number of constraints, some of which were not, or could not have been, anticipated. For one thing, the progress of the Programme was slowed down by the inadequate budgetary provision for materials, equipment and other inputs necessary to achieve the SEREP targets. Even where the funds were available, the spare parts and inputs were not delivered on time. This was, in fact, a carryover from the 100 days Programme. In addition, the time span for some of the SEREP activities was too short, and that is why some of its anticipated targets were not realised.

Staff motivation and morale also fell short of the SEREP expectations due to inadequate training, facilitation and remuneration. As has been mentioned already, the voluntary retirement scheme had created a sense of panic, insecurity and despondence among the workers, some of whom feared for their jobs and whether they would get their benefits quickly if and when they retired. Besides, workers due for retirement had no motivation to work as diligently as would be expected. They simply marked time pending their exit from the Corporation. Despite the management's efforts and promises to settle retirement benefits as quickly as possible, many of the workers were not reassured. Our revenue improvement efforts during SEREP were to a large extent neutralised by the government's inability to settle its bills, including arrears. The tariff review, which we had submitted to the Ministry of Water, Lands and Environment was not approved as we had hoped, and this invariably constrained SEREP performance and outcomes.

Job well done, but more challenges

All in all, during SEREP, we did very well in some aspects of water production, water distribution, customer care, revenue enhancement and cost reduction, but not so well in other aspects. For example, we achieved our targets in response time to blockages, leaks and bursts and in staff productivity. However, we did not meet our targets in water production, NRW, billing and collection efficiencies and the reduction of arrears. Nonetheless, we did well enough to sustain the momentum of the 100 days Programme by focusing on our customers and striving to improve corporate financial viability. By the end of the Programme, the NWSC had gained sufficient strength and confidence to pursue and fulfil its corporate mandate on a continuous and financially sustainable basis. Personally, I was happy and proud of the SEREP achievements. I was proud of my whole team, from the top management and lead champions down to the ordinary members of the staff in the Corporation, for their invaluable contribution to yet another successful programme. We had the courage to move on.

Although I was satisfied with what SEREP had achieved and with our growing corporate momentum through concerted teamwork, I had no illusions about the enormity of the challenges ahead of us. One of the immediate challenges was the future of the Corporation in the context of government policy of divestiture and privatisation of public enterprises. Was NWSC a candidate for divestiture and privatisation?

Indeed, in view of the progress that the Corporation had made during the 100 days and SEREP, the government and the donors were prepared to defer the privatisation debate and give NWSC a chance to continue implementing the change management programmes, which were beginning to show remarkable success. However, even with the goodwill from the government and the donors, there was no guarantee that the Corporation was going to continue on the same positive performance trend, especially given the pitfalls encountered during SEREP. Similarly the Corporation was also not certain about the government's commitment to supporting the internal reforms without any interference and paying its water bills on time. There was therefore a need for a clear framework through which both parties could commit themselves towards fulfilling certain obligations.

In this respect, the government, the Board and the management of NWSC chose to enter into a performance contract. This contract would bind NWSC to deliver services to the satisfaction of the customers at a level negotiated and agreed by both parties. On the other hand, the government would in return support the internal reform initiatives of the Corporation and guarantee us a certain level of autonomy. Since we had proved our performance potential, we were ready to take on the new challenge of the government performance contract. This is the subject of the next two chapters.

SEREP was another learning experience, very much like the 100 days Programme and some of the lessons drawn from it are listed in the following box.

Lessons learnt

- Performance improvements can be compared to ascending a ladder, the steps are interconnected and need to be taken one at a time – programme-by-programme – for an organisation to achieve its goals and objectives.
- Managers must continuously think of new ways – of new programmes – to sustain the momentum, knowing very well that every success is pregnant with new challenges and expectations.
- Managers must be proactive in building on past achievements and turning constraints into opportunities.
- Customers are the geese that lay the golden eggs. Make it your ultimate responsibility to know their needs, preferences and expectations, and cultivate their support for a continuous and sustainable performance improvement.
- If properly done, shaking up organisations through restructuring and right sizing results in efficiency gains, enhanced productivity and improved financial performance.
- Always look out for new incentives (like the introduction of trophies, for example) to recognise and reward outstanding performance.
- Many efficiency gains can be achieved by considering strong cost-containment measures.

Down to work! NWSC staff of Lira Area replacing worn-out parts of a service line. Through such activities, the Corporation was able to address the high levels of NRW.

Improved maintenance practices: regular maintenance of the network and all installations resulted in provision of reliable services to customers. Right: staff of Mbarara Area servicing a valve in the network

Below: staff of Entebbe Area flushing a transmission main to ensure customers get quality water.

Focusing on the customer: publicity and public awareness campaigns formed a big part of the 100 days and SEREP. Above: Entebbe Area staff using innovative means to sensitise and communicate with customers.

Customer sensitisation through strategic alliance meetings. Above: the management of Masaka Area in one of the strategic alliance meetings with a local community.

Left: the branch manager of Najjanankumbi Branch, Mr Richard Muhangi, addressing customers during customer sensitisation drive.

Performance reviews and competition were among the main channels to drive and transform the Corporation.

Left: the Area Manager of Lira, Mr Charles Ekure, receiving the overall winner's trophy from the Board Chairman, Mr Sam Okec (second left), as the MD (third left) cheering him on.

Right: the winners of the Area competitions in a jolly mood as they celebrate and proudly show off their trophies.

Above: the winners for the Area competitions posing for a group photo with the management, Board members and key stakeholders.

Recognition did not stop with the staff; Right: a customer of Jinja Area riding off happily with a brand new bicycle in recognition of his support to the Corporation.

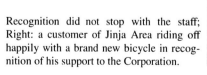

Focusing on the Customer and Financial Viability | 69

Celebrating every achievement: the Corporation used the evaluation ceremonies to celebrate its achievements and interact with different stakeholders. Dr William Muhairwe (centre) with NWSC staff and other stakeholders during bull roasting as part of the performance-evaluation ceremony.

Celebrations were part of the 100 days Programme and SEREP. Above: staff celebrating with a dance after hard work.

Chapter Four

A Pat on the Back

First Government Performance Contract

On 31 August 2000, the Government of Uganda and [the] National Water and Sewerage Corporation (NWSC) signed a performance contract for three years permitting NWSC to function as an autonomous parastatal. It is the first parastatal to enjoy this status in recognition of its good performance...

The Monitor, 22 October 2000

The transformation of NWSC's operations by the 100 days Programme and SEREP in less than two years created a positive atmosphere of good feeling among the urban water sector stakeholders both inside and outside the Corporation. Under these programmes, the NWSC leadership had turned around 'one of the most dysfunctional corporations' in Uganda, a fact that was recognised by the African Almanac when I was rated among the top 100 Africans, together with three other Ugandan managers and a high court judge, for good performance in 2001. *The New Vision* of Monday, 5 March 2001 carried a story which, in part, read

> '... four Ugandan Directors and a High Court Judge have been rated among the top 100 Africans of the year. ... the criterion for selection was based on concrete achievements, especially those with an impact on a large number of people ... under the leadership of Dr Muhairwe, what has been an inefficient state-owned corporation for more than 25 years, was turned into a profit-making agency, something almost unheard of in the East African country...'.

By the end of SEREP, the NWSC Board and management had demonstrated beyond any shadow of doubt that, with a clear vision and a sense of purpose, as well as focused programming, it was possible to turn the Corporation into a commercially viable institution. The radical improvement in the performance of NWSC had proved that it was possible to 'make money out of water' in a developing country, as *The New Vision* columnist, Gawaya Tegulle, wrote in the issue of 1 January 2001.

Our staff not only became more confident but also more customer friendly by internalising the culture of putting the customer first. On their part, our

© 2009 William T. Muhairwe. *Making Public Enterprises Work*, by William T. Muhairwe.
ISBN: 9781843393245. Published by IWA Publishing, London, UK.

customers were beginning to trust us a little more with the improvements we had registered, in the quality and delivery of water and sewerage services. They also became more assertive in presenting their complaints, comments and suggestions. The government and the donors were pleased that, at last, we were on the right course and as such we deserved 'a pat on the back' in the form of financial and policy support. The role of the government in the recovery of the Corporation is discussed in detail in Chapter 13, while the role of the donors we partnered with is dealt with in Chapter 12.

In this chapter, I demonstrate that, after a good start of a turnaround process, key stakeholders such as the government can give you a pat on the back by facilitating a better enabling environment to perform. In our case, the government entered into a performance contract with the Corporation, the details of which are the focus of discussion in this chapter.

A nod to continue

In appreciation of the Corporation's good performance, the government, as the sole shareholder, lender of last resort and policy maker, entered a new relationship with us. This relationship was defined by a performance contract – the first of its kind in Uganda's public enterprise history. The Corporation was given operational freedom to pursue its programmes geared at enhancing efficiency, cost-effectiveness and delivery of quality service in all our operational Areas. The performance contract, commonly referred to as the first performance contract or PC I, triggered the concept of 'contractualisation' of the management and governance of water and sewerage operations. Hitherto the conventional thinking had been that performance contracts between government and public enterprises could never work. The introduction and eventual successful implementation of the performance contract, as demonstrated later on in this chapter, proved that contractualisation with well-thought-out objectives can lead to improved performance, irrespective of the contracting parties.

This performance contract was most welcome for several reasons. First, it demonstrated that the government had confidence in the leadership of the Corporation and that this was a way of saying 'thank you' for a job well done in turning NWSC around. This view was independently expressed by the media. An article published in *The Monitor* of 22 October 2000 entitled 'Gov't gives NWSC a big thank you', quoted at the beginning of this chapter, captures the mood and attitude the public had towards our performance.

Secondly, it was a challenge to the Corporation's leadership to keep the change management momentum gained during the 100 days Programme and SEREP on the right course in the drive towards full-fledged commercialisation. Thirdly, it signified the redefinition of government–NWSC management relationship and commitment by both parties to achieve incremental and sustainable improvement of the delivery of water and sewerage services in the medium and long terms.

The preparation and negotiation of the performance contract took about four months. The Ministry of Finance, Planning and Economic Development drafted the contract and it was subsequently discussed in detail by all parties involved. On our side, we proactively established a core negotiation team in April 2000, which comprised managers from the finance, commercial, human resources, engineering and corporate planning departments. By establishing this team, I wanted to ensure that the eventual terms of the contract were agreeable and binding on all the departments. Each department was consequently asked to make an input as far as their obligations in the contract were concerned.

During the negotiations, it was evident that the Finance Ministry was ready for a programme that would ensure incremental performance. After the negotiation team had thoroughly analysed and discussed the contract terms, the performance contract was presented to the Board in July 2000. Because of the importance attached to the contract and the novelty of the whole concept, the Board approved the contract within the same month despite its busy schedule. The performance contract was then sent to the Ministry of Water, Lands and Environment for comments and approval. The Ministry consented to the performance contract, referring to it as ambitious and a step in the right direction.

Under the contract, the government granted the Corporation full commercial freedom in all its operations, including capital expenditure, commercial borrowing for investment, determining staffing and salary levels and proposing appropriate tariffs. On its part, the Corporation undertook to use its managerial, professional, technical and commercial skills 'to oversee the development of policies, strategies and plans' to fast track full commercialisation of the Corporation. The NWSC was to work closely with the Ministry of Water, Lands and Environment and that of Finance to ensure the realisation of the performance contract objectives. Box 4.1 below presents a summary of the objectives that the performance contract aimed to achieve.

Box 4.1: The PC I objectives

> The objectives of the PC I were to
>
> - Sustain the NWSC's financial and commercial viability;
> - Grant NWSC operational and commercial freedom in the quest to fast track full commercialisation;
> - Make the Corporation more attractive to potential investors;
> - Maximise the utilisation of the managerial, professional and technical skills of the staff of the Corporation;
> - Enhance productivity, cost-effectiveness, accountability and customer care and
> - Extend water and sewerage services to underserved areas, especially the urban poor.

Performance contract targets

The PC I spelt out a detailed and comprehensive list of performance targets that the Corporation had to achieve between July 2000 and June 2003. These targets required internal reorganisation of the Corporation to make it more efficient, cost-effective and more responsive to customer demands and expectations. The reorganisation included, among others, strengthening management efficiency in water production and distribution, financial management and accountability, human resources management and development, and customer service delivery. In each of these key areas, the government and the Corporation agreed on specific targets and deadlines to be achieved. Although we knew that achieving these targets was a tall order, we were, nevertheless, determined to deliver in order to consolidate the newly won government support and confidence.

The terms of the performance contract required NWSC to deliver water and sewerage services to a target population of over two million people in 12 towns. The key performance targets for each year are shown in Table 4.1. The operational efficiency gains were envisaged to translate into an increase in revenue generation and a reduction in operating expenses and hence an increase in the operating surplus of the Corporation.

In addition, it was agreed that, during the contract period, the NWSC would initiate a major distribution mains network programme to extend water services to the urban population, especially the underserved urban poor. The agreed target was to connect a minimum of 18,800 new consumers by 30 June 2003. The total number of NWSC staff was to be reduced from 1,200 in June 2000 to 800 by 30 June 2003, targeting a staff productivity ratio of 10 staff per 1,000 connections by the end of the contract.

Table 4.1: Performance contract targets for 2001 to 2003

Performance indicator	Year 2000 Base Year	Year 1 2001	Year 2 2002	Year 3 2003
NRW (%)	43	39	37	35
Staff Productivity (staff/ 1,000 connections)	18	16	13	10
Proportion of Inactive Accounts to Total Accounts (%)	26	20	17	15
Collection Ratio (%)	89	89	89	92
Metering Efficiency (%)	85	85	90	97
Capacity Utilisation (%)	65	63	65	66
Operating Surplus (UGX million/month)	**250**	**394**	**848**	1,090

Apart from the quantitative targets, the performance contract also contained a wide range of ambitious qualitative internal reforms to be carried out by the end of the contract period. These included the design and implementation of a water distribution network plan, the appointment of a distribution network extension team and the establishment of a marketing/customer liaison group. The Corporation also hoped to develop and implement a coherent customer policy to promote customer awareness and to work out a customer charter containing guidelines regarding our relationship with customers. Furthermore, the Corporation was required to prepare a master plan and an integrated information system focusing on incremental improvements in the quality of services delivered, as well as to review and improve fault reporting and data collection procedures to address customer complaints promptly.

In the area of operations management, the NWSC was expected to update the assets register and to ensure effective management of its facilities and infrastructure, and the optimisation of financial and human resources. In the same regard, the Corporation agreed to refine the materials and works management systems in order to improve utilisation. The Corporation was expected to ensure that its staff internalised the planning processes and procedures aimed at improving water distribution efficiency, cost-effectiveness and customer responsiveness. The Corporation was also required to improve the regularity and reliability of the delivery of water and sewerage services to the satisfaction of its customers.

The Corporation undertook to initiate and implement a series of internal reforms in human resources management, information flow, financial controls and regulations, and compliance with the best international accounting standards and practices. In regard to human resources management and development, we decided to introduce a bottom-up approach to increase staff productivity and participation. We also undertook to review and rationalise the existing remuneration structure and to introduce a new performance appraisal scheme covering all members of staff, including top managers. As part of capacity building, we committed ourselves to the introduction of a corporation-wide staff training and staff development programme to enhance the workers' managerial and supervisory competencies.

Both the government and the Corporation leadership recognised the importance of developing and using effective management support tools and systems. The NWSC was required to introduce new information technologies, especially computer hard- and software applications to collect and analyse data to inform the Corporation's ongoing decisions, strategies and policies.

We were also required to review and, where necessary, overhaul the existing financial controls and procedures in budgeting, revenue collection and expenditure to ensure optimal and cost effective utilisation of available financial resources.

Financing and monitoring of PC I

Realisation of the NWSC contractual commitments would have far-reaching financial implications for us as an organisation. To fulfil its contractual obligations, NWSC required additional financial resources over and above its recurrent and capital expenditure provisions. Neither the government nor the Corporation had the requisite money to support the implementation of the performance contract. At the same time, commercial borrowing was out of the question because our indebtedness was still very high. Therefore, it became necessary to devise other ways by which to finance the performance contract. As a first step, the government agreed to maintain a freeze on the NWSC debt-servicing obligations for the duration of the contract. Secondly, the government undertook to subsidise our activities in the contract, which were primarily driven by a 'social-mission' objective rather than commercial considerations. Thirdly, the government committed itself to clearing all arrears incurred after 30 June 1999 and to ensure that subsequent bills were settled on demand. Both the government and the Corporation's management were optimistic that a combination of internally generated resources, savings from debt servicing and grants from development partners would suffice to finance the performance contract.

With effect from the 2000/2001 financial year, NWSC was required to submit regular performance reports to the government in accordance with the terms of the contract. These reports included the three-year strategic corporate plan, annual budget, annual corporate performance reports, audited accounts, monthly returns, quarterly management accounts and any other information required by government to assess our performance in light of the contract.

The government and the Corporation agreed to establish a four-member Performance Contract Review Committee (PCRC) to monitor and evaluate its implementation. The Ministry of Finance, Planning and Economic Development on the one hand and that of Water, Lands and Environment on the other were each represented by one officer while two external directors represented NWSC. The Ministry of Water, Lands and Environment was assigned the duty of working out the specific terms of reference for the committee. Although the PCRC was not established until October 2002, this did not in any way affect the progress of the implementation of the performance contract. In actual fact, the PCRC was constituted in time to evaluate actual NWSC achievements against the contract targets.

The contract gave us an opportunity to continue with the change management initiatives which had enabled us turn the Corporation around. As mentioned earlier, this contract was a vote of confidence in the Corporation's efforts to deliver on its mandate and an endorsement of what had been achieved by the internal reform initiatives. In fact, many of the activities that the Corporation undertook to accomplish under the contract period were already under way, in some form, by the time the contract was signed on 31 August 2000.

It is not surprising, therefore, that the entire NWSC community embraced the performance contract with single-minded enthusiasm and determination. At the signing of the contract, I confidently told the press that the Corporation was prepared to face the challenging targets with resolve so as to succeed in spite of the operational obstacles we were likely to encounter along the way. We therefore initiated Area Performance Contracts (APCs), which are the subject of Chapter 5, as a tool by which the objectives of the first performance contract would be achieved. The APCs were to cover all NWSC Areas except Kampala, which was under the management of a private operator. In this connection the operationalisation of the performance contract was to follow two paths; the first was to be through the introduction of APCs and the other would be through private sector participation (PSP) in Kampala.

These APCs were to be effected between the Corporation, as the employer, and the Areas as the implementing quasi-autonomous business units. The Areas were the operational microcosms of the performance contract. The APCs were intended to transform NWSC operational Areas into autonomous business-like commercially oriented units. However, they were also designed to overcome the shortfalls of SEREP and to eradicate bottlenecks that had impeded the full realisation of the SEREP objectives. The Corporation signed 11 APCs, excluding Kampala.

PC I implementation

The implementation of the one-year renewable APCs was concurrent but not conterminous with the performance contract. While the performance contract ran from July 2000 to 30 June 2003, the effective date of APCs was October 2000. Initially, the first APCs (APC I) were supposed to end in September 2001, but were later extended to 30 November 2001 to allow ample time to develop a new, comprehensive batch of APCs (APC II), which ran from 1 December 2001 to the end of November 2002. The third and last of these APCs (APC III) ran from January to December 2003, after which this change management approach was replaced by a more radical approach, the Internally Delegated Area Management Contracts (IDAMCs), which are discussed in detail in Chapter 8. The performance contract and APCs were inseparable sides of the same coin. The success of the performance contract was dependent on the corresponding success of the APCs. I regarded the signing of the performance contract as a win-winsituation for both government and NWSC. Chapter 5 describes how the performance contract was implemented. The box below highlights some reflections on the signing of the performance contract.

Reflections on the signing of PC I

- Contractualisation with well-thought-out objectives and incentives, irrespective of the contracting parties, can lead to improved service delivery.
- A contract is a gesture of goodwill between two parties. It does not necessarily have to be legalistic. If it is, long and disappointing divorce processes become unavoidable.
- Fairness and transparency between the parties are key ingredients of the success of any contract.
- Contracting parties need to participate actively in the design process of the contract to be able to internalise and appreciate its objectives and the respective roles and responsibilities of each party.
- A framework for operationalisation of the contract should be put in place by both parties to ensure its successful implementation. Even then, the framework is not an end in itself! Flexibility, transparency and fair dealing in implementation are key factors.

Chapter Five

Swinging the Pendulum from the Centre to the Grassroots

Area Performance Contracts

The government agenda of modernisation sets challenging new performance objectives for organisations, from the delivery of high-quality services that meet the needs of their customers in doing more within the constraints of available resources through continuous improvements in the ways organisations operate. Performance management underpins the operations and processes within a strategic change management programme framework. Sound practices and targets, which are both flexible and reactive to change, are needed to achieve performance improvement.

Author's observation in managing a public enterprise

After signing of the first performance contract with the government in August 2000, we were faced with the challenge of implementing it without losing focus of the direction and achievements we had recorded by the end of SEREP II. As pointed out at the end of Chapter 4, we were keen to realise the performance contract targets at Area level using the Area Performance Contracts (APCs).

Before the advent of the APCs, the Area operations had suffered from too much centralisation, compartmentalisation, sluggishness and bureaucratic red tape, among numerous other constraints. For example, top managers at the headquarters made trivial decisions regarding personnel, procurement, vehicle repairs and, above all, expenditure, which could have been carried out meaningfully at the Area level. This excessive centralisation of decision making and programme initiation left the Area managers little room and hardly any chance to manoeuvre and, as a result, were suffocated of their initiative. In addition, Area managers could not easily be held accountable for good or poor performance. There were no rewards for good performers or penalties for underachievers. All in all, Area managers, at that time, tended to behave like traditional civil servants rather than commercially oriented managers whose challenge was to improve performance and pass on efficiency gains to customers. We agreed that it was no longer desirable and, most importantly, sustainable, for NWSC Area managers to continue walking in the shadow of the centre. Time had come for the NWSC headquarters to delegate many of the

© 2009 William T. Muhairwe. *Making Public Enterprises Work*, by William T. Muhairwe. ISBN: 9781843393245. Published by IWA Publishing, London, UK.

responsibilities to Area managers in the interest of enhancing effectiveness and efficiency. This called for the redefinition of the relationship between the NWSC headquarters and the Area managers. Therefore, this chapter goes into details of how the pendulum had to be swung from the centre to the grassroots through a series of Area Performance Contracts (APCs).

Background of the APCs

Although the APCs were the operational arms of the first performance contract, their history goes far back to 1999, when I met and first discussed the idea of introducing contractual arrangements within the NWSC with a team of officials from the World Bank, led by Mr Alain Locussol. The Bank had been one of the key development partners in the urban water sector since the 1980s. During our discussions, we shared mutual concerns about the shortfalls of the change management programmes implemented under the 100 days Programme and SEREP, and the impending contract with the government. We concluded that it was important to devise new strategies and to revitalise our management efforts to sustain the momentum we had gained and to push forward towards the goal of full commercialisation. It was during the meeting that we contemplated the possibility of empowering the Area managers to run their own shows within an agreed contractual framework by which the managers would be held accountable for their performance.

At about the same time, negotiations to establish a performance contract commenced between the government and the NWSC management. The negotiations that culminated in the signing of the PC I in August 2000 gave our contemplation about the future of the Corporation after SEREP a new sense of urgency and direction. The fact that the preparation of the performance contract was a participatory process made it possible for us to have a good idea of what was happening and, therefore, we were able to realign our plans and activities in anticipation of the materialisation of the contract. In an effort to keep or even speed up the change in the management momentum that had begun under the 100 days Programme and sustain through SEREP, we had to discuss the APCs alongside the PC I. This enabled us to harmonise the performance objectives of the APCs with those of the PC I and eliminate the possibility of any performance gaps that would be created by delayed implementation of the APCs. During discussion of the APCs, we decided to set higher APC targets than those set under the performance contract.

Drafting the APC framework

After meeting the World Bank officials, the idea of empowering the Areas preoccupied most of our discussions and planning for whole of the year 2000.

I appointed a committee comprising senior managers from the Head Office, including the Corporation secretary, and some Area managers. The committee was expected to design a contract framework which would form the basis of our discussions, taking into account the need to address SEREP shortfalls and the objectives of the PC I whose negotiations were in progress at that time.

The committee submitted its first draft for discussion in March 2000. When I received it, I sought advice and comments from many people both inside and outside the Corporation, including Dr Richard Franceys, at that time an associate professor of Sector Utility Management at the Institute of Infrastructure and Hydraulic Engineering (IHE), Delft, the Netherlands. Dr Franceys, a renowned institutional development expert and a friend of the NWSC, had already done a lot of research work in utility management and also trained the Corporation's senior managers in 1994.

I have singled out Dr Franceys because his comments on the first draft were very instructive in shaping the subsequent revision and refinement of the APCs. In summary, he welcomed the idea of introducing APCs between the Area managers and the NWSC management as a radical innovation, but warned against the dangers of making the contracts too legalistic. He recommended that, given the interdependence of the Areas and the Corporation's management, and the limited development of management systems and procedures at the Area level, the proposed relationship between the Areas and the Head Office should take the partnering mode rather than the one based on very strict technical and legalistic constructs. He suggested that the draft be revised to tone down the exaggerated legalese and to produce a new document in a more practical, realistic and clear understandable language. He further advised that the APC drafting committee should benchmark with the United Kingdom partnering contracts, which were undergoing pretesting with reasonable success at the time.

Armed with the comments and suggestions, the drafting committee went back to the drawing board to fine-tune the APCs. Emphasis was put on aspects that could help improve performance rather than those that would open the gates of obstructive and self-defeating litigation. Accordingly, the committee worked in the direction of removing unnecessary legal technicalities and to spell out clearly the obligations of the Corporation's management on one hand and Area managers on the other hand, to achieve common objectives. The final draft framework of the APC was ready by May 2000.

APC secure staff support

Having finalised the APC framework, it was important to secure the confidence, support and cooperation of Area managers and the entire staff, both in the field and at the headquarters. From our experience during the 100 days Programme and SEREP, we had learned that new ideas, innovations and approaches invariably provoked resistance unless they were successfully sold to all the stakeholders. These fears were alluded to by Mr Joseph Kaamu who stated in his recollections of the APC experience that

> Many people were at first sceptical, calling it a veiled attempt to privatise the organisation, with the attendant threats of lay-offs and downsizing. Fortunately the Union was involved right from the start and the benefits to the staff and the Corporation as a whole were self-evident.

In order to avoid any unforeseen resistance, our initial strategy was to try out the implementation of the APC in one or two pilot Areas before extending it to all Areas. For this purpose, we chose Kasese and Fort Portal Areas, which were operated jointly by one manager, Mr Harrison Mutikanga. He was and he is still among the Corporation's hard working and enthusiastic managers, open to new ideas. He had earlier indicated that the performance of the two Areas could have been improved through reduction of what he referred to as 'very high and unnecessary operating costs', but nothing could be done, owing to the centralised decision making. As a manager, he was well disposed to experimenting with new management methods and innovations. He was the ideal man through whom the APC would be sold.

On the other hand, we had another experienced, calm and open-minded Area manager called Mr Joel Wandera, well known for his great skills in mobilising staff for the implementation of new initiatives. He had just been transferred from Jinja to Masaka. During the short time that Mr Wandera had been in Masaka, he had already come up with innovative ideas to improve performance, for example, his ideas on how to rationalise staff and medical costs and to increase revenues drastically through a number of demand-driven water extensions were very impressive. According to him, the only perceived missing link to his proposed improvement ideas was independent decision-making power. Because of this, Mr Wandera was another potential champion for the implementation of the APC.

In May 2000, I called the two managers for a meeting during which I attempted to sell the APC idea to them. It was attended by some World Bank officials on a mission in Kampala. When I introduced the idea, the World Bank officials gave their strong endorsement since they had already bought the concept. I highlighted the objectives and expected benefits of the APCs and stressed that they were a panacea to the constraints observed in the previous performance enhancement programmes.

Mr Wandera was clearly reluctant; he indicated that he would need more time to think about the subject and to consult with colleagues in his Area before giving a definitive response. We decided not to push him against his wishes.

Mr Mutikanga was more enthusiastic, though guarded in his response. On the one hand, the prospect of exercising autonomy in his Areas was too attractive to ignore because it would give him scope and space to innovate and make decisions without fear of being overruled by the Head Office. On the other hand, he did not relish playing the 'guinea pig' in this APC experiment. Why did the MD pick on Kasese and Fort Portal of all Areas? What if the experiment did not work? Despite his reputed charisma and propensity to stand out in the management crowd, on this occasion Mr Mutikanga was rather hesitant to grab the opportunity of being a pacesetter in the transformation of management across the Corporation. That is why I taunted him as the 'coward manager', a man who, for all his talents and leadership qualities, appeared too scared to play the pilot role in kick-starting the APC. He recalls that

> ...business planning and performance contracts were first introduced using Fort Portal Area as pacesetters (guinea pig). It is under this programme that I got my nickname of "Coward Manager". I clearly remember the day I was baptised with a new name of "Coward Manager" by the Reverend MD and Godfather Ato Brown at Church-MD's office. I was too scared of change by then especially accepting more autonomy and getting ready for PSP-like ...

Jokes aside, I assured the 'coward manager' that there were sound reasons for choosing him to swing the management pendulum from the centre to the grassroots. I urged him to consider the piloting challenge as an opportunity for Kasese and Fort Portal to be at the forefront of the Corporation's quest for management excellence. In the end, my exhortation did the trick. By the time he left the meeting, he had not only accepted the challenge but had also agreed to do the preparatory groundwork within the next two months.

Taking on the APCs mantle

In June 2000, as a follow-up of the May meeting, we invited Mr Mutikanga and his management team to Kampala for a two-day introductory and harmonisation workshop. The headquarters team, which had been spearheading the APC drafting process, discussed the idea of the APC with Mr Mutikanga and his team in greater detail. In this workshop, I specifically pointed out that one of the major shortcomings of SEREP was lack of sufficient, deeper and broader participation by Areas in the Corporation's decision-making and management processes. I also pointed out that the main purpose of the APC was to empower Area managers to handle the affairs of their own Areas in close partnership with the management at the Head Office. The presentations in the workshop made it clear that the contracts would separate and specify the obligations of the headquarters, on the one hand, and those of Area managers, on the other, in the delivery of water and sewerage services. The design of the APC was such that both parties would agree on performance indicators and set SMART targets for the realisation of the contract objectives. I also explained that the contracts would carry the 'carrot and the stick' instrument in the form of incentives and penalty mechanisms; outstanding performers would be rewarded with trophies, cash prizes and performance bonuses, while underachievers would be penalised in accordance with the penalty provisions of the contract.

Mr Mutikanga and his team were quick to embrace the incentive side of the contract but did not relish the proposed penalties, which included the dismissal of the Area manager who showed consistent failure to live up to the contract expectations. According to him *'The disincentive element* in the contract was so scaring as we had to lose part of our salaries and be reprimanded (withdrawn) for non-performance.'

During the tea break, having realised that our manner of introducing the subject had perhaps been too blunt and abrasive for comfort, I guided the team from the headquarters to tone down the language at the next stage of the discussion. We agreed that we should strategically, as part of the change management process, play down the penalties and emphasise what was at stake. Since performance improvement was the ultimate objective, we reached a consensus that 'blame storming' by imposing penalties for its own sake was not the desired goal, but rather that the purpose of the APC was to put in place systems that would henceforth encourage Areas to improve performance and enhance customer satisfaction.

When we resumed, the Corporation secretary and I assured Mr Mutikanga and his team that the penalty would be used as a last resort when everything else had been tried and failed to deliver the desired results. I remember the Corporation secretary, for example, using reassuring words like, 'don't fear, management will not do anything arbitrary or be highhanded to penalise Areas'. This was done to diffuse whatever apprehensions had crossed the minds of Mr Mutikanga and his team. Moreover, the arbitration clause in the proposed contract stipulated that any disputes between the Areas and management at the headquarters would be handled by an 'independent' party, namely, the chairman of the Technical Committee of the NWSC Board of Directors.

These assurances persuaded Mr Mutikanga and his team to try out the idea of the APC as a pilot project. He then asked the drafting committee to have an open mind and revise the APC framework to incorporate all additional issues which had arisen and had been discussed and agreed upon during the workshop. Such issues included the need to provide the start-up capital and to give Area managers more autonomy in staff management issues by, for instance, having

Area disciplinary committees (ADCs). Other issues related to easing of penalty ceilings from a failure level of 90% to 85% and the inclusion of social-mission objectives in the APC framework. I reassured Mutikanga's team that by pioneering the APC, Kasese and Fort Portal Areas had nothing to lose, but everything to gain.

As a result of the harmonisation workshop, he and his team agreed to draw up a business plan in consultation with the Corporation's management. The plan incorporated the SWOT approach to analyse the strengths, weaknesses, opportunities and threats of the existing water and sewerage services delivery system, and to show how the APC targets could be achieved. On its part, the Corporation's management undertook to facilitate the successful preparation of the business plan. Besides, KfW, a German agency that was already funding the water projects in Kasese and Fort Portal Areas, immediately came on board to assist in the preparation of the business plan. The process for Kasese and Fort Portal Areas was elaborate and took one and a half months to finalise, with several drafts moving back and forth from the Area manager to the NWSC Head Office, and KfW for comments, suggestions, revision and refinement.

In particular, KfW insisted that an Area must be given maximum autonomy, with as little interference from the Head Office as possible. We had no objection to these KfW suggestions. By August 2000, the Kasese and Fort Portal APC pilot project was ready for implementation.

The signing of the pilot APCs was scheduled to coincide with the SEREP evaluation workshop in August 2000. During that meeting, we intended to finalise the Area business plans before signing the APCs. We also planned, during the course of the workshop, to sell the APC strategy to other NWSC Areas. To our astonishment, the final touches to the Kasese and Fort Portal business plans triggered unexpected enthusiasm for the APC strategy among the participants, to the extent that it was not really necessary for the management to belabour the point. Most Area managers were determined to take on the APCs if given a month for preparatory work on the basis of the Kasese and Fort Portal Area business plans. We realised that our initial cautious pilot project approach was after all not needed and we therefore accepted the challenge to operationalise the APCs in all our Areas except Kampala, which was still under the management of a private operator. As a result, all Areas were ready with their business plans by the end of September 2000. The APCs were signed and took effect at the beginning of October 2000.

The APC on the ground

The APCs were signed by the Area managers on behalf of the Areas and the director, Technical Services on behalf of the NWSC as the employer, and endorsed by me and the chairman of the Uganda Public Employees' Union (UPEU). This endorsement did not only demonstrate the importance management attached to the success of the APCs but also signified the resolve to work in close partnership with the NWSC workers through their union. This was a historic event in NWSC, as attested to by many staff. For example, Mr Joel Wandera, Masaka Area manager, recalls that

> For the first time in the Corporation history, a legal document was prepared and both players, Head office and Area managers, had to append signatures to it. Serious negotiations on targets and other aspects of the contract had to take place and a common position agreed on.

Under the APCs, all NWSC Areas were, in practice, to operate as autonomous commercial entities whose primary objective was to improve service delivery, achieve financial viability and ensure operational sustainability. The salient features of the APCs were plain and simple:

- give autonomy to the Area managers in running the affairs of the respective Areas,
- enhance commercial orientation of the Area management team,
- create result and output-oriented management, and
- increase accountability and have clear separation of responsibilities between Areas and headquarters.

According to Mr Wandera, *'the real increase of autonomy to National Water and Sewerage Corporation Areas started with Area Performance Contracts.'*

The APCs clearly defined the separate and distinct contractual obligations of the Area managers on one hand, and the NWSC as the employer, on the other. According to Mr Mutikanga,

> Area Managers had never been empowered before like in the APC, where we were allowed to decide on optimum staffing levels, we use operational funds with less interference from headquarters as long as the expenditure ceiling was never exceeded. This flexibility freed Area Managers and made Areas more accountable to achieve agreed-upon targets.

Roles and responsibilities of the Area managers: Area managers assumed full responsibility for the implementation of the APCs. The APCs granted full autonomy to Area managers in most of the operational matters, including the maintenance of the Corporation's facilities and equipment, operation and maintenance of vehicles and motorcycles, procurement, personnel management and the incremental improvement of water and sewerage service delivery to the satisfaction of customers. Specifically, the contracts spelt out the Areas' performance targets and expected outputs in water production and sewerage services, water distribution and customer care and personnel management. Area managers were required to identify the requirements in each of these links in the water service delivery chain and to develop appropriate and effective systems and mechanisms to ensure optimal operation. The contracts also emphasised the centrality of planning, maintenance, cleanliness, problem solving and management in all aspects of water and sewerage services.

Area managers undertook to set up Area procurement committees to handle all Area procurement-related matters, including bidding and tendering. Area managers became accounting officers for their respective Areas under delegated authority, subject to their employer's conditions and regulations of service. They became responsible for setting financial milestones in their business plans, collecting revenue, recovering arrears, preparing recurrent and capital budgets and controlling expenditure. They also became responsible for paying staff salaries and creditors' outstanding bills; including electricity, telephone and ground rent and rates/bills from monthly collections, and general management of Area finances and books of accounts, including bank accounts.

Within the APC framework, Area managers enjoyed considerable freedom to hire, discipline, promote, demote and dismiss the employees under their charge. They were free to determine the optimum number of staff required to perform their contractual obligations. Excess employees were referred to the headquarters for redeployment or retirement. Area managers could 'hire and fire' staff on contract terms of employment or outsource service, if and when necessary. However, hiring of staff by the Areas was restricted to staff members in scales 5 to 10. The other scales, 1 to 4, required NWSC Board approval. Area managers were given the autonomy to regularly review staff levels and to advise the management at the headquarters about increasing or reducing the number of employees in their respective Areas.

They were allowed to carry out performance appraisals of members of staff and to recommend promotion for good performers and demotion for chronic underachievers. They were also allowed to discipline their staff, provided such disciplinary measures were taken in accordance with the employer's terms and conditions of service. For this purpose, Area managers were empowered to set up ADCs to advise them. In exercising their powers, Area managers had to be guided by high standards of professionalism and best management practices, as well as principles of fair play, equity and natural justice.

The autonomy granted to the Areas was of course not absolute or irreversible. There were limits to what they could or could not do, as mentioned earlier. The

Areas did not have power either to fix salaries and allowances or to determine the terms and conditions of service for permanent staff. Under the APC, these responsibilities were retained by the management at the headquarters. The promotion or demotion of staff was also subject to consent from the headquarters.

Furthermore, they could not exceed budgeted expenditure ceilings without approval from the headquarters. In any case, Area managers could not buy materials above UGX 3 million without the approval of the Corporation's management. The headquarters reserved the right to dismiss or transfer these managers under circumstances that were clearly set out in the APCs. The invisible but restraining hand of the headquarters was still considered necessary to guide the Areas towards achieving the desired objectives and targets, which would in turn lead to realisation of the goals of the PC I contract.

Roles and responsibilities of the Head Office: The APCs were as specific about Head Office obligations as they were about those of the Area managers and their management teams. The most important contractual obligation of the Corporation was to facilitate the work of the Area managers to achieve the APC objectives and ultimately those of the government performance contract. The Corporation undertook to hand over its office premises to the Area managers or to rent such premises in those Areas where the Corporation did not own office accommodation. All the NWSC's physical premises, equipment and other assets, including vehicles, motorcycles and office furniture, were vested in Area managers for operational purposes. The Corporation also undertook to provide facilitating loans to Area managers on application as start-up capital to cater for additional requirements specified in the Area business plans. These loans were subject to a 10% interest charge.

In order to maximise economies of scale and to effect savings, all bulk purchases of chemicals, pipes and fittings remained centralised for the duration of the APC period. However, the Corporation was duty bound to stock enough input materials for water production and distribution and to guarantee supplies to Areas on demand. Besides, the Corporation undertook to revise the prices of materials in its stores to ensure that the prices were competitive with the prevailing local market prices for the same quality materials. Area managers were not, however, obliged to buy items from the central stores if the prices were higher than local market ones. The Corporation also retained financial responsibility to provide transport to all Areas, raise capital for new investments, including those of a social nature, coordinate donor activities in the urban water sector and insure all assets, equipment and employees in all NWSC Areas.

To perform its monitoring and evaluation roles effectively, the Head Office set up a task force with five committees, dealing with billing, collection, arrears recovery, cost control and water and sanitation services to oversee the operations on the ground. The role of each sub-committee was to carry out evaluation of the APCs and to oversee their day-to-day implementation. Evaluation for each Area was to be carried out monthly to determine performance improvements, constraints experienced and to moot remedies. In addition, the APCs were to be jointly evaluated on a quarterly basis. This was to show the progress in performance *vis-à-vis* the performance contract.

The stick and the carrot

In order to maintain staff enthusiasm, the APCs also accommodated the tradition of competition, with trophies and cash prizes being given to star performers after each quarterly evaluation in areas of billing, revenue collection, arrears recovery, cost efficiency and water and sanitation services. All in all, the cash prizes ranged from UGX 1.5 million for the best overall performer and UGX 1 million for the overall runner-up to UGX 300,000 for the winners in each of the individual operational Areas. In addition, all employees and the management of an Area which achieved at least 95% of the targets were entitled to performance bonuses equivalent to 25% of their monthly basic salary. With the introduction of the Stretch Programme, which is presented in Chapter 6, the performance bonus increased to as high as 50% of an individual's monthly basic salary. Also, Area managers were eligible for promotion in recognition of exemplary performance.

On the other hand, the APCs spelt out penalties for Area managers whose performance consistently fell short of expectations. Managers whose Areas performed below 85% of the targets for three consecutive months would forfeit the bonus. And for managers whose poor performance could not be justified, demotion or dismissal would be the prize. The penalties were, however, to be used as a last resort. From the outset, emphasis was to be more on incentives than on penalties. Fortunately, these penalty clauses were rarely invoked; managers performed satisfactorily.

Support Services Contracts (SSCs)

From the outset, the management recognised that it was important to back up and give complementary support to the APCs by 'commercialising' support services that were to be provided by the Head Office. These services were earmarked as procurement and stores, block mapping, water quality control, static plant, information technology (IT) and Gaba Water Treatment Works and Bugolobi Sewerage Works. As part of the strategy to facilitate the implementation and success of the APCs, the Corporation's management initiated complementary SSCs to enhance result-oriented management, commercialisation, accountability and respective departmental responsibilities. The SSCs were designed deliberately, targeting the departments and sections at the headquarters, whose services, in light of the significant role they played in the overall performance of the Areas, were considered critical to the success of the APCs. Furthermore, the overall goal of introducing the SSCs was to ensure a holistic approach in the implementation of the performance contract. The heads of department of these support services were accordingly given performance targets to be achieved within the contract period. The SSCs ran alongside the APCs and were subject to the same quarterly performance evaluation criteria.

The main objective of the SSCs was to provide timely backup support and logistics that the Areas required for effective and smooth implementation of the APCs. The specific objectives of the SSCs, as summarised in Box 5.1, focused on the realisation of the aims of the APCs and eventually the success of the government performance contract.

Box 5.1: Specific objectives of SSCs

- Creation of synergies to enhance staff commitment, simplification and coordination of multiple sectional and departmental tasks to boost performance;
- Develop a more focused and coordinated approach for effective execution of Area operations and tasks aimed at creating a favourable environment for job satisfaction;
- Inculcate among staff the spirit of sectional and departmental team spirit and hard work with emphasis on output-oriented approaches and practices and
- Initiate problem-solving and performance-oriented business plans, and develop bulk material delivery systems based on hired transport.

Therefore, besides ensuring successful implementation of the APCs, the SSCs aimed at increasing efficiency in the operations management of these departments/sections. In short, the SSCs were designed to alleviate existing operational inefficiencies that were likely to impede the success of the APCs.

The Procurement Department, for example, had no motivation to bargain for lower prices as the procured items would always be drawn for use by the Areas at any cost. The SSCs were therefore designed in such a way that if the Procurement and Stores Department procured materials at a price higher than the prevailing market prices, the Areas would be under no obligation to use them. The Procurement and Stores Department was also given a time interval within which to procure materials and was given a target to provide 60% of Areas' procurement needs at a price lower than the local prices in the respective Areas. Otherwise, from an economic point of view, it was not prudent for an Area manager to spend his or her time and money in travelling to central stores, for example, to get a ream of paper procured at UGX 8,000 when the same item could be purchased locally at UGX 7,000. Such an action would be tantamount to double loss to the Corporation.

Under the SSCs, the departments/sections were not only commercialised but were also empowered to deal directly with the Area managers as equal business partners. Although the objectives of the SSCs were welcome in theory, they were hard to implement in practice and encountered a number of organisation-related constraints. The main constraints are summarised in Box 5.2.

Box 5.2: Constraints of SSCs

- Reporting bottlenecks between various sections and departments that hindered day-to-day operations of support services;
- Conflicts over centres of power as a result of restructuring;
- Some restructuring activities under SSCs needed approval from top management, which some people shied away from because of misconceptions;
- Inadequate utilisation of the IT services and functions across all the support services units and
- Inadequate participation in the design, formulation and implementation of SSCs by water quality, training, research, development, internal audit and consultancy units.

The above constrained the realisation of the envisaged SSCs objectives and targets. The way the SSCs were introduced was another major impediment to their successful implementation. It involved a hasty transformational mode rather than the cautious evolutionary mode. This approach turned out to be a mistake and I was made to understand that lack of focus and cohesion in the SSCs could derail the APCs. Out of anxiety to keep on track, it seems I put excessive pressure on the managers and staff to speed up the clogged planning and implementation process thus leading to this hard-to-implement contract. I recall a comment made by one of the affected managers to the effect that '*It was like taking the horse to the river but of course it was not possible to force it to drink the water!*'

However, I have never regretted taking the risk of testing the SSCs. In fact, the attempt to combine the APCs and the SSCs was a useful learning experience that proved to me that support services are vital to the overall performance and success of any organisation, which no good manager can afford to ignore.

Gauging progress and re-energising the APCs

The APC era marked the beginning of systematic quarterly evaluations. In the previous programmes, the evaluations were planned at the end of the Programme; but because the APC duration was rather long, it was deemed necessary to incorporate strategic quarterly performance reviews within the APC implementation framework. This was one way of continuously rejuvenating the programmes.

The actual performance of Areas assessed against the targets was evaluated in a series of workshops on a quarterly basis between February 2001 and July 2002. These fora were attended by Area managers and members of the APC task force and its committees, as well as by NWSC directors and senior managers. In some cases, ministers and other senior government officials, civic leaders and prominent customers also attended. The purpose of the workshops was to review the progress of the APCs in terms of achievements, constraints and challenges; to share experiences; to compile quarterly evaluation reports for the Corporation's management to consider and to map out a way forward for the next quarter. These fora also helped us to keep in focus with the government performance contract.

All in all, six APC evaluation workshops, hosted in different NWSC Areas across the country – Mbarara, Mbale, Jinja, Kabale, Masaka and again Mbale. The aims of these workshops, which were hosted and spearheaded by the respective Area managers, were to highlight qualitative and quantitative achievements of the APCs, to reward good performers, pinpoint constraints/ shortfalls and to involve local dignitaries and customers in the festivities and celebrations of the Corporation. During these fora, the participants were duty bound to make resolutions to inform future policy directions regarding field operations of the APCs in the next quarter. Throughout these workshops, emphasis was put on the virtues of hard work, team spirit and confidence, competition, innovation and improved service delivery and sustainability. There was a lot of enthusiasm among the staff, and the results of the APC I evaluations showed positive performance trends in most of the Areas.

However, the first and second quarterly evaluations of the second series of APCs (APC II) showed clearly that performance in some of the Areas had slackened and there were clear signs of complacency among some managers, which was affecting performance negatively in some Areas. It also impacted negatively on the Corporation's efforts to achieve the government performance contract targets. Management, therefore, decided to re-energise the APCs through a new radical approach, the Stretch-out Programme, which is the subject of Chapter 6. This programme was later complemented and refined by the One-Minute Management programme presented in Chapter 7, which focused on individual responsibility, performance and incentives in order to boost productivity, service delivery and, most importantly, to enable achievement of the APC objectives. In essence, the introduction of Stretch-out and One-Minute Management programmes augmented the implementation of the last part of the APC II and the whole of the APC III.

APCs achievements

The success of the APCs ensured that the Areas invariably gained local prestige and a sense of achievement, monetary incentives and, of course, enhanced customer satisfaction, which was amplified by consecutive customer surveys. During the periodic evaluation workshops, actual performance of each Area in various operational aspects of water and sewerage service delivery, including billing, NRW, active accounts, new connections, revenue collection, arrears recovery, water production and distribution and cost efficiency was measured against the APC (I, II and III) targets.

Table 5.1 shows some of the key achievements at the end of the three phases of the APCs. The achievements are structured into Kampala, other Areas and overall achievements, to show the impact of the APCs in the Areas and on the overall performance of the Corporation. The table shows significant improvements in performance on all indicators in the other Areas where the APCs were implemented. Specifically, NRW reduced considerably from 37% before the introduction of the APCs to 22% by the end of the Programme.

It is fair to say that the APCs represented a great leap for NWSC towards the unpredictable future. It triggered staff enthusiasm in all the Corporation's operational Areas. The APC also proved to be a suitable performance improvement tool in the eyes of key stakeholders. The APCs were seen as a good start-up activity to implement the wider reform strategy of separating operations from monitoring/oversight responsibilities. In principle, the Programme received strong support from the government and development partners.

Bringing more towns on board

I cannot wind up this narrative on the APCs without making a brief reference to the seemingly unrelated issue of bringing more towns on board during the APC period. The initiative to raise the number from 12 to 17 came from President Yoweri Museveni's Presidential Election Manifesto of 1996. Although, for financial reasons, we did not relish overseeing more towns, the Corporation accepted the challenge when the government undertook to subsidise those that failed to break even, as was provided for in the performance contract. Originally,

Table 5.1: Achievements of APCs

Indicator	Before APCs (2000)			After APCs (2003)		
	Kampala	Others	Overall	Kampala	Others	Overall
NRW (%)	46	37	43	45	22	39
Staff Productivity (staff/1,000 connections)	15	25	18	8	15	11
Proportion of Inactive Accounts to Total Accounts (%)	30	21	26	22	18	21
Collection Ratio (%)	94	81	89.3	93	98	95
Metering Efficiency (%)	85	84	85	93	97	95
Capacity Utilisation (%)	74	54	65	83	65	75
No. of Towns covering operating costs except depreciation (out of 15)		7			10	

owing to political pressure, it had been agreed that Bushenyi/Ishaka, Soroti, Arua, Kitgum and Kisoro would be brought under the NWSC operational mandate in 2002. However, in due course, Kitgum and Kisoro benefited from an Austria-funded water rehabilitation project and as such their water systems remained outside the NWSC orbit of operations. Accordingly, in 2002, only three towns – Bushenyi/Ishaka, Soroti and Arua – became part of the NWSC family, bringing the total number of Areas managed by the Corporation to 15.

When the Corporation took over Bushenyi/Ishaka and Soroti, there was little to boast about in terms of physical facilities or revenue generation in those Areas. In Bushenyi/Ishaka, for example, revenue collection was a mere UGX 2.5 million per month. And out of a mere 300 accounts, only 250 were active. Thanks to NWSC's intervention, water service delivery and revenue collection in Ishaka/Bushenyi improved significantly within a short period. By October 2002, average monthly revenue collection increased to over UGX 10 million (460% increase), while NRW reduced from 50 to 35%. Water production rose from 8,000 to 15,000 cubic meters per month. Although this remarkable progress was too late to include Ishaka/Bushenyi in the second APC, the Area performed well enough to get weaned off the conditional grants. This outstanding achievement within a short time ensured that Bushenyi/Ishaka was qualified to join the APC III and the Stretch-out Programme.

Similarly, Soroti Area rose from the doldrums to become one of the most promising performers. Its problems ranged from chronic political interference to being completely 'dry' of water for two to three days a week. In order to turn it around, we appointed Mr Charles Ekure, known for his boundless energy, resourcefulness, fearlessness and determination, as Area manager. By the end of 2002, he had revitalised Soroti, stabilised its finances, motivated its staff, enhanced its water service delivery (production, distribution, response time to complaints, water bursts and leaks) and boosted its customer satisfaction. For example, by October 2002, revenue collection had improved from about UGX 10 million to over UGX 20 million per month. On its part, the Corporation gave consistent backup support to the Area in an effort to cope with meddlesome politicians, consumers who habitually defaulted on payment or who were connected illegally. Although Soroti has generally been on the mend since the NWSC assumed responsibility, it has not yet broken even and continues to depend on a subsidy for investments.

Of the three towns that the Corporation brought on board in 2002, Arua had unique problems relating to war and of internal displacement that stretched way back to the 1980s. Arua is also unique in that it was the pioneer in indigenous private sector participation under the auspices of the Arua Municipality. In 1989, the Arua Municipality hired Mr Fred Irumba, an experienced water engineer, to operate its water and sewerage services. However, being a small private operator, he lacked working capital to cover operating costs and investment to meet the demands of the Area on a sustainable basis. By asking the Corporation to take over, the government had in effect come to the conclusion that Arua was not yet ready for private sector management. When NWSC took over the Area, it retained Mr Irumba on a six-month contract pending the appointment of a suitable alternative. To our pleasant surprise, Mr Irumba and his staff did a wonderful job and put Arua on the right track. And, accordingly, there was no need to replace him after his contract expired. The Area's collections shot up to UGX 30 million by April 2003 (from UGX 10 million). By the end of that year, customers had access to water on a 24-hour basis, unlike before. It also registered 35 new connections per month, which was almost equal to new connections in Jinja, a more established and larger Area. In spite of this good performance, Arua was still far from breaking even; it required a subsidy to cover part of the operational costs and new capital development projects.

Taking over Arua town was an experiment in indigenous private sector participation that we have since endeavoured to replicate through the ambitious Internally Delegated Area Management Contracts (IDAMCs), which are discussed in Chapter 9. The Corporation's successful takeover of Bushenyi/Ishaka, Soroti and Arua clearly demonstrated our growing capacity and confidence to deliver safe, clean and affordable water to millions of Ugandans living in urban areas. At the same time, it was a timely reminder that the task of water service delivery was still unfinished business, requiring innovative strategies and approaches in order to accomplish new tasks and challenges. It also needed consistent and sustainable ways of providing fresh impetus to change management momentum.

Impact of the APCs on the PC I performance

The implementation of the APCs and the strengthening of PSP in Kampala had a profound impact on the achievement of the PC I objectives and targets. By the end of June 2003, the Corporation had achieved or even surpassed most of the PC I targets. For one thing, all the qualitative targets, as presented in Box 5.3, were implemented within the contract time frame. For example, we implemented internal reforms with regard to human resource management and development, financial management, corporate management and private sector participation within the stipulated period. Similarly, the development and implementation of the water network expansion plan and the establishment of a project management team and a market/customer liaison group were accomplished within contract deadlines. We were also able to introduce a management information system, which also involved computerisation of the Corporation's accounting function on time.

Box 5.3: Key performance contract milestones

Milestone	Target Date	Achievement Date
■ Development of network expansion plan	February 2001	July 2000
■ Establishment of project management team	December 2000	August 2000
■ Establishment of market/ customer liaison group	January 2001	January 2001
■ Establishment of Management Information System	March 2001	May 2001
■ Updating of asset register	March 2001	June 2001
■ Performance Improvement action plan	September 2001	September 2001
■ Rationalisation of remuneration structure	October 2000	January 2001
■ Performance appraisal scheme	October 2000	January 2001
■ Customer Policy	November 2000	November 2000
■ Customer Charter	End of January 2001	January 2001
■ Focused staff training programme	December 2000	December 2000

The Corporation's quantitative performance during the contract period was equally impressive. Table 5.2 shows the actual performance of Kampala, other Areas and the Corporation as a whole for some of the key performance indicators, between 1 July 2000 and 30 June 2003 against the contract targets. The figures show that, by the end of the contract in June 2003, the Corporation had registered significant improvement in all the performance indicators compared to the performance before signing the contract, though in some cases the contractual targets had not been achieved. The performance was a clear demonstration that the two-pronged approach: APCs in Areas and PSP in Kampala, contributed significantly to the realisation of performance contract objectives.

There was significant improvement in the customer base, largely due to the expansion of the distribution network, mainly funded from our own resources. For example, 50% of the funds were used to extend the network by 154 kilometres in the financial year 2001/2002 and 77.4% of the funds that we used to do a further 354 kilometres in 2002/2003 came from internally generated

Table 5.2: NWSC performance against PC I targets (2001 to 2003)

Performance indicator	Year 2000 Base Year	Year 1 – 2001 Target	Year 1 – 2001 Actual	Year 2 – 2002 Target	Year 2 – 2002 Actual	Year 3 – 2003 Target	Year 3 – 2003 Actual
NRW (%) – Overall	43	39	43	37	40	35	39
Kampala	46	41	47.5	40	44	37	44.5
Other Areas	37	33	32	33	30	29	26.7
Staff Productivity (staff/ 1,000 connections)	18	16	16	13	12	10	11
Proportion of Inactive Accounts to Total Accounts (%)	26	20	26	17	24	15	21
Kampala	30	21	25	16	24	13	22
Other Areas	21	19	27	18	24	16	18
Collection Ratio (%)	89	89	85	89	92	92	95
Metering Efficiency (%)	85	85	86	90	92	97	95
Kampala	85	84	88	88	91	97	93
Other Areas	84	86	82	93	93	96	97
Operating Surplus (UGX million/month)	250	394	364	848	580	1,090	670

sources. Furthermore, the Corporation increased the share of internally generated funds for overall capital expenditure from 21% in the year 2000/2001 to 39% in the year 2001/2002. This percentage was sustained in the third financial year. What these figures demonstrate is that we were slowly but surely moving towards financial viability, not only in terms of recurrent expenditure but also in terms of resources for capital development in line with the performance contract expectations. The improvement in staff productivity was attributed to the substantial reduction in the number of staff from 1,213 to 950 employees and the increase in the total number of connections by the end of the contract. However, staff numbers were not reduced to the expected contract target levels, partly because the increased number of connections necessitated retention of more staff. Also, additional employees were inevitably brought on board when we assumed responsibility for water and sewerage services in Arua, Soroti and Bushenyi/Ishaka. Without these two eventualities, we would certainly have achieved staff reduction and staff productivity targets. Irrespective of whether we reached the contract targets or not, the most important thing was that the Corporation was moving in the right direction towards customer satisfaction and operating as a commercial entity.

With regard to the financial viability, annual turnover improved by over 43% from UGX 25.8 billion to UGX 37 billion by the end of the contract. Similarly, there was a consistent increase in the operating profit during the contract period, despite the Corporation's failure to achieve performance contract profitability targets.

The profitability underperformance in respect to the contractual targets during the contract period can be explained by a number of factors, including an unanticipated increase in the private operator's management fees for the Kampala Area, the 'write down' of overvalued stocks, the increased wage bill, increased electricity costs and increased expenditure on water production inputs. If these operational costs had not gone up significantly during the contract period, we would have comfortably achieved or even surpassed the contract profitability targets. In any case, the fact that there was a consistent improvement in operating profits during the contract period was an important move in the right direction, towards the realisation of the contract objectives.

Constraints: That NWSC failed to achieve some of the targets of PC I raises a number of questions that must be answered. Did we miss some of them because the architects of the PC I had set the targets too high or was the Corporation a poor marksman? In other words, did the contract set unrealistic targets without taking into account the bottlenecks that were bound to slow down our performance? Alternatively, did we miss the targets due to internal technical, financial, resource and administrative deficiencies? Were there any unforeseen external factors that impinged on our performance during the contract period and, by so doing, diminished the Corporation's efforts to realise and even surpass the contract targets?

As demonstrated in Table 5.2, our failure to meet some of the targets was as a result of Kampala Area's poor performance. Kampala is a unique Area for NWSC. It is the Corporation's greatest asset, and at the same time, its greatest liability. The Area is the greatest asset because it has the highest number of customers and generates at least 70% of the total revenue. On the other hand, Kampala has a lion's share of the Corporation's challenges, including NRW, illegal connections, suppressed accounts and arrears. Kampala generally determines the overall fortunes of the organisation. When Kampala Area excels, for example, in revenue collection, the Corporation does very well. But if Kampala Area falters, for example, in reducing NRW, it overshadows the exemplary performance in other Areas.

During the contract period – indeed since 1997 – the management of Kampala Area was under the control of private operators. As already mentioned, when I became the Corporation's chief executive in 1998, Kampala Area was under the management of Gauff, a German private operator, under the auspices of the KRIP, which meant that we had limited control and influence on its affairs. However, given the importance of Kampala Area to the overall performance of the Corporation, I took the initiative to ensure that it was part of our internal reforms, and as such the 100 days Programme and SEREP involved Kampala too. For example, the branch manager for KRIP, Mr Joshua Kibirige, was made the Task Force chairman for Customer Care during the 100 days Programme, and the KRIP contract was renegotiated to make it more cost-effective.

When the KRIP contract expired in 2001, ONDEO Services Uganda Ltd (OSUL), a French international operator, took over Kampala and ran it from 2002 to 2004. The rationale for engaging international private operators was to tap into the benefits of international expertise and exposure.

On our part, we had reservations about the efficacy of retaining a private operator when we had an ample pool of local expertise to do the job, as events eventually proved. The international experts were not only too expensive for the Corporation but were also unfamiliar with the local working environment. It took them almost a year to learn the ropes, leading to frustration and resulting in performance way below expectation.

Besides, the dual allegiance of the workers to the Corporation and to the private operator was bound to be a source of tension and friction that blurred the lines of responsibility and accountability. In the case of OSUL, we had no choice because the government and the donors had insisted on its involvement as a condition in the PC I. By the time our misgivings were attended to, much damage had already been done to the performance of Kampala Area, in particular, and the Corporation in general.

This shows that private sector participation *per se* is not always a panacea for poor performance. In fact, for eight months (2001–2002) between Gauff and OSUL management contracts, when we were directly in charge, Kampala Area performed much better in terms of operating profits than it did under Gauff and subsequently OSUL. If Kampala is excluded from the performance contract equation, the quantitative data show that other Areas performed very well. For example, Table 5.2 shows that by June 2003, all NWSC Areas (excluding Kampala) achieved a NRW level of 26.7% in comparison with the contract target of 35%.

Secondly, public enterprises such as the NWSC, or any business organisation for that matter, are not islands operating in isolation. They are part and parcel of a wider society and economy. The business fortunes of public enterprises are – for better or for worse – invariably tied up with what happens elsewhere in the economy and society at large. In the case of NWSC, our performance from 2000 to 2003 was to a large extent dictated by the social and economic trends in the country. In particular, several factors, including, but not limited to water production input costs, customer attitudes and behaviour, privatisation policy and procurement laws, contributed to overall PC I underperformance. Box 5.4 summarises the constraints encountered.

Box 5.4: Constraints

- Erratic electricity supply and high costs
- External macroeconomic factors
- Resistance to change
- Tariff review restrictions
- The burden of 'social mission' activities
- Poor private sector participation performance in Kampala Area
- Bureaucratic procurement laws

Since 1998, one of the biggest constraints on the NWSC's water production operations has been the high price and unreliability of electricity supply. As one of the biggest electricity consumers in the country, NWSC is very sensitive and vulnerable to any electricity price increases and interruptions. During the contract period, the Uganda Electricity Distribution Company Ltd (UEDCL) increased its tariff significantly. This increase not only pushed up our overall water production costs but also impacted adversely on our profitability performance. This is partly why the Corporation's profitability performance during the contract period was substantially off target.

Thirdly, like most public and private enterprises in Uganda, NWSC imports most of its essential water production inputs, such as chemicals and equipment, and therefore requires foreign currency. Since the Corporation is not paid in dollars, pound sterling, euros or in whatever other foreign currency, it has to buy the required forex at commercial rates. Unfortunately, the Ugandan Shilling depreciated substantially during the contract period. As a result, our production costs increased more than anticipated and inevitably diminished our chances of achieving the contract profitability targets.

Despite the implementation of a customer-friendly policy and establishing a customer charter, there were some black sheep among our customers, especially in the Kampala Area. There is truth in the saying that old habits die hard. Some of the people in our operational Areas have been bent on consuming free water. They continued to engineer illegal connections or to tamper with meters in order to manipulate the readings. They never seemed to appreciate that the production and supply of water costs money. Some of them stubbornly refused to pay their bills, forcing the Corporation to disconnect them. These negative attitudes explain why NRW and inactive connections' actual performance fell short of the contract targets.

Although the performance contract granted us commercial freedom in all our operations, there were still some restrictions regarding what we could or could not do. For example, we could not raise the water tariff to realistic commercial levels without the approval of the government, for fear of far-reaching social and political repercussions. During the contract period, a full cost-recovery tariff would have required an increase of 135%. Such a drastic increase would have been politically and socially unacceptable. And yet the Corporation could not realise its full commercialisation dreams without the freedom to determine the water tariff levels to recover the full costs.

As a provider of water and sewerage services, NWSC is a unique enterprise whose operations cannot be driven by the profit motive alone. Throughout its history, the Corporation was obliged to take into account other considerations, notably 'social-mission' objectives. Indeed, until the advent of privatisation, the social-mission objectives were perhaps far more important than making money or profits. Although the quest for commercialisation was the driving force behind the performance contract, we could not dare ignore our social responsibilities. For example, the Corporation could not pack its bags and pull out of non-viable towns, which a private exclusively 'for profit' operator would not have hesitated to do. On the contrary, NWSC took over Arua, Soroti and Bushenyi/Ishaka, not because of their promising commercial potential, but at the behest of the government.

To its credit, the government recognised that the NWSC could not carry all the costs of the social-mission objectives single-handedly. Whenever it requested the Corporation to extend services to Areas that were considered commercially unviable, it subsidised the operations. In the year 2002/2003, for example, it provided a subsidy of just over UGX 680 million towards operations, and invested in the three towns mentioned above. The subsidy provision for the three towns in the year 2003/2004 was UGX 390 million. In addition, the government provided subsidies of UGX 400 million and UGX 436 million in the 2002/2003 and 2003/2004 financial years to enable us extend water services to the urban poor. These subsidies helped us meet the 'social-mission' objective, although not all the operational costs were covered.

Nevertheless, the Corporation reached or surpassed most of the qualitative and quantitative contract targets. Even where actual performance fell short of the targets, the improvement trends were upwards and generally satisfactory. Both the government and the NWSC leadership were satisfied that the PC I had served a useful purpose in realising the dream of full commercialisation of NWSC services. More importantly, the Corporation's internal reviews and the Performance Contract Review Committee evaluation process provided vital information to feed into the subsequent programmes.

The results of the performance contract proved that, if properly managed, public utilities had the potential to become 'cash cows' for the government. In recognition of a job well done, the government did not only award a 25% bonus to the NWSC managers but also entered into a second performance contract (PC II), which was signed on 17 December 2003, to last another three-year period: from 1 July 2003 to 30 June 2006.

But whatever all our achievements, there was no room for complacency and self-satisfaction. It was prudent to keep thinking ahead in order to re-energise ourselves, to look out for and seize new opportunities and to prepare ourselves for likely uncertainties. In Chapter 9, I explain how the APCs soon grew into the refined and more purposeful IDAMCs. But, in the meantime, let us immediately turn to the fascinating story of 'Stretch-out', a new, revolutionary management experiment, which was designed to revitalise the APCs, and which transformed the NWSC almost beyond recognition in an unbelievably short period of time.

Lessons learnt

- Government policies and strategies are never set in stone. Policy makers are flexible and can therefore embrace sound and viable policy options. What public enterprise managers need to do is to win over such officials through the force of argument and performance rather than through complaints and incessant memoranda asking for subsidies. Policy makers will support ideas that work, even when it is not in tandem with declared official policy.
- Public enterprises are not islands operating in isolation. They exist and operate in a specific social and economic environment that invariably impacts on their performance. So, one should not be discouraged if they are not able to reach set targets. Targets are simply a means to an end; they are not ends in themselves. What matters is the confidence that your organisation gains once it is established that it is moving in the right direction.
- Managers of public enterprises need to define their working relationships and expectations from the outset. Each side must clearly understand its rights and obligations. These should then be subjected to independent and periodic appraisal in light of changing objectives, goals and targets.
- Since most people, including policy makers and customers, move at different speeds, emanate from different directions and have different motivations, it is important to be patient and resilient as one strives to win the confidence of staff members and rally them to support the organisation's destination.
- Privatisation *per se* does not necessarily translate into star corporate performance. Go for what works rather than rigidly conforming to policy positions of the day.
- Corporate performance springs largely from not only the expertise of policy makers and head office management directors but also from having the right strategy, human resources and operational activities. Optimal autonomy should therefore be granted to operational units.
- Keep celebrating success whenever possible, in a bid to remind yourselves of where you have come from, where you are and where you would like to go.
- Expect failure at one point or another; catch it early by keeping communication channels open.

Chapter Six

The Group Incentive Mechanism

Stretch-out Programme

I'm nature's greatest miracle. I'm not on this earth by chance. I'm here for a purpose and that purpose is to grow into a mountain not to shrink into a grain of sand. Henceforth will I apply all my efforts to become the highest mountain of all and I will strain my potential until it cries for mercy. I will increase my knowledge of mankind, myself, and the goods I sell, thus my sales will multiply. I will practise and improve, and polish the words I utter to sell my goods for this is the foundation on which I will build my career. Also, I will seek, constantly, to improve my manners and graces for they are the sugar to which all are attracted.

Og Mandino
The Greatest Salesman in the World, p. 70

As pointed out in Chapter 5, all the preceding managerial and administrative interventions, such as the APCs, were geared to enhancing attainment of the objectives and targets of the Performance Contract between the government and the NWSC.

From the 100 days Programme to APC, as part of our operational change management strategy, we used SMART (specific, measurable, achievable, realistic and time bound) criteria to set targets and milestones in order to enhance performance, productivity and commercial viability. Since the launching of the change management programmes in 1998, we had made significant progress towards the realisation of the Corporation's commercial and 'social-mission' objectives. However, the struggle to reach the best practice standards in the delivery of water and sewerage services in Africa and to realise the Corporation's vision of becoming 'the best service utility provider in Africa', were still 'a bridge too far'. The quarterly evaluation of APC II had also revealed slackened performance, with some Areas failing to deliver on their APC II expectations. For this reason, the NWSC management decided to embark on a radical new approach – the Stretch-out Programme, which I came across during one of my official travels abroad, to speed up the transformation of the Corporation.

The Stretch-out Programme, which was launched on 25 August 2002, just before the end of APC II, was designed to achieve two objectives – one general and one specific. The general objective was to enhance and sustain the achievements of the previous change management programmes. The specific objective was to complement and energise the APCs by adopting a radical new approach to address the mindset and operational obstacles that stood in the way of achieving and even surpassing APC SMART targets.

Beginning with Jinja in September 2002, the Stretch-out Programme was gradually extended to all NWSC Areas. Eventually, with the consent of OSUL, it was implemented in Kampala. By the end of 2003, all our Areas had embraced Stretch-out and a new change management dispensation had begun.

The subsequent sections explore the various perspectives and actions taken to incentivise the group through a Stretch-out Programme. I demonstrate that sometimes good ideas may exist somewhere – it is just a question of courage and enthusiasm to discover these good ideas and put them to the right use.

In praise of Jack Welch and Spencer Johnson

In July 2002, I went to the United States to attend a World Bank-sponsored short course on 'Privatisation and Infrastructure Development in a Market Economy' at the John F. Kennedy School of Government of Harvard University. The course, held from 14 to 26 July 2002, targeted senior executives from the private and public sectors across the world. It focused on the pros and cons of privatisation, and was facilitated by eminent professors and consultants in the US. Another participant from Uganda was Mr Michael Opagi, the head of the Privatisation Unit of the Ministry of Finance, Planning and Economic Development.

I had read many books by professors at this prestigious Harvard School, some of which were compulsory reading material, including those of Mr Michael Porter on competitive strategy and advantage, when I was a graduate student at the Munich, Ludwig Maxmilian University, in Germany. I therefore looked forward to the Harvard course to gain more knowledge and business management insights, which would positively impact on my work at NWSC. As expected, when I arrived at Harvard, the atmosphere was congenial for reading and reflecting on what we were doing in Uganda. This reinforced my anticipation for stimulating lectures and shared professional experiences over the days that the course was to last. (Incidentally, while at Harvard, I traced Prof. Porter, who was kind enough to autograph two of his famous books for me.)

During the course, I met many people with long and varied experiences in the field of management and the privatisation of public utilities. As expected, we exchanged ideas and shared practical experiences about the pros and cons of privatisation around the world. Unfortunately, what I learnt from the course was neither impressive nor instructive in as far as the challenges facing NWSC were concerned. The experiences drawn from the privatisation of public utilities around the world, including those of developing countries such as Uganda, were mixed. For example, we learned that privatisation experiments in Argentina, Bolivia and Chile had not yielded expected results. Indeed, during the course, we hardly encountered a single privatisation case that was not controversial. Even in the United States, the proud homeland of private enterprise, one professor, who was one of the course facilitators, told us that he failed to get re-elected as mayor of one of USA's large cities, partly because of his enthusiasm for privatisation of the city's water and other services. These experiences left many people wondering whether privatisation was the panacea for the inefficient and ineffective water service delivery in developing countries.

Although, to me the course in many ways did not have much to write home about, my visit nevertheless turned out to be extremely useful. I visited several bookstores in and around Harvard looking for books on strategic management. In one such store, I saw a best seller prominently displayed on the shelves of the management section. Since it is unusual for management books to hit best seller lists, I was naturally curious to find out what that best seller was about. This book was about *Jack Welch: The GE Way* by Robert Slater. I bought it immediately, not knowing that it would radically change my perceptions and soon transform the nature and content of our change management programmes.

Once I started reading the book, which centred on Jack Welch, a star performing chief executive of General Electric, I could not put it down. I read it from cover to cover, over and over again, taking notes. What a great relief it was from the rigours of the privatisation course! I was amazed by Welch's stretching concept, which was simplified and concisely expounded. In my experience, no other book had ever affected me the way this one did. I could not believe that Welch had done wonders at General Electric, as detailed by author Slater, through stretching. I went back to the bookshop and bought Welch's autobiography to learn more about the man who made General Electric tick. It was the same success story. I talked to some of the professors and other acquaintances at the John F. Kennedy School of Government about Welch. They confirmed unanimously that he had worked miracles at General Electric and transformed it into one of the most successful multinationals since he became its chief executive at the relatively youthful age of 44. But how did he do it? The answer was stretching, one of the most revolutionary management concepts to date.

Welch begins his compelling case for stretching to reach for the stars with an indictment of old-fashioned management, which tends to be driven by monitoring, supervision and control. For him, a manager of yesteryear 'controls rather than facilitates, complicates rather than simplifies, acts more like a governor than an accelerator'. He argues that old management practices stifle the energy of the workers, block the flow of information, stifle or, worse still, kill initiative and innovation and slow down business operations to a snail's pace.

For these reasons, he argues that the old-fashioned overbearing, insensitive, remote and manipulative manager should be discarded in favour of a leader who has the clarity of vision to inspire, facilitate and empower the workers to get on with their work without interference.

According to Welch, the three key secrets of good leadership are simplicity, speed and self-confidence. Simplicity means focus and clarity, as well as flexibility and responsiveness to change. The virtue of speed is to make decisions in minutes, not days letalone weeks or months. Self-confidence is the key to security, receptivity to new ideas and methods of work, and a clear sense of responsibility. These secrets call for the removal of bureaucracy, the creation of an open working and decision-making environment devoid of boundaries, the empowerment of workers to make quick, on-the-spot decisions and, above all, fast and direct communication. In short, his world of work knows no bureaucracy, boundaries, and boss element; it is distinguished by simplicity, speed and worker participation.

Welch argues that given the centrality of the people in any business enterprise, it is important for leaders to foster new ideas, worker participation and creativity in programme design and implementation, and to provide the needed resources to reach specified goals. That is why, at General Electric, he introduced workout programmes to pick the brains and unleash the energy and enthusiasm of the workers to ensure the success of the organisation. He also eliminated the 'boss element' to encourage a sense of equality and informality at the work place and to give the workers the freedom to come out with new ideas and to confront 'the boss face-to-face'. In this environment of simplicity, speed and self-confidence, what was at stake was not what the workers did in General Electric but their contribution to the 'battle of ideas' in the organisation.

Box 6.1: Jack Welch's key secrets of good leadership

- Speed, accuracy and clarity
- Simplicity
- Self-confidence
- Boundless decision-making mechanism
- Direct communication across the organisation (no memos, no intrigue, no cliques, etc.)
- No boss element in decision making
- Full-fledged worker participation

For me the most inspiring of his ideas was stretching, by which he meant exceeding goals. Most business leaders are content to set and achieve SMART targets. For Welch, SMART targets are simply a point of departure – a means to an end. While it is fine to work out SMART targets, stretching aims much higher and the sky is the limit. Stretching is a dream to strive to achieve seemingly impossible targets.

According to Welch, although stretching calls for superhuman effort and commitment at all levels of an organisation beyond the call of duty, it is nevertheless rewarding even if actual performance is not dead on target. From his own experience at General Electric, he found that 'by aiming for what appears to be impossible, we often actually do the impossible, and even when we don't quite make it, we inevitably wind up doing much better than we would have done.'

On my study trip to the US, I travelled with my wife, Vicky. She had been invited to visit her elder sister and brother-in-law who lived in Ohio. After the Harvard course, I joined her and her hosts. While there, I could not help mentioning the exciting story of Welch to whoever cared to listen, both in Ohio and later on in Washington. I was already bubbling with enthusiasm about the concept of stretching. Vicky, who is a more avid and faster reader than I, quickly went through Robert Slater's book and she, too, shared my excitement. Then, on our way back to Entebbe via London, she bought a small book: Spencer Johnson's *Who Moved My Cheese?* En route from Washington to Heathrow and in the transit lounge, I noticed that Vicky was completely absorbed in reading the little book. As soon as she finished, she said: 'You must read this book. You will like it as much as you liked Jack Welch,' as she handed the book over to me.

I picked the book and looked at the drawings of cheese slices on the cover and thought that Vicky was making fun of my newly acquired enthusiasm for stretching. I told her to stop pulling my leg and to leave me in peace to enjoy my nap. But Vicky looked dead serious. So when I got on the plane to Entebbe, I read the 60-page book. Within an hour I had read it cover to cover. What an enjoyable treasure I discovered! I was lost for words and did not know how to thank her for drawing my attention to this wonderful little book. She could tell that Spencer Johnson had had as much intellectual impact on me as Jack Welch. With a telling smile, Vicky said she had bought the book for me. It was one of her best gifts ever, which I have treasured ever since. Even as we talked, I could not stop reading the book again and again and thinking about its contents until we reached home. As I read *Who Moved My Cheese?* numerous questions flooded my mind. I started to analyse myself, my wife, my friends and, above all, my colleagues in relation to the four principal characters in *Who Moved My Cheese?* – Hem, Haw, Sniff and Scurry.

In *Who Moved My Cheese?* Spencer Johnson presents a simple but compelling argument. Since change is inevitable for individuals and organisations alike, it is imperative to anticipate and monitor change, to adapt to change quickly before it is too late and to keep changing with change. For Johnson, individuals and business organisations tend to get stuck in the mud because they fear change. They do not dare plunge into the unknown. Blinded by complacency and false security, they do not see the writing on the wall. Since they cannot anticipate and monitor change, they are incapable of jumping onto the bandwagon of change to exploit and enjoy opportunities around the corner. So when, sooner or later, change overtakes them, they sulk, mourn and feel sorry for themselves. Instead of blaming themselves for self-inflicted misfortunes, they ask: 'Who moved my cheese?'

Johnson gives useful tips about changing with change. To start with, do not be complacent – do not take things for granted. Do not be a slave of the fear of change. Instead be bold; overcome fear to embrace change. Abandon outdated habits, ways of thinking and methods of work. Be vigilant, look out for and detect the signs of change. Quickly look for and exploit opportunities that may present themselves on the change trajectory. Individuals and organisations that notice and embrace change invariably find and exploit new opportunities and, ultimately, reap the fruits of change.

Eye openers: For me Spencer Johnson and Jack Welch were welcome and inspiring eye openers. As I read the two books over again, it dawned on me that despite the success of NWSC's change management programmes, there was still plenty of room for improvement. I realised that the use of the SMART criteria to set our programme targets was in many ways self-constraining. They were too cautious and deliberately set low to ensure success and to avoid the possibility of failure. Jack Welch reminded me of one of my Area managers who used to set his targets very low in order to achieve more than 100% success and, by so doing, win trophies and cash prizes. In this sense, the fear of failing to reach SMART targets tended to take the upper hand in the minds of our change management programme designers and implementers.

Accordingly, the SMART targets were not bold and daring enough. Furthermore, at NWSC we were easily satisfied with SMART targets. This became a fertile ground for complacency and self-satisfaction. NWSC managers were not exerting themselves hard enough to exceed SMART targets. As one senior manager aptly observed at the time, we seemed to be content to stay in the 'comfort zone of SMART targets'. Reading Jack Welch convinced me that it was time to throw caution to the wind and embrace the stretch-out concept. The Corporation is heavily indebted to Jack Welch and Spencer Johnson for the origination of its Stretch-out Programme.

Although our change management programmes had improved our work culture and performance immensely, Jack Welch opened my eyes to the fact that we were still shackled in old, bad habits. The Corporation was still suffering from bureaucratic red tape, resulting in un-called-for delays in decision making and implementation. The upward and downward movement of memos at the Head Office and between headquarters and the Areas negated the principles of speed and direct communication. NWSC managers seemed to attach more importance to cumbersome paperwork and procedural details than getting work done. The old habits of intrigue, cliques, backbiting, rumour mongering and infighting still pervaded the Corporation's work environment.

In spite of our deliberate efforts to bring senior management closer to their subordinate staff and to encourage informality, 'the boss element' still kept management and the NWSC 'foot soldiers' miles apart. The manager – worker distance seemed to be anchored on the old English saying that 'familiarity breeds contempt'. Reading Robert Slater's book on Jack Welch taught me that nothing could be further from the truth. Bridging the distance between managers and their workers was the only way to generate teamwork, group enthusiasm, confidence and creativity at the place of work.

Moreover, there was limited worker participation in the initiation and implementation of our change management programmes, including the APCs. This meant that the Corporation was not doing enough to use the brains of all the workers at the headquarters and in the Areas and to capture the best ideas during programme design, implementation and evaluation. Lack of total worker involvement also meant that members of staff did not have 'a sense of ownership' of the change management programmes. Without a sense of ownership, there could be no total commitment and urge to excel. Of course worker empowerment goes with responsibility. Once the workers feel that they 'own' the Programme, they are part of the decision-making process and have a stake in the success and future of their organisation; they will stretch themselves to the limit. It was, therefore, imperative to inculcate the 'spirit of common purpose' and to unleash staff enthusiasm, initiative and innovation through worker involvement.

The psychological disposition of the managers and their staff was still far from perfect. The Corporation had not yet outgrown the mentality of looking at its workforce as an asset to pamper, regardless of the level of performance, rather than a cost to cut down to optimal levels. The Corporation also tended to attach more importance to workers being present – the 8.00 a.m. to 5.00 p.m. mentality – than being productive. Furthermore, management tended to assume the posture of a benevolent employer whose duty is to provide for the welfare of its loyal and hardworking employees.

On their part, the workers tended to take their jobs – their cheese – for granted. They felt they were entitled to job security until their retirement. They did not seem to realise that there was a direct link between the security of their jobs and increased productivity and the Corporation's ability to make money. Indeed, their attitude reminded me of a manager who lost his job in one of the Areas because he had not cared to notice that 'his cheese had moved'. Instead of blaming himself, he had resorted to writing anonymous letters attacking the management. There was a dire need for attitudinal change in the NWSC management and the workforce.

Reading Johnson's *Who Moved My Cheese?* made me realise that the change management programmes so far had been initiated from above rather than below. I wondered how many of us at the NWSC anticipated, appreciated and enjoyed the opportunities as well as the challenges that came with the change management programmes. To what extent could we initiate change from below? In what ways could we turn the challenges of inevitable change into stretching opportunities to reach for the stars? How could the ideas of Johnson be combined with those of Jack Welch to stretch the ongoing APCs to the limit? I decided to become a Jack Welch disciple in Uganda. After all, if stretching did wonders for General Electric, why would it not work for NWSC? The next step was to win over the Board of Directors and my colleagues to my stretching enthusiasm.

Selling the stretch-out concept

It took several hectic weeks of hard bargaining and negotiation to sell the concept and to get it off the ground. It is one thing to be converted to an idea but it is quite another to persuade colleagues to do likewise. As I nursed my new idea of moving beyond SMART, I was not sure whether the Board members would share my stretching enthusiasm and endorse the implementation of the Stretch-out Programme. I knew from experience that my chairman would not object to any ideas or programmes aimed at enhancing performance and service delivery provided they conformed to government policies and regulations and did not compromise the good name of NWSC. He had consistently supported the previous change management programmes and I had no reason to doubt his backing of the Stretch-out Programme. Since I was not sure how the rest of the members, many of whom were new, would react, I decided to float the idea selectively before selling it to the entire Board.

Apart from the chairman, I informally introduced Jack Welch's stretching concept, as well as Spencer Johnson's idea of anticipating, monitoring, accepting and taking advantage of inevitable change, to Mr Gabriel Opio, the chairman of the Board's Finance Committee, to Mr Yorakamu Katwiremu, the vice chairman and also the chairman of the Technical Committee of the Board and to Mr Abdulahi Shire, the chairman of the Administration Committee of the Board. I stressed that the board's endorsement of these ideas would not only catalyse the APCs and subsequently accelerate the performance of the government – NWSC contract, but it would, more importantly, revolutionise the methods of work and decision-making processes at the headquarters for better performance. I gave them copies of the two books and urged them to reflect on the relevance and applicability of the ideas to the Corporation.

To my pleasant surprise, the board members readily agreed that changing the workers' mentality and attitudes was just as important to the success of the change management programmes as was the mobilisation of material and financial resources. Indeed, Mr Katwiremu spoke for many of his colleagues when he pointed out that changing the mindset was even more fundamental than providing the physical and financial resources. Although buying and installing pumps at the waterworks was an indispensable component of our work operationally, it is a simple, straightforward task, but on the other hand, changing the attitudes of workers towards their work and towards each other is a much more problematic case, requiring careful human resource management skills and practices that could only be nurtured and inculcated over a long period of cultural and social transformation.

He was so enthusiastic about stretching that he offered to give a brief favourable presentation at one of the APC quarterly review workshops in Mbale, which was planned for mid-August 2002. I became confident that given these informal and positive exchanges, Board members would endorse the idea.

The next hurdle was to win over management colleagues to what may have looked to them like an outlandish, fanciful dream. Stretching was a revolutionary concept; it called for simplicity, openness and good ideas to inform decision making. I was not sure that my colleagues would willingly abandon the safe world of rules and regulations, controls and procedures, tight budgets, hierarchy and authority in favour of the risks and uncertainties of stretching into the unknown. Wouldn't stretch, devoid of bureaucracy, controls and defined boundaries, be a recipe for anarchy?

Most of the NWSC senior managers are down-to-earth professional engineers dealing with practical engineering and management problems of delivering water and sewerage services as effectively and efficiently as possible. Would such practical, no-nonsense managers embrace stretching with all its implied dreaming and fantasising about setting seemingly impossible targets, which demanded superhuman effort and commitment? Dreaming and fantasising were luxuries of poets and creative writers, not for down-to-earth professional engineers! Why abandon our dependable SMART criteria for devising and implementing our change management programmes that had stood the test of time? In any case, why initiate a new programme well before the conclusion of APC II, whose implementation was still in progress?

I set the ball rolling through colleagues who were most receptive to change and who had been instrumental in the design, implementation and evaluation of the previous and ongoing change management programmes. I started with four of my senior managers – Messrs Johnson Amayo, Silver Mugisha, Edmond Okaranon and John Bosco Otema – whom I knew were always positively disposed to new ideas and methods of work. These managers were chosen for several reasons. All of them had mastered and internalised the essence of change management programming. Messrs Mugisha and Okaranon were invaluable assets owing to their intellectual rigour and analytical skills.

Mr Amayo was not only the manager of operations but was also an effective behind-the-scenes operator who had played a leading role in the previous change management programmes. Mr Otema, a sociologist and the senior marketing officer, was chosen for his public relations skills – he could serve as a bridge between the management and the employees in the new stretch environment. These senior managers became the pioneer stretch champions.

When I introduced the idea of stretching, I emphasised the need to reduce bureaucracy and the boss element to foster speed and simplicity, as well as worker involvement and self-confidence. I also explained why, in my opinion, the idea was very important. I called on them to be candid and earnest while discussing with me the pros and cons of stretching. I wanted them to be frank about the relevance, suitability and application of this new programme. I gave them copies of *Who Moved My Cheese?* and Robert Slater's book on Jack Welch to read, after which we would meet again to have a detailed brainstorming session on the subject. I also requested Mr Mugisha, who had completed his masters' degree in the Netherlands and was pursuing a PhD, to study the two books and identify the key stretch-out concepts regarding good leadership that would inform our subsequent meeting.

During the next meeting, Mr Mugisha presented the 'Stretch-out Programme: Good Leadership Brief Notes' and the 'Stretch-out Programme: Inception Report' for brainstorming. He also advised that we marry the ideas about change in *Who Moved My Cheese?* and Jack Welch's stretch-out concepts to design and implement the Stretch-out Programme. My colleagues agreed with me that stretching would be timely to catalyse the APCs to aim much higher than the SMART targets. They, however, stressed that the idea should be sold to as many managers and staff as possible so that its significance would be understood by the entire NWSC family before formal introduction. We therefore agreed to distribute copies of the two books to all the managers and staff. We also agreed that my colleagues would henceforth become the champions of selling and developing the Stretch-out Programme. With the four senior managers on board, I knew that stretch was bound to succeed.

The managers and the rest of the staff were required not only to read and understand the ideas in the two books but were also obliged to think about how those ideas could be used to overcome operational drawbacks in the ongoing APCs. What needed to be done to reduce bureaucracy and the 'boss element', to empower staff, and to dismantle the boundaries that hampered speed, simplicity, worker involvement and receptivity to change? In order to ensure that the managers and the rest of the staff read and internalise the two books, we took the unusual step of setting tests, at the risk of being accused of treating grown-ups like school kids. Senior managers were asked to make presentations about their understanding of stretching, to determine its applicability to the Corporation and to explain how stretching could help exceed SMART targets. The rest of the staff simply took tests on what they had read in order to establish how much they understood of this new idea.

To develop and popularise the Stretch-out Programme, we adopted a two-pronged strategy. On the one hand the strategy was to choose a pilot Area where stretching would be put into practice and has its operational efficacy tested before bringing other Areas on board. The pilot Area was expected to develop the content and the implementation strategy with the assistance of officers from headquarters whose role would be that of a facilitator. All staff of the pilot Area were to be involved in the Programme design and implementation. On the other hand, the strategy required a series of workshops to pick the brains of key NWSC actors, including Board members and leaders of the workers' union, about the inputs and modalities required for the implementation of the Programme.

We chose Jinja, the largest Area after Kampala as the pilot area. In Mr Andrew Sekayizzi, the Jinja Area manager, we had a receptive, open-minded and enterprising officer. He was one of the few fire-fighting managers in NWSC whom I could count on any time. He commands authority and knows how to move and motivate his staff in a simple but sure way. Any member of the staff who could not getalong with Mr Sekayizzi was usually beyond redemption. I had no doubt in my mind that he was the ideal manager to pioneer the Stretch-out Programme.

In 1999, after winning the chairman's trophy at the end of the 100 days Programme, we had posted him to the troubled Jinja and he had turned it around in record time. Jinja became one of the best performers. It is easily accessible, conveniently located 80 km from the NWSC headquarters, which meant that senior managers and stretch-out champions could go to Jinja in the morning and return to Kampala in the evening. It was therefore possible for the champions to facilitate stretch-out working sessions at minimum cost in terms of transport and overnight subsistence allowances. For these reasons, Jinja was the ideal pilot area for this programme.

I subsequently spoke to Mr Sekayizzi about the idea of stretching. In my previous communications with my managers and staff, I used to assume a formal, ivory tower mentality. In this case I chose to tone down the 'boss element' and treat him as my colleague and partner rather than a subordinate. From the onset of stretch, formality and distance had to give way to informality and cordiality. So when I called Mr Sekayizzi, I deliberately asked: 'Is that Andrew?' instead of the usual: 'Is that the Area manager?' This was my first step towards breaking down the 'boss element' boundaries that had kept managers and their subordinates apart since NWSC came into existence in 1972.

I talked to him as a colleague rather than a boss imposing solutions or giving mighty, unquestionable directives from on high. I chose my words carefully and I remember saying something along the following lines:

> I have confidence in you and the workforce of Jinja as a whole. I want you and your team to determine your destiny, that is, in conformity with our corporate goals. We should strive to build a corporate culture that empowers and encourages teamwork, transparency, free of unnecessary bureaucracy and the boss element. I want to empower you and reward your performance. Why shouldn't headquarters allow you to stretch yourselves rather than curtailing your potential?

I told him about the management's desire for him to become one of the chief stretch champions in the Corporation and to pioneer it in Jinja. When I detected some hesitation in his tone, I assured him that there was no need to rush. I told him to think about it and to consult with his colleagues.

Much later, Mr Sekayizzi told me that our telephone conversation was puzzling for he did not understand why I had deliberately introduced stretch with unusual humility, modesty and persuasion. But it did not take him long to know why. Stretching was a revolutionary departure from the past.

To wind up the conversation, I promised to visit Jinja the following weekend to have a more detailed discussion on stretch-out. I also promised to bring him copies of *Who Moved My Cheese?* and Robert Slater's book on Jack Welch, as well as the 'Stretch-out Programme: Good Leadership Brief Notes' and the 'Stretch-out Programme: Inception Report' for him to study and share with his colleagues in Jinja. I was optimistic that whatever reservations they entertained, Mr Sekayizzi and his management team were likely to accept our proposal for Jinja to pioneer stretch-out once they grasped its essence.

The following weekend, I went to Jinja as I had promised to promote my pet stretch-out subject. Indeed, before leaving for Jinja, I invited Mr Sekayizzi and his management team for dinner at Sunset Hotel where I was booked to stay. Over dinner, I was informal and casual as possible. I told my guests to relax.

I assured them that I had come to pick their brains and to involve them in what I was proposing to do rather than to lecture or give them directives. I expressed the hope that Mr Sekayizzi had already briefed them about the purpose of the dinner-cum-meeting. I then briefly introduced the stretch-out concept and stressed its importance as a management tool to energise the APCs.

I briefly recounted my Harvard experience and brought in the story of Jack Welch and how he had made General Electric succeed through cultivating simplicity, speed, confidence and worker involvement, and by getting rid of bureaucracy and the boss element – in short, stretch-out. Secondly, I told them about *Who Moved My Cheese?* I noted that at NWSC, we had a choice to make; either to be complacent and fail to see change, which would culminate in collapse, self-pity and blaming other people for our own shortcomings, or to anticipate, monitor, adapt to and benefit from the changes I was calling on them to initiate. I gave my audience the relevant literature to study and discuss the stretch merits, opportunities and likely constraints.

Having explained how the ideas of Jack Welch and Spencer Johnson had inspired me, I summarised what I considered to be the stretch-out objectives and the expected outputs. I informed them that Jinja had been proposed as the stretch-out pilot Area subject to their voluntary consent. I emphasised that this proposal was the management's expression of confidence in Jinja staff and leadership, and recognition of their distinguished record in the previous change management programmes. I sought their candid opinions about the proposal and urged them not to mince their words or hide their feelings since they would be required to implement the Programme. Deep down I was a bit scared that my initiative could be turned down, a prospect that was too ghastly to contemplate.

The initial response of my audience seemed to confirm my worst fears. It did not take me long to sense their scepticism and reservations. They were blunt enough to say that while stretching may have done wonders at General Electric, it might not be appropriate for NWSC, where the circumstances were completely different. They called for a sense of realism and cautioned me against urging them to bite off more than we could chew. They argued that since APC II was still in progress, it was premature to evaluate its performance or to call for its revamping through the initiation of yet another change management programme. Given these reservations, we agreed that the Jinja team would study the relevant literature I had given them, to help them think through the concept, and to meet later on for a more definitive discussion.

A week later, I went back with the stretch champions, Messrs Amayo, the operations manager, Mugisha and Okaronon, the principal engineers in charge of operations, and Otema, the senior marketing officer. I also enlisted the support of one of our senior officers in the audit department, Mr Paul Mugoya, a Jinja man, to help us explain the complexities of stretching and its revolutionary possibilities for the future of the Corporation. Before going to Jinja, Mr Sekayizzi had assured me that his management team had read the stretch concept literature and had reached a general consensus that it could yield positive results for Jinja and the Corporation as a whole.

By the time we travelled to Jinja, Mr Sekayizzi and his colleagues had, in principle, accepted the stretch-out initiative. The challenge facing the champions from the headquarters and the Jinja management team was therefore to formulate stretch-out targets over and above the SMART targets under implementation in APC II. The meeting took the whole day and much of the night. This protracted meeting was an exhausting foretaste of stretching but it was worth it. By 1.00 a.m. the champions and their Jinja colleagues had reached consensus on stretch-out targets with emphasis on operating margins, and Jinja had agreed to pilot this adventure.

I hardly slept that night. I was anxious to forge ahead with the Programme without delay. The thought of dilly-dallying letalone failing was a frightening prospect. The first thing I did the next morning was to telephone Jinja and find out what had happened. There was good news. The Jinja Area management had agreed to pilot stretch-out and was eager to get started. I was thrilled that we were making progress. However, before launching the Programme, it was imperative to bring the entire Jinja workforce on board. The success of stretch-out required the support and participation of the workers from the onset to the end. It was important for the Jinja Area workers to play a leading role in working out the strategies aimed at agreed stretch-out targets and to identify constraints that were likely to be encountered along the way. It was therefore agreed that another meeting would be held soon to involve the workers in preparation of the Programme. The headquarters champions returned to Kampala with their heads held high; stretch was on course.

Before the Jinja stretch-out workshop, we went to Mbale to sell our new programme to a wider NWSC audience. The Mbale workshop was planned to kill two birds with one stone – to evaluate the second quarter of APC II and to formally introduce the stretch-out concept to Board members, senior managers and all Area managers. In my opening remarks, I commended the Area managers for the APC II progress so far but observed that this progress was still constrained by bureaucracy, stifling boundaries and 'I am the boss' mentality.

I assured them that through stretch-out, the Corporation's management would give the Area managers more power and freedom than they already had under APC II and would ensure that all employees – high or low – would be fully involved in the design and implementation of all future programmes. I pointed out that the management, in consultation with all Area managers, had already identified constraints impeding the progress of APC II. That is why, starting with Jinja, the Stretch-out Programme had been initiated to address the APC II constraints.

All Board members spoke in favour of the stretch-out initiative. They pledged their full support for its implementation and welcomed the progress that had already been made in Jinja. They encouraged other Areas to embrace it since it was likely to accelerate achievement of ongoing APC II targets. The Area managers needed little encouragement. They were keen to take on stretch-out as soon as they received the green light from the headquarters. The Mbale workshop turned out to revolve around how and when the Areas would start on this new programme rather than whether or not the stretch initiative should be accepted as a management tool. Therefore, under the guidance of the stretch champions, the Mbale workshop worked out a concrete timetable to implement the Stretch-out Programme, starting with Jinja as a pilot area. Unlike previous change management programmes, each Area was to take on stretch when it was ready to do so.

A subsequent follow-up workshop was held at our training centre in Kampala. It was attended by all heads of department to explore ways and means of assisting Jinja in a bureaucracy-free, boundary-less and boss element-free environment, underpinned by enhanced area management empowerment and full-scale worker participation. On its part, the Jinja Area management was required to spell out what sort of material and technical assistance would be needed for realisation of stretch-out targets.

Mr Martin Ssekibala of Group To Consult, assisted by the in-house stretch champions, facilitated the workshop. Board members followed the workshop proceedings with keen interest. The forum was very successful, with all the participants freely participating in the deliberations and giving all sorts of ideas, including the suggestion that Areas should be empowered to recruit staff up to Scale 8. The stage was thus set for the final push to launch stretch-out in Jinja Area through deliberate and concerted worker participation.

Modus operandi

Effective understanding and conceptualisation of stretch among the employees was undertaken using a process known as 'workout'. It was a process of building trust and self-confidence among employees by re-defining the relationship between the boss and the subordinate. The aim was to create a new approach and attitude to work by instilling values of speed, simplicity and self-confidence at all levels of organisational hierarchy. Employees in each Area were taken on a three-day workout workshop in an off-the-site venue.

The schedule started with setting out an Area's stretch-out vision, mission and targets. The Area staff, facilitated by the stretch champions, would chart out these elements, which reflected the Area's ambition. After this, members of staff, on individual basis and without any interference, wrote down all the issues they felt were affecting their performance. These were then compiled and categorised according to function. The staff would then be put into three groups, chosen according to their functional competence — technical, commercial and finance, and administration and personnel — to discuss the issues raised about that particular functional area. They would then formulate strategies to address the constraints that might impede the realisation of the agreed stretch targets.

During the strategy formulation session, both 'hardware' and 'software' issues envisaged to affect the achievement of the targets were discussed exhaustively. Software issues were qualitative in nature and were related to constraints impeding speed, simplicity, self-confidence and worker involvement. Hardware issues, on the other hand, were those regarding lack of tangible materials and resources and, therefore, which were likely to significantly hamper effective and efficient Area operations. Both issues were discussed comprehensively and all staff were encouraged to contribute ideas and make recommendations freely on how best to address them.

Due to the openness of the discussions, a number of software issues concerning the behaviour of their managers were brought to the fore.

Because software issues critically influence organisational behaviour and eventually staff attitude at work, they had their separate sessions, which were aimed at cultivating harmony among staff.

All the strategies – both hardware and software – formulated during the group sessions were precise, unambiguous and clearly spelt out the offices or individuals responsible for their implementation, time by which those strategies were to be executed and the cost implications.

After formulating the strategies, all the groups met in a plenary session. A member from each group presented the issues that were raised and the corresponding resolutions or strategies that had been formulated. The facilitators guided the session in harmonising all the issues raised. In this session, the Area manager was required to say either 'Yes' or 'No' on behalf of the Area management team. In some cases, he had the option of requesting for more time, especially if there was a need for more consultations. As a result of the bottom-up approach adopted in the workout and the high level of consensus at the plenary sessions, it was rare for the Area manager to say 'No'. The champions were at hand to ensure objectivity in the entire exercise. In the case of software issues, if a solution/strategy concerned behaviour and/or conduct, an individual or the manager gave 'Yes' or 'No' if they had the capacity to influence the intended person's behaviour through the normal managerial procedures.

The workout sessions required total commitment from all participants. All staff were fully involved and in most cases worked for long hours, from

8.00 a.m. to midnight, sometimes going to as late as 2.00 a.m. This was in itself a show of commitment and desire to see the Programme succeed. The staff were energised and confident of achieving their targets.

Jinja Area workout

In order to ensure the full participation of the Jinja Area workforce without any distraction, we decided to hold an uninterrupted, around-the-clock workshop at the Jinja YMCA hostel to launch this programme. This facility was chosen not only to minimise expense but also for its spartan setup. I personally invited each participant. The idea was to emphasise the equality of all participants regardless of their individual positions. The participants, including senior managers and even Board members, had to share rooms and eat the same food. We bought T-shirts and caps for each participant to wear. Suits and ties were banished from the workshop. The first rule of the Jinja meeting was for participants to call each other by the first name instead of Mr or by official titles. The champions and Mr Ssekibala of Group To Consult were on hand to facilitate the workshop deliberations, but the workers were encouraged to be the principal actors and thinkers. This was done to enhance self-confidence, self-worth and a sense of ownership of the Stretch-out Programme.

This workshop started on 23 August 2002 and lasted three days. The Corporation's Board vice chairman opened it by stressing the importance of stretching. After giving an introductory overview of the background, purpose, the modalities and expected outcomes of the workshop, I urged participants to make their contribution freely without any inhibitions. To encourage free discussion, the workers met on their own, without their Area manager, in two working groups. One of the groups was assigned the task of dealing with the technical aspects of water and sewerage services, notably water production and water quality, water distribution and NRW, plant maintenance, block mapping and the extension of the distribution network.

The second group was required to deal with the commercial, financial, customer care and personnel dimensions of the Corporation's mandate, such as billing and revenue collection, budgeting and auditing, procurement, personnel management, the degree of Area autonomy, the commercialisation of operations and customer satisfaction.

After the working group discussions, all the participants came together in what we called a harmonisation session. I personally attended this session to get a feel of the discussions and to make my own contribution where necessary. The methods of work of this session were both confrontational and conciliatory. The Area manager and the workers were on opposite sides and the champions were in between. The groups presented their stretch-out strategies to the Area manager who, on his part, accepted or rejected any proposed strategy there and then or asked for more time to evaluate any of the proposed strategies. On their part, the champions discreetly intervened from time to time to moderate the lively debate or to provide some information for purposes of clarification and moderation.

The workshop went much more smoothly than expected. No hitches were serious enough to disrupt or bog it down. By the time it ended on 25 August 2002, all participants were upbeat and in good cheer. The scope and intensity of the workshop deliberations, the creative energy and enthusiasm of the participants, the proposals of diverse strategies and the offer of practical solutions were certainly impressive. The workshop was sufficient proof of the value of worker involvement, if any reassurance was needed.

Accordingly, I praised the enthusiasm with which the Jinja workforce had embraced stretch-out, regardless of the uncertainties and risks involved. I was optimistic that the implementation of stretch-out would dramatically improve Jinja's performance by reducing red tape and empowering the workers. I assured the workers that better performance and increased productivity would not only culminate in improved operating margins but would also make it possible for the Corporation to pay performance bonuses. When the Jinja Stretch-out Programme was formally launched on 25 August 2002, I was delighted beyond imagination. My dream had come true. Of all our change management programmes, stretch-out was the most revolutionary breakthrough. It was a refreshing wind of change that was about to change our methods of work for the better. Between September 2002 and January 2003, other Areas, including Gaba and Bugolobi in Kampala, voluntarily embraced stretch-out at their own pace and timing.

As mentioned earlier, the Stretch-out Programme was aimed at enhancing the APCs. Since the Areas were already operating under performance contracts, the Stretch-out Programme was considered as an amendment rather than a replacement, and was operationalised through a Memorandum of Understanding (MoU), between the headquarters and the Area management teams. The signing of the memorandum usually took place at the end of the workout sessions, which usually took three to four days. The Board and Union representatives witnessed or endorsed the memoranda to signify their respective support and commitment. The champions and Area managers then supervised the implementation of the stretch workout recommendations.

The MoU clearly spelt out each party's role and obligations within the Stretch-out Programme and its implications for the ongoing APCs. Based on the principles of stretch highlighted earlier, a special incentive package in addition to the APC incentives was included in the MoU. This was aimed at motivating and rewarding the workers' superhuman efforts in improving performance, even when they did not achieve the stretch targets.

As part of the MoU, headquarters was required to provide all the agreed requisite financial and technical support to the Area. This meant establishing an environment free of bureaucracy, which would enable the Area to get the required information, equipment and finances in time. On the other hand, the Area undertook to implement fully all the agreed strategies and strived to improve performance and service delivery in accordance with agreed stretch-out targets.

MoU covered four categories of services, namely, personnel and administration, finance, commercial and customer care, and technical. The issues were already catered for under APCs, the main qualitative difference was that, in this case, staff identified problems affecting each category during the workout sessions and listed their possible solutions. As such, there was more worker participation and ownership of strategies.

For example, in the personnel and administration category, the problems listed included lack of facilities such as telephones, transport, protective gear such as gloves, masks and overalls, bureaucratic communication bottlenecks, delays in decision making and work monotony due to lack of rotation. On the other hand the problems that were listed under finance, commercial and customer care ranged from delayed billing and poor debt management to data updating bottlenecks and flawed customer handling resulting in lack of adequate customer satisfaction. Furthermore, the problems in the technical category included delays in attending to bursts and leaks, inadequate tools and materials inputs, defective equipment and inadequate planned preventive maintenance.

The Jinja MoU also contained solutions to the numerous and diverse problems that were identified during the workout sessions. With regard to personnel and administration, some of the proposed solutions included introduction of transport management procedures, getting rid of selfish practices in the use of facilities such as telephones and refining the upward and downward flow of information between staff and management. Other solutions included requisitioning materials and protective wear in time; being sensitive and responsive to subordinate staff and putting in place mechanisms to ensure speed, simplicity and continuous worker involvement at all levels of stretching in order to enhance performance improvement, productivity and service delivery.

In finance, commercial and customer care, the solutions offered were, among other things: streamlining accountability mechanisms and procedures, expeditious settlement of bills, keeping accurate customer information and updating it regularly, and rigorously enforcing customer agreements.

Solutions to technical problems ranged from fault detection and mobile motorcycle patrols and surveillance to comprehensive block mapping and round-the-clock emergency preparedness to respond to leaks, bursts and overflows.

The Jinja stretch-out offered us a lot of experience, which was used in the planning and implementation of stretch in other Areas. Under the facilitation of the headquarters stretch champions, all Areas worked out their vision, mission and three-month stretch-out targets and signed memoranda.

The new arrangement, where the Areas formulated their own visions and missions, was completely new and inspiring in the history of NWSC. More interesting is the fact that most Areas formulated their visions over and above that of the Corporation. For example, the vision for Masaka Area was 'to be number one in the provision of water and sewerage services in Uganda.' The one for Soroti Area was 'to be a role model for the provision of efficient and cost-effective water and sewerage services in Africa', stretching beyond Uganda!

However, it was Lira and Mbarara Areas which stretched their visions and dreams farthest to the world stage. The vision of Lira Area was 'To be a role model in the provision of water and sewerage services in the world.' Similarly, the vision for Mbarara was 'To be the best utility in the world in effective and efficient provision of water and sewerage services to a delighted customer.' Not even the NWSC vision had dared to stretch that far! What all these visions signified was that stretch-out had, by the beginning of 2003, become an exciting programme to an extent that everything was achievable in the minds of the workers.

The stretch-out mental revolution

The Stretch-out Programme marked the beginning of a mental revolution. Initially it was implemented in two phases. The first lasted three months and the second six months. Ever since life has never been the same in NWSC and all the subsequent programmes have adopted the stretch concepts. It was a revolutionary departure from previous change management programmes in NWSC. Stretch introduced new ways of thinking, of doing things and of closer worker-to-manager relationships. To monitor the progress of stretch, a series of evaluation workshops were conducted by the champions from the headquarters in close collaboration with staff and management of the individual Areas. As each Area completed Stretch I, it embarked on a six-month Stretch II programme to sustain the revolutionary momentum that had been gained. For example, Stretch II in Jinja began in December 2002 and was completed at the end of May 2003. Other Areas followed the Jinja example. The lessons learnt from Stretch I were used to inform the design and implementation of Stretch II. The achievements registered by stretching were both quantitative and qualitative. During the Programme, emphasis was put on setting stretch targets that were far above those of SMART, which had been set under APC II. This was done with the objective of catalysing and motivating Areas to ensure that at the end of the day the APC targets were achieved or even exceeded. As expected, Stretch resulted in performance improvement that enabled most of the Areas to exceed their APC targets.

Of all the stretch-out achievements, none was more fundamental than the qualitative 'mental revolution' that swept through the Corporation like a hurricane in 2002–03. This revolution entailed profound changes in attitudes and work methods across the Corporation. Gone were the traditional bureaucratic barriers that had hitherto impeded open and free flow of information. Gone was the distance that had previously kept the managers apart from their subordinates. Stretch-out was, above all, a revolution that shifted emphasis from hierarchy, rules and procedures to the primacy of ideas, speed, simplicity, and worker participation. In less than a year, stretch-out transformed the Corporation's organisational behaviour beyond imagination, to the astonishment of other public utilities and the satisfaction of our customers. It generated unprecedented staff enthusiasm and enhanced performance at all levels.

Matching words with deeds

One of the fundamental changes brought about by stretch-out was that henceforth, the management at the headquarters as well as in the Areas honoured the commitments spelt out in the MoU – matching words with deeds. At the beginning of stretch-out, many Area managers were sceptical about headquarters' resolve to comply expeditiously with its MoU obligations and let go of its powers and functions.

The Corporation's top management kept its side of the bargain throughout the implementation of stretch-out. For example, immediately after signing the MoU with the headquarters, Jinja Area received the funds and other required inputs to kick-start the implementation of stretch-out without any delay. Once the required resources had been released, Jinja had no excuse to dilly-dally under the pretext of lack of facilitation.

Similarly, as stretch-out gradually got under way, headquarters did not hesitate to devolve power and responsibility, in order to break bureaucracy and speed up decision making and programme implementation. Areas were empowered to manage their own affairs and take responsibility for their actions. A good example was Mr David Opoka, the Mbale Area manager, who seized the opportunity to run media advertisements and promotions and to devise new ways of speeding up water business operations without seeking prior clearance from the headquarters. Indeed, all Area managers were free to implement any decision provided headquarters was kept informed. Thus, the devolution of power and functions speeded up the process of decision making drastically and culminated in enhanced performance.

From their business to our business

Before stretch-out, members of staff tended to see provision of water and sewerage services as the business of their employer because they were not directly involved in decision making, or in programme design and implementation. Employees did not appreciate that the destiny of their employer was closely tied up with their own. They did not seem to recognise the fact that the success or failure of their employer reflected their own success or failure. Stretch-out transformed this mindset.

NWSC became 'our' business; employees transformed their work attitudes, habits and behaviour. Most employees were prepared to invest more time, effort and energy to meet stretch-out demands and expectations. Workers reported to work on time and worked beyond the call of duty throughout working hours. Many workers, including senior managers, worked from 7.00 a.m. to 8.00 p.m. going beyond the normal working hours without overtime pay. Some of the technicians willingly answered emergency calls to repair leaks or bursts even when they were off duty. Absenteeism declined drastically in all Areas. Sister public utilities were astonished by this 'wind of change' in NWSC, speculating that its workers worked that hard because they had received more pay, which was not the case.

Stretch-out transformed the working relationship between supervisors and their subordinates. Managers and their staff discussed ideas and activities without fear of victimisation or recrimination. Stretch-out introduced a revolutionary culture of open and positive criticism to banish the boss element from our operations. This attitudinal change and worker involvement enhanced self-esteem and self-confidence. Stretch-out also revolutionised the language of discourse. Words such as best practice, innovation, mission, vision, ingenuity, benchmarking, performance indicators, break-even, cost reduction and operating margin became commonplace lingua franca.

Stretch memories and experiences

Nothing illustrates the revolutionary nature of stretch-out better than the memories and experiences of some of the Areas. According to Mr Kenneth Mushabe, the Kabale Area manager, stretch-out gave him the opportunity to know his workers better, to tap their ideas and to energise them into dreaming and stretching their abilities. Furthermore, communication and feedback from headquarters to Kabale improved greatly. Kabale was facilitated to repair faults and defective lines, extend the distribution network, get new vehicles and motorcycles as well as install new connections at maximum speed. For Mr Mushabe, Kabale had not experienced such intense activities before stretch-out.

Mr Mutikanga, the Area Manager Fort Portal/Kasese, praised stretch for creating a free atmosphere for discussion, which left no stone unturned and allowed for criticism of anything that hampered good performance. According to him, stretch-out had unleashed boundless energy and talent that could move mountains to achieve stretch targets.

Mrs Sylvia Tumuheirwe Arinaitwe, the Entebbe Area manager, vividly recalls that, initially, the whole idea of stretch-out had sounded 'strange and crazy'. For her, stretch-out was a bitter/sweet concept. It had good ideas but the realisation of those ideas seemed to impose demands that went beyond what was humanly possible. She was later to describe the fundamental behavioural change in her Area as follows:

> Everybody had been reformed, [the] boss element gone and each one's door open to everybody in a bid to implement the open-door policy that came with the programme. The concept of everybody wearing T-shirts on Fridays and at all NWSC gatherings was another experience ... By the end of the programme the Area had been greatly facilitated and great performance improvements registered. For the first time, the Areas had their own visions and mission statements. Behavioural change had been registered in almost all personnel. A culture of open positive criticism had emerged in the personnel to the extent that we no longer need small discussion groups excluding supervisors to generate ideas ...

According to Mr Okwir, the Tororo Area manager, 'Stretch-out can be rated as the most beneficial and successful programme in NWSC.' Mr Ameda, the Arua Area engineer, recalls it thus:

> Without the stretch-out programmes, Arua Area would be worse off or nowhere. The Area thus owes the NWSC top management and the MD all the credit for this great innovation that has had significant impact on the Area and Arua town. With stretch, NWSC will never be the same again.

Mr Emudong, the Lira Area manager, summed it up in the words:

> ... the big animal called bureaucracy had been killed. The freedom which was given to Areas to implement their strategies was unimaginable. Every task which required resource allocation in terms of money from headquarters was quickly acted upon.

The above words clearly capture the extent and intensity of the qualitative 'mental revolution' that had occurred in all Areas, thanks to the implementation of stretch-out.

Summing up the gist of stretch-out gains

Undoubtedly, stretch-out was a revolutionary experience without precedence in the history of NWSC. Apart from the quantitative achievements and the qualitative 'mental revolution', the tangible gains from the Programme start with improved service delivery and customer care, way beyond what had been achieved by the previous change management programmes.

Secondly, by creating a sense of ownership through worker participation, stretch-out inculcated responsibility and accountability at all levels. Thirdly, unlike its 'predecessors' it ushered in speed, simplicity and worker involvement in all our activities and operations.

Above all, it led to a new spirit of renewal and optimism about the future prospects of NWSC with or without privatisation. At a practical level, stretch-out marked the beginning of a new phase in the modernisation of the Corporation through the refurbishment of its offices, the upgrading of information, communication and accounting technologies, and the acquisition of new operational vehicles and motorcycles.

Refining the reward system

Stretch-out also involved trophies, cash prizes and performance bonuses introduced earlier on, to recognise exemplary performance. Since the stretch targets were much more difficult to achieve than the SMART ones, the stretch-out memorandum introduced the *pro rata* formula into the reward system that put special emphasis on operating margin performance.

If an Area achieved 5% improvement over and above APC SMART targets, it was paid 25% of monthly basic pay, even if it did not reach the stretch targets. On the other hand, if an Area did not meet the said target, it still received a *pro rata* performance pay. In addition, an Area that achieved the set stretch operating margin target received another 25% of monthly basic pay. Therefore, some of the Areas that performed well enough to meet or exceed the minimum stretch memorandum expectations won 50% of monthly basic pay, in addition to the traditional trophies.

Apart from the cash incentives, presents and salary increments, tests were administered in different Areas to find out whether employees had understood and internalised the stretch concepts. Members of staff who excelled in those tests received extra points, which would count in the race for promotion and other fringe benefits such as loans. A casual worker in Fort Portal, who emerged best performer in a stretch test, won himself full employment in the Area and was subsequently promoted from a sewage attendant to a meter reader. In Kasese, at the discretion of the management, field workers were given 'stretch motorcycles' and hardship allowances to facilitate their work in their highly mountainous area. Such incentives were designed to demonstrate that workers matter, and that their contribution is central to the success of the Stretch-out Programme.

Stretch shortcomings

Some of the lessons from the Stretch-out Programme are highlighted at the end of this chapter. However, just like the previous change management programmes, it had its own weaknesses, which had to be addressed in order to realise its full potential. Firstly, the Jack Welch stretch model was applied rigidly without modification. This model tended to encourage confrontation rather than consensus. Despite the moderating and mediating influence of the champions in workout sessions, some workers tended to concentrate on emotional and personal issues and to point accusing fingers at their supervisors.

Instead of raising issues pertinent to work by pinpointing performance constraints and proposing practical solutions, some workers tended to present a long list of unhelpful or unsubstantiated complaints against their supervisors. On their part, managers tended to assume a defensive position and to dismiss legitimate complaints and suggestions. To avoid the adversarial approach of the Jack Welch model, we shifted to what we code-named the Entebbe model, which encouraged open discussion and consensus.

Secondly, the 'Yes. No. Let me think about it' approach to staff proposals during the workout plenary sessions, whose purpose was to ensure speed and simplicity in decision making, left Area managers with little room to manoeuvre. On-the-spot 'Yes or No' decision making presupposed that headquarters would automatically endorse whatever the Areas had decided. On one occasion, headquarters reversed an Area manager's 'Yes' commitments and this threatened to undermine his authority and confidence. Fortunately, headquarters used its veto powers sparingly and Area managers did not have 'to eat their own words' or 'lose face' before their own staff too often. In order to avoid embarrassing climb downs, the limits within which Area managers had complete freedom to say irreversible 'yes or no' to their staff were clearly specified.

Thirdly, the three-month period for Stretch I, and even the subsequent six-month period for Stretch II, was too short to implement all aspects of the Stretch-out Programme. Although Jinja was chosen as the pilot area, other Areas jumped on to the Stretch-out bandwagon too early, before lessons had been drawn from that operational experience. The fact that stretch started in different Areas at different times meant that they could not be evaluated at the same venue and time. Individual Area evaluation exercises turned out to be too costly and time consuming for those involved. In addition, separate stretch evaluation exercises meant that Areas would not get the opportunity to share experiences.

It goes without saying that behavioural change, by its nature, is a protracted process. It cannot occur overnight. In the case of stretch-out, some managers at the headquarters were reluctant to relinquish their powers and functions because they believed that removing bureaucratic controls and approval procedures would culminate in abuse of office or misuse of the Corporation's resources. Some managers also wanted to maintain the boss element on the ground that it shielded them from the dangers of familiarity-induced insubordination. They also argued that on-the-spot decisions were detrimental to careful deliberation and consultation over various issues that were raised in the stretch workout sessions. Fortunately, the voices of the old mindset – of those who feared change or failed to accept and adapt to change – were not strong enough to derail the Stretch-out Programme or to slow down its momentum. The few cases of misconduct that were attributed to reduced bureaucratic control were quickly and effectively nipped in the bud.

The Stretch-out Programme focused on group performance, participation and incentives. The programme did not pay much attention to the role of individuals in the transformation of behaviour or in underscoring organisational success. As stretch-out progressed, we began to realise that maximising individual potential was as important as catalysing group participation and performance. This meant that, in addition to stretch, it was imperative to focus on individual responsibility, performance and incentives in order to boost productivity and service delivery. That is why we decided to introduce the 'one-minute management concept', which is the subject of the next chapter.

Lessons learnt

- Do not be content or satisfied with success, however great it may be. Shop around for new ideas from whatever source and use them to sustain the change momentum.
- Having a cooperative, confident and self-respecting workforce is the key to organisational change and good performance. Involve your workers in decision making to give them confidence and self-worth.
- Behavioural change is a protracted process requiring patience, deliberation, all-round participation and consultation.
- Motivate your workers to release their energies, enthusiasm and ambitions to effect change and reward them accordingly. Employees are quick to learn and appreciate new ideas and techniques to effect change provided you give them a sense of ownership of the organisation.
- Go for stretch to make what seems impossible easily achievable. You may not reach the stars but you will outshine the SMART guys of yesteryear.
- Instead of a remote, top down and autocratic management style, go for horizontal management, which knows no bosses and no boundaries. Always remember that simplicity, speed, clarity, self-confidence and direct face-to-face encounters oblivious of status are the secrets to organisational success.
- Since change is inevitable and the only constant in organisations, as in life, accept, anticipate and initiate change; monitor change, and keep changing. Do not fear change. Do not take things, including your jobs and businesses, for granted. Always remember that changing behaviour, attitudes – in short, an organisational work culture – is even more important than providing abundant physical and human resources.

Breaking the Boss element and empowering staff: The stretch-out concept emphasises no boss element and full staff participation.

Left: Head Office Staff smartly dressed in T-shirts listening attentively during the Head Office Stretch Workout session.

Workouts taken to the Areas. Right: staff of Mbarara Area smartly dressed in T-shirts posing for a group photo during the Stretch Workout session.

Below: staff of Mbale Area after their Stretch Workout.

The Group Incentive Mechanism

Breaking the boss element.

Above: the Area manager of Entebbe, Mrs Sylvia Tumuhairwe (right), and the representative of head office, Mr Odonga Chalres – chief manager engineering services – showing off the signed Entebbe Stretch document.

Left: Dr William Muhairwe is joined by Mr Charles Odonga as he shares a joke with the staff during the Stretch workout session for Entebbe Area.

Below: In the true spirit of Stretch; Dr William Muhairwe (right) practicing the Stretch-out concept of no boss element as he is serving his staff during the stretch workout session.

Chapter Seven

The Individual Incentive Mechanism

One-Minute Management Programme

... I would like to extend a One Minute Praise to you and your team for the innovations you put in place in Gaba, that have contributed to the reduction of power costs. It has always been the conviction of management that the only way for the Corporation to be more productive and keep on track to realising its vision is to create an environment that fosters innovation, self-confidence and taps creativity of all staff. Management appreciates and takes a lot of pride in your efforts and would like to encourage you to share these innovations with the other Areas....

Internal memo from the Kampala Water authorised representative to Water Production manager, 28 May 2008

Following the recent joint technical audits that were conducted in your Areas, of which you received a copy, several anomalies were noted. Unfortunately, you have failed to furnish me with a response within seven days as was required. This serves as a One Minute Reprimand to you and to request for a response within three days latest... we will carry out another audit without any prior notification and if you fail to put in place proper systems to rectify all the anomalies... thereby attracting three consecutive reprimands, my office will have no alternative but...

Internal memo from the chief internal auditor to the Area managers of Entebbe and Mbale, 9 November 2006

Before I discuss the individual incentive mechanism exhaustively, which is the basis of this chapter, it is important to point out at the onset why it was deemed important as an intervention tool. This unique management and pro-people development initiative was introduced after some colleagues and I realised that it would address the shortfalls of the APCs, SMART and the 'group' stretch interventions. It was aimed at enhancing individual staff confidence and productivity. I would also like to point out that this chapter is to a large extent informed by the following publications: *One Minute Manager* by Kenneth

© 2009 William T. Muhairwe. *Making Public Enterprises Work*, by William T. Muhairwe.
ISBN: 9781843393245. Published by IWA Publishing, London, UK.

Blanchard and Spencer Johnson, *Leadership and the One Minute Manager* by Kenneth Blanchard, Patricia Zigarmi and Drea Zigarmi and *The One Minute Manager Builds High Performing Teams* by Kenneth Blanchard, Donald Carew and Eunice Parisi-Carew.

Consequently, this chapter presents how NWSC realized that it was not enough to look at the group alone but also important to further look at individual staff productivity. We benchmarked best practice on "One Minute Management" Concept elsewhere and adapted it to NWSC operations. Let us now look at the various dimensions of change management and how the one minute management concept was made a reality.

Group contribution: Tapping into individual participation

Stretching was a wonderful and momentous experiment that made the entire NWSC family – from the Board to the rank and file – proud. Indeed, it demonstrated that the sky was the limit. Our initial quarterly stretch reviews showed that performance in all Areas, at all levels, was encouraging.

The experiment was beginning to bear fruit. All Area managers reported that the team spirit, which steered performance, was buoyant. Even our customers and the public at large were beginning to sense a refreshing wind of change sweeping through the Corporation, thanks to stretching. The success of stretching was reflected in the bonuses and rewards handed out to the Area teams on equal basis, regardless of individual input or contribution.

But despite the dynamism and triumph of stretching, my sixth sense told me that we needed to go beyond group or team success and assess individual contributions to the Programme. The stretch appraisal reports raised many questions in my mind: to what extent was group stretch performance a sum total of individual effort and contribution? Were all individuals in the Areas stretching equally? Did they deserve to be part of group success and to share the bonuses and rewards on an equal basis? While some staff worked beyond the call of duty, others worked to the clock; some did not even bother to work, except under close supervision. Was it really true that with stretching, all managers and staff were working beyond the call of duty? Had the lazy ones abandoned their bad old habits or were they harvesting where they had not sowed? I decided to put my ear on the ground to find answers to these troubling questions.

I began talking to all categories of staff, managers, secretaries, plumbers, sweepers, drivers, etc. about their experiences of stretching in their respective Areas. My questions were always the same: How is your Area performing since the introduction of stretch? How are you personally doing? Have the guys who used to be lazy or thieving changed their habits for the better? What do you think we should do in order to enhance stretching in your Area?

The answers to these questions were always the same. Yes, stretching was doing wonders, thanks to reduced bureaucracy, worker involvement, simplicity, self-confidence-building measures and teamwork. But while some individuals had abandoned their old habits and faults to become key players in stretching, others had not changed at all and were slowing down or even pulling down their colleagues. What was troubling was that the managers did not seem to recognise this problem, letalone correct it through action against the culprits.

One morning, I talked to a member of the staff from one of our Areas. He complained bitterly that despite his hard work and sleepless nights to complete stretch tasks, his efforts were neither recognised nor rewarded by the Area manager. He was treated on par with his more sluggish and work-shy colleagues who had to be pushed to do even the simplest of tasks. I challenged him that his complaint could not be true because statistics at headquarters showed that his Area was among the best performers. He retorted that, yes, his Area was doing very well but that some people were not stretching hard enough to deliver the results. 'They are sharing [the fruits of] our sweat,' he complained. This and other conversations with a cross-section of workers across the NWSC spectrum brought home some truths that stretching had operational shortcomings, which needed to be addressed if the Programme was to remain on course.

The first shortcoming was that it emphasised the group at the expense of individual recognition. Our programme was stretching the group, but not all individuals. As a result, some individuals tended to hide behind group success. Deo Kyambadde recalls that

> ... after successful implementation of the stretch concept, it was realised that there were people who were still hiding under the cover of group performance and collective success at the cost of a few that were actually doing the job.

Secondly, the reward system was flawed because incentives and bonuses were given out on equal basis without taking into account individual contributions. This meant that individuals in the group got equal benefits for unequal work, thus offending the best performers. The latter felt that they were being short-changed. This led to internal grumbling. The best performers slowed down, feeling they had nothing to gain from their individual effort.

Thirdly, stretch lacked the mechanism to recognise individual contribution to group achievements and an appropriate formula to reward star performers. It also lacked a self-correcting mechanism to address these shortcomings. Clearly, this had to be rectified. The only problem was to figure out how to do it.

The dilemma of stretching at headquarters

While all NWSC Areas had accepted the stretching concept, the same was not true of headquarters managers and staff. Yes, headquarters had agreed to propagate and spread the stretch spirit, to supervise its implementation, to monitor its progress and to evaluate its performance. But, ironically, many headquarters staff seemed to think that stretch did not apply to them. On the contrary, these headquarters managers seemed to think that they were above stretching.

Although there were some exceptions, headquarters managers generally tended to treat stretch with indifference. Besides, some of the previous change management programmes at headquarters, such as the Support Services Contracts, had not worked according to our expectations and were, in fact, as good as dead. The question was whether stretch-out at headquarters would work where the Support Services Contracts had failed.

For once, the Board members were getting impatient that headquarters was not stretching. The chairman, in particular, was breathing down my neck, insisting that we must introduce it at the headquarters as we had done in the Areas. 'What are you waiting for?' he asked whenever he met me. I always dithered and dodged his questions because, honestly, I did not yet know what to do. I was not yet sure how stretch-out at headquarters could be implemented and would work in practice. Then I came across the *One-Minute Management* concept and my prayers were answered.

A timely idea: The one-minute manager

In early March 2003, accompanied by my colleague, Mr Charles Odonga, the chief manager, Engineering Services, I attended the Water Week in Washington, DC, at the invitation of the World Bank. I was invited to present a paper entitled: 'Improving Performance through Internal Reforms by the Public Sector: A Case of NWSC'.

As is the norm while on my business trips, I looked up some literature. While in transit at Heathrow Airport, we wandered around and visited the management section in one of the bookstores. My attention was drawn to *The One Minute Manager* by Kenneth Blanchard and Spencer Johnson. This book was in fact one of many 'One-Minute Manager' titles. In addition to *The One Minute Manager*, I bought two more books in the series – *The One Minute Manager Builds High Performing Teams* by Kenneth Blanchard, Donald Carew and Eunice Parisi-Carew, and *Leadership and the One Minute Manager* by Kenneth Blanchard, Patricia Zigarmi and Drea Zigarmi. I also bought Colin Powell's autobiography for my colleague in appreciation of his support and good services.

On the plane to Washington, while Mr Odonga was busy reading Powell's autobiography, I quickly perused the three One-Minute Manager books to figure out what they had to offer. The force of the ideas in those books and the compelling simplicity with which it was presented impressed me beyond words. In fact, no other book had impressed me so much since I read *Who Moved My Cheese?* and *Jack Welch: the GE Way*.

Throughout the flight to Washington I was so deeply absorbed by the books that I did not say a single word to Charles or anyone else. Fortunately, the books were easy to read and master the gist of the simple but most profound management ideas I had ever come across. It was only much later, after I had gone through the books that I told Charles that the one-minute management concept was the only way to give fresh impetus to our ongoing Stretch-out Programme. As we exchanged ideas, I was pleasantly surprised to learn that the same concept had been articulated in Powell's autobiography. From then I began to ask myself: If the one-minute management concept had worked so well for Powell, surely could it not also work for the NWSC?

Coincidentally, my wife, Vicky had travelled to South Africa where she too bought a whole range of One-Minute Manager books. Being a quick reader, she had gone through all the books by the time I returned to Kampala. So when I arrived home, I found her bubbling with enthusiasm about her intellectual discoveries. After exchanging greetings, she immediately plunged into the ideas and merits of the one-minute manager and its relevance to our Stretch-out Programme.

Vicky's powerful arguments in favour of one-minute manager as an indispensable compliment to stretch-out reinforced my conviction that we had discovered the answers to the stretch-out shortcomings. With a telling smile, I told her that she was preaching to the converted. A new change management programme was about to shift stretching from the group to the individual.

The basics of the one-minute management concept

The one-minute management concept is based on the premise that of all the productive resources in any organisation – large or small – none is more important than its human resources. Any organisation that ignores the centrality of its human capacity in its operations does so at its own peril. The one-minute management concept is built on several basics that are briefly described in this section.

Autocratic and democratic managers: In their *One Minute Manager,* Blanchard and Johnson begin their case in favour of people-centred management by shooting down the traditional notions of management. According to the two authors, these ranged from the 'autocratic' at one end of the management spectrum to the 'democratic' at the other end.

The 'autocratic' manager is depicted as hard-nosed, high-handed, realistic and profit-minded. The trouble is, he puts the organisation first and the people last. In autocratic management the approach is top-down, directive, bottom-line and is devoid of worker involvement in the decision-making processes. An autocratic manager strives to please superiors but ignores the views and interests of subordinates.

Inevitably, there is a sense of alienation among workers who see the organisation as 'theirs' rather than as 'ours'. This, in the long run, is detrimental to worker satisfaction and productivity. In contrast, the 'democratic' manager is seen as pro-people, supportive, considerate, consultative and humanistic. This type of manager puts the people first at the expense of the organisation.

According to Blanchard and Johnson, none of these managers is good enough. A good and effective manager is one who combines and balances the interests of the organisation and its workforce for the mutual benefit of the two parties. Effective managers make their staff 'feel good about themselves' and this in turn increases productivity, both in quality and in quantity.

The three secrets: So what should managers do to make their staff feel good in order to produce the desired results? In *One Minute Manager*, the authors spell out three basic principles that are instrumental in one-minute management. These are goal *setting*, *praising* and *reprimand*. In one minute management, all the members of the staff in an organisation, from management to rank and file, are required to set time-bound goals in a clear and concise manner, to periodically review it and to determine whether the goals are matched by individual behaviour. Since research has shown that '80% of really important results are from 20% of the goals', goal setting should be limited to say three to five goals within a specific time frame. Managers should never assume that their workers know exactly what to do.

In goal setting, it is the responsibility of the managers in conjunction with their staff to spell out specific goals against which performance can be measured. Goal setting is indispensable because it motivates people; it gives them a sense of purpose and achievement. It also gives managers feedback on performance for 'feedback is the breakfast of champions'.

Secondly, in one-minute management, it is important to praise individuals when they do something right in order to encourage them to aim even higher. As Blanchard and Johnson graphically put it, 'Help people reach their full potential, catch them doing something right.' They advise managers to praise people immediately in specific terms, in a friendly and sensitive manner about what has been done right and its positive impact on the organisation and to encourage them to continue the good work. In this regard, individuals should be

The Individual Incentive Mechanism

praised for doing something right even when their unit, or organisation for that matter, is not doing well.

In one-minute management, praising is extremely useful because all individuals love to be recognised and appreciated for what they have achieved – in other words, to be winners. Praising encourages people to improve their behaviour and performance, and its purpose is to catch people when they are doing 'something approximately right and move them towards the desired behaviour until they learn to do it exactly right'.

The old-fashioned tendency of praising people only when they do something 'exactly right' can never transform individuals into best performers. In any case, praising is part of a continuous performance review process. Instead of waiting for annual performance appraisals, individual performance is assessed and praised on a continuous basis, which is certainly fairer, scientific and transparent.

Thirdly, like one-minute praise, one-minute reprimand must be used immediately in specific and clear terms when an individual does something wrong. The individual must be clearly told what wrong has been done. Managers must make it clear that wrongdoing is not acceptable because it is neither good for the organisation nor good for the individual's career development. Managers must be candid and honest in their reprimand(s) and should be careful not to offend individual sensibilities.

One-minute reprimands must be fair, targeting behaviour not individuals. While poor performance may not be acceptable and deserves criticism, the value of the individual should not be in doubt. Managers should also realise that 'when a reprimand is over, it is over'. In other words, it is counterproductive to reprimand individuals for the same incident of wrongdoing over and over again.

The purpose of a reprimand is to give feedback and to correct behaviour, but not to punish or rebuke people for the sake of it. For this reason, reprimand should only be used against individuals who have the skills and the knowledge to perform specific tasks but fail to do so for some inexplicable reasons. If an individual fails to perform due to lack of experience or skills or understanding of the task at hand, then the one-minute reprimand is not the answer.

In such a case, the individual and the one-minute manager should go back to the drawing board and set goals afresh. Apart from feedback, reprimands have the virtue of being open, direct and reassuring. They help to avoid rumour mongering, suspicion, persecution, mistrust and intrigue, which have been the curse of many organisations around the world.

The individual skills – the driver: However, on their own, one-minute goal setting, praising and reprimand cannot work in a vacuum where individuals do not know what they are worth, what they want and how to reach the desired performance. For individuals to know where they are coming from, where they are and where they are going, they must have the skills and the knowledge to do so.

According to Blanchard and Johnston, investing in people is the key to the success of one-minute management. The two authors point out that most organisations devote 50–70% of their money to paying salaries and a mere 1% on staff development and training. They also spend more money maintaining buildings and equipment than on developing the skills and capacities of the work force. According to Blanchard and Johnson, 'The Best Minute Spent Is The One Invested in People.' Investing in people is one of the principal sources of worker motivation, commitment and productivity.

Organisations that neglect staff development and training do so at their own peril. Every individual in an organisation is a potential winner although some may be disguised as losers. To build competence, commitment and confidence, individuals must be systematically nurtured and trained so that they master the art of self-management and self-evaluation, setting goals, and measuring behaviour against goals.

As Blanchard and Johnson put it, 'We are not just our behaviour, we are the persons managing our behaviour'. They also stress that while 'Goals begin behaviours, consequences maintain behaviours'. That is why it is in the interest of managers to train and develop individuals to acquire appropriate behaviour to accomplish organisational goals.

Situational leadership: In *Leadership and the One Minute Manager*, the authors demonstrate the relationship between the one-minute management concept and 'situational leadership styles'. Situational leadership refers to the actions and behaviours between managers and the people they manage in an organisation. It demands that instead of being a boss in the traditional mode where subordinates labour for their superior, a situational manager must be responsive to the needs of the subordinates in the organisation, and provides the resources and suitable conditions for the accomplishment of their goals. A situational leader is a facilitator, not a master, whose principal function is to lead but not to dominate.

Situational leadership revolves round two basic ideas. To begin with, working hard *per se* is no guarantee to success. Therefore, it is better to 'work smarter rather than harder'. In other words, putting more thought into work is superior to working long hours, day in and day out. The second basic idea is that in any organisation, individuals have different backgrounds, skills and competencies and operate in different situations from time to time.

This means that managers must understand what makes different individuals in their organisations tick and the different situations in which they work in order

The Individual Incentive Mechanism | 145

to apply the dictum of 'different strokes for different folks' or even different strokes for the same individual in different situations.

In order to be an effective situational leader, the manager must do three things. The first of these is to diagnose the needs and capabilities of the subordinates in the unit, division or organisation and the circumstances in which they operate. The second step is to apply appropriate leadership styles in accordance with the diagnosis in a flexible manner. The third step is partnering, which means that the manager and the subordinate staff must agree on the appropriate leadership style to be used under specific circumstances. Blanchard and his colleagues also pinpoint the basic leadership styles in one-minute management. These are directing, coaching, supporting and delegating.

Leadership styles: In a directing leadership style (as though a person was in Senior 1), the manager supervises closely the work of a subordinate to accomplish a specific agreed task or goal. This style is suitable where decision making is of utmost urgency and the organisational stakes are very high. It is also appropriate to individuals who are either new in the organisation or are yet to develop the requisite skills and confidence so as to work without close supervision. In this case, the manager structures and organises the task, and teaches the individual how to do the work and supervises performance closely. This style of leadership tends to be 'autocratic' and is characterised by one-way communication in which the manager tells the individual what, when, where and how to do a specific task.

As the individual gains experience and confidence (as though in Senior 2), the manager shifts from the directing to the coaching leadership style, which combines task supervision and monitoring with explaining decisions and seeking opinions and suggestions and supporting progress.

In the coaching leadership style, the manager provides a lot of support, but also introduces the element of two-way communication in which the subordinate's ideas and suggestions are sought in order to cultivate confidence, initiative and risk-taking. Coaching also introduces the element of encouragement and praise when the individual makes progress. If, through monitoring and supervision, the shift to coaching is found to have been premature, then the manager quickly reverts to directing before it is too late.

Once the individual is firmly on the ground and by and large knows what to do, the manager shifts the leadership style gear from coaching to supporting. In supporting leadership style (as though in Senior 3), the manager facilitates and supports the individual's efforts towards task accomplishment and the two begin to share responsibility for decision making and accountability. In this leadership style, the emphasis is on consultation, encouragement and praising. If it is found that the individual is not yet up to the mark, the manager shifts back to coaching.

When the individual is finally on top of the job by depicting that he has gained the requisite skills and the initiative to accomplish tasks independently, devoid of guidance, then the manager shifts from supporting to delegating.

This means that the manager hands over the day-to-day decision-making, problem-solving, trouble-shooting and risk-taking and accountability responsibilities to the individual concerned. At that point, the manager occupies a reserve and backup position, only to be consulted when and if necessary.

Indeed, the ultimate goal of an effective one-minute manager is to reach the delegating leadership style so that 'when the best leader's work is done, people say "we did it ourselves"'. If an individual falters in exercising his delegated responsibilities, the manager reverts to supporting.

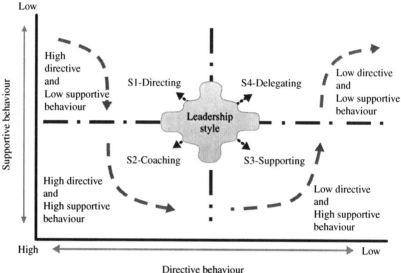

Model of Leadership Style

Adapted from Blanchard and Hersey: *Leadership and the One Minute Manager.*

Blanchard and his colleagues advise managers to adopt a gradual, step-by-step approach in shifting from one leadership style to another, whether moving up or down the leadership style ladder. There are no shortcuts to progress. Abrupt shifts back and forth, from one leadership style to another, can be disconcerting and frustrating for individuals and their managers alike. The manager must always ensure that the individual is ready and capable of moving to the next leadership style level. If there is an error of judgement in shifting from one leadership style, this should be rectified immediately, but without offending individual sensibilities. Managers should vary their leadership styles according to the situation and the people being managed.

For Blanchard and his colleagues, the key to the selection and application of a particular leadership style is flexibility. Managers should be pragmatic in their leadership styles by taking into account not only the differences between individuals but also the differences between situations. Even for the same individuals, the manager can apply different leadership styles. For example, managers can delegate responsibility to an individual to perform a particular task but choose supporting for the same individual when it comes to a different task. Similarly, the transformation of any given situation may dictate a shift from one leadership style to another or even a combination of different leadership styles.

The individual development levels: Given the differences between individuals and situations, it is imperative that managers determine what Blanchard and his colleagues have called the individual development level in order to choose the appropriate skills and to apply the one-minute management principles of goal setting, praising and reprimands. As Blanchard and his colleagues point out, 'there is nothing as unequal as the equal treatment of unequals'.

According to Blanchard and his colleagues, there are four different individual development levels code-named D1, D2, D3 and D4 that match different leadership styles, namely S1 to S4. The four individual development levels are low competence but high commitment (D1), high commitment and low competence (D2), moderate to high competence but variable commitment (D3) and a combination of high competence and high commitment (D4). Blanchard and his colleagues show which leadership style is appropriate for which individual development level.

Matrix of development levels: Matching the leadership style to the development level: Directing (S1) is suitable for D1 individuals who are enthusiastic and committed because they are new at their jobs but are not competent because they lack experience and confidence to accomplish tasks or goals.

Coaching (S2) is appropriate for D2 individuals who have gained some competence but lack commitment because they are disillusioned due to inexperience and lack of confidence. They require lots of support and encouragement to revitalise their self-esteem and commitment.

Supporting (S3) is for competent D3 individuals who have gained lots of experience and skills but are not yet fully motivated because they lack confidence and, accordingly, they still need support from their managers to rise to the occasion.

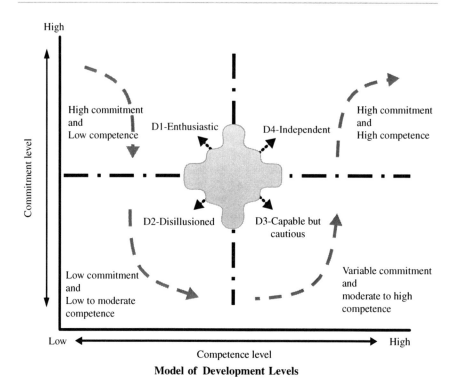

Model of Development Levels

Adapted from Blanchard and Hersey: *Leadership and the One Minute Manager*.

Delegating (S4) is for highly motivated, committed and competent D4 individuals who are willing and capable of working independently at their own initiative and as such do not need much supervision.

Knowing the individuals and the different (and constantly changing) situations in which they work at all levels of the organisation is one of the most important secrets of effective one-minute management. In setting individual goals and tasks, one-minute managers always take into account past experience, performance and, above all, competence. One-minute managers do not require individuals to punch above their weight. A mismatch between individual experience and competence and goal setting invariably leads to frustration, poor performance and disillusionment.

So a situational one-minute manager uses not only different strokes for different folks but also different strokes for the same folks in different situations.

Partnering: Another secret of situational leadership in the context of one-minute management is partnership between managers and their staff. According to Blanchard and his colleagues, 'situational leadership is not something you do to people but something you do with people'. This means that individuals must be partners in planning, implementing, monitoring and evaluating all the activities in which they are directly involved, from the beginning to the end.

Partnering requires setting performance standards to realise agreed-upon individual goals. The goals must be specific and measurable, motivating, attainable, relevant and trackable – in short, SMART.

Turning the pyramid upside down

For me, reading the *One Minute Manager* and *Leadership and the One Minute Manager* was very revealing. The two books forced me to rethink the existing management styles in public and private organisations. It forced me to reassess the way we had been managing the staff and affairs of NWSC since 1998.

The two books revealed to me what we had been doing right, what we had been doing wrong, and what we needed to do to improve management and performance. In particular, the ideas and concepts articulated in the two books convinced me that we had to adopt the one-minute manager and situational leadership techniques to close the operational loopholes in our ongoing Stretch-out Programme.

Traditional management styles in Uganda, which we inherited from the British class system, resembled a pyramid. The typical style was top-down, dictatorial and often high-handed. Bosses were masters and subordinates were obedient servants, at the beck and call of their superiors.

Bosses were remote from their grassroots staff. The higher up the manager was, the more remote he was from ordinary staff in the organisation. It was not uncommon for individuals to spend their entire working lives without meeting the boss of their organisation. The bosses believed in keeping maximum distance from the subordinates on the outdated pretext that 'familiarity breeds contempt'.

Subordinate members of the staff were required to serve the interests or even the whims of their superiors. Managers summoned their staff to give them instructions or directives – but not to seek their opinions or suggestions. A good example was the practice of managers summoning their secretaries to dictate letters in an officious and unfriendly manner.

Thus, managers set the goals and their staff have to struggle to accomplish those goals without raising any questions. Managers set out to find their staff doing something wrong rather than doing something right. Managers are quick to claim credit for any success and to pass on the blame for all the failures or shortcomings to their subordinates.

Also, they tend to be quick to reprimand or tell off the workers but they rarely praise individuals in recognition of good performance. Managers are insensitive and indifferent to the needs and aspirations of their subordinate staff. People are expected to know what to do without being trained or coached and without regard for their individual development levels.

Compared to other public enterprises and, indeed, by Ugandan standards, the NWSC had made substantial progress in adopting a more sensitive and responsive style to bridge the distance between the Board, management and staff since the 100 days Programme in 1998.

We had encouraged Board members, management and staff to mingle freely, for example, on occasions where trophies and cash bonuses were awarded. On all these occasions, all the members of the NWSC family used to eat roast meat and dance together in an informal and relaxed atmosphere.

Despite the progress we had made since 1998, we had not gone far enough to turn the management pyramid upside down, to transform managers into servants rather than into masters. For one thing, it was not easy for managers to outgrow outmoded management styles.

The case for one-minute management in NWSC: Many of our managers, both at headquarters and in the Areas, were still stuck in the traditional mode of thinking and management. They tended to apply the same management style for all people and for all occasions regardless of experience, past performance, competence or circumstances.

Secondly, the group incentive mechanism left little room for the recognition and appreciation of individual performance. Even the praises and (more often) the reprimands tended to target groups rather than individuals. As a result, individuals were not given the opportunity to take responsibility for the outcome of their actions. The time had come for NWSC managers to change their approach: from the group to the individual.

All I wanted to know was when we were going to implement the one-minute management concept in NWSC! The concept was exciting; however, we adapted the approach slightly on the basis of our experience. Instead of directing first, supporting second and finally delegating, we realised that strong support and direction go hand in hand from the beginning if we were to lead staff from dependence to independence (see graphs below). A person who is new at a job, with limited competence levels, requires significant support and directing to get acclimatised with the realities of the task. Support can be given in the form of encouragement, courtship and moral buttress, among others. Low competence employees need this approach or else they get disillusioned and leave the job prematurely. Blanchard and colleagues point out that the

The Individual Incentive Mechanism

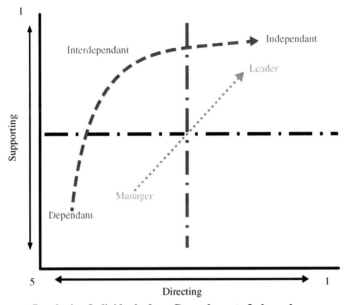

Developing Individuals from Dependence to Independence

Source: Developed by Paul Reiter, Jan Jansen and William T. Muhairwe during a capacity-building workshop for utility managers in Dubai, March 2008.

one-minute manager, for any given task or situation, must move from being a manager on whom subordinates depend to being a leader of whom subordinates are independent and from whom they require minimum direction and support.

Leading by example and setting my own goals: This shift in approach had to start from the top. Since 'charity begins at home', I had to set the example by setting my own goals for the realisation of the corporate vision and mission. I had to evaluate my own performance against my goals on a regular basis.

I had to push myself to the limit in order to push others to realise their own goals. I also had to start practising to praise those who excelled in their work and reprimand those who, without good reason, failed to live up to the goals they had set themselves.

The results of this new management style were truly amazing. People began to be candid in what they said and did. People readily accepted criticism whenever they fell short of their goals, without taking offence. The working atmosphere at the NWSC headquarters became more relaxed, cordial and results oriented. The traditional saying that 'familiarity breeds contempt' was torn down.

Putting one-minute management in action

Before implementing the one-minute concept in the Corporation, a lot of groundwork had to be done to bring everyone on board and make sure that we were all speaking the same language. This process was not as easy as we expected and we encountered resistance from some staff who thought that we were rushing into too many change management programmes. But the experience gained from the previous change management programmes helped us to deal with the resistance and the Programme was implemented on pilot basis.

Breaking the news to top management: On 25 March 2003, soon after my return from Washington, I called a special top management meeting during which I announced my enthusiasm for the one-minute management concept. I summarised the main ideas: one-minute goal setting, praising and reprimands. I argued that the adoption of one-minute management was what the NWSC needed to overcome the shortfalls of stretching.

In retrospect, this method of work would not only enhance stretching but also galvanise the management's initial intervention tools, the APC. And looking at the wider picture, all these interventions were aimed at helping the Corporation achieve its objectives and targets that had been enshrined in the performance contract.

All individuals in the Corporation had to set measurable and motivating goals that were relevant to their day-to-day assignments. Since none of the senior managers present had read the one-minute manager books yet, I distributed copies so that they could study and internalise thoroughly the key concepts it contained, before circulating the books to other managers and staff in their divisions and departments.

I requested Dr Silver Mugisha and Ms Evelyn Otim to read the books and prepare summary notes for discussion at the next top management meeting.

Selecting the champion team: Soon after the top management meeting, I asked Odonga, Amayo and Mugisha to spearhead the implementation of the one-minute management programme. I chose these managers as champions because they had been my most reliable lieutenants since the introduction of the 100 days Programme.

They had not only mastered the complexities and technicalities of change management programmes but had also followed through the implementation of those programmes. These colleagues were clearheaded, highly motivated, receptive and open to ideas. Over the years, they had demonstrated their unyielding enthusiasm in pursuit of their assignments, from the beginning to the end. I had no doubt in my mind that these colleagues were the right champions – they were the right men for the right job.

Incidentally, one of the great enthusiasts of the one-minute management concept and someone who became its champion in his own right even before the rest of the Corporation was brought on board was the late Mr Charles Ochieng. At the time when the one-minute management programme was introduced, he was the Corporation's chief internal auditor.

He not only read and mastered the one-minute concepts and articulated them vigorously at one of the senior management meetings but, more importantly, he also immediately began to require people in his department to read and internalise the books, long before the scheme officially got off the ground. I was so impressed by his mastery of one-minute management concepts that I decided to add his name to the list of my lead champions.

Unfortunately, he passed away a few days later. I was so upset when his life was cut short that today I am yet to overcome the sense of loss, personally, for his department and the entire Corporation.

At a top management meeting on 31 March 2003, I encouraged managers not to lose heart and quoted from the Bible: John 14.1, which reads: 'Do not be worried and upset', Jesus told them. 'Believe in God and believe also in me ...' I told colleagues that Charles' death served as a challenge to all of us to implement the success of one-minute management in his memory.

Dealing with resistance: Notwithstanding Charles' enthusiasm, I had somehow underestimated resistance to the implementation of the one-minute management concept. Some of the managers and members of the staff were suffering from 'change management programme fatigue'.

They grumbled that the programmes were being introduced too quickly for the good of the Corporation. The first phase of stretch-out had barely ended and yet, here the MD was rushing into another change management programme!

One member of the staff objected with the following words: 'How many management concepts do we have to learn and for what reason? What are their benefits to the workers? How about stretch?' Some managers saw the one-minute management concept as an academic exercise of dubious relevance to the challenges that the NWSC was facing.

Some members of the staff complained behind my back that they were sick and tired of reading books and doing tests as if they were back in a classroom. The MD, so the argument went, was becoming obsessed with change management and forcing staff to bite off more than they could chew.

Every time he went abroad he came back with exotic books containing outlandish ideas. It was about time for someone, somehow, to break his change management fanaticism. So, I was surprised but not discouraged, in mid-April 2003, when I discussed Mr Mugisha's summary of the one-minute concepts, to find that some of the senior managers opposed to the idea.

Despite these subterranean objections, I was determined to ensure that one-minute management became a reality in the NWSC as soon as possible. Between April and June, I did everything in my power to kick-start one-minute management.

Whenever I met my managers in the corridor, at a meeting or wherever, I always asked them: 'Have you completed reading the one-minute manager books? Have you mastered and internalised the concepts? Have you circulated the copies to your colleagues and staff?' I was particularly determined to implement the Programme at the headquarters. On one occasion, I cautioned the top managers that those who dared 'to come near my firing line, to frustrate the programme ...' would face dire consequences.

On 20 May 2003, I called another top management meeting during which I expressed concern about the slow progress of one-minute management and set a deadline of 23 May, by which date I expected all divisions, departments and sections to have been brought on board. It was clear to every manager and member of the staff that I was not joking and, therefore, they had to do their best to avoid crossing me.

Some of the people found the speed at which we were moving daunting, humbling and even frightening. But we had no room for laggards. The manager Commercial/Billing, Mr George Okol, in his recollections attests to this:

> The Managing Director was particularly firm and very determined to see the programme implemented ... The MD was on top of the whole process by being available to staff all the time ... moving from floor to floor checking on progress of the goal setting process ...

Ensuring understanding of the concept: The groundwork for one-minute management started with the circulation of the one-minute manager books to all managers and members of the staff. All the managers were required to read the books and internalise the concepts. As with stretch-out, the members of the junior staff were required to take tests to find out whether they had read the books and understood the concepts. Surprisingly, the performance of some of the rank and file (secretaries, drivers, assistants and messengers) was quite outstanding.

Once the members of the staff had read the books, they were required to put the ideas into practice. In other words, they had to begin setting their visions, missions and goals in line with those of the Corporation, and in accordance with the 2003–2006 Government of Uganda/NWSC Performance Contract targets. Of course, setting visions, missions and goals, which was a new and novel experience for most people, was no easy task. However, the champions were on hand to provide guidance. In framing goals, individuals were assisted to spell out milestones.

According to George Okol, the MD himself was very involved in refocusing the goal-setting process. He recalls that:

> Noticing that the Divisions were experiencing problems in precisely formulating their goals as required, the MD wasted no time to correct the process ... He personally reviewed a number of goals from the managers and guided them on how to develop their own goals ...

Piloting the one-minute management process: The implementation of one-minute management started simultaneously with the first Stretch-out Programme for the headquarters. The implementation strategy was to begin with a one pilot division and to extend it gradually to other divisions and departments, drawing on the confidence and experience gained and the lessons learned from the pioneers. We chose the Engineering Division, where most of the lead champions were based, to spearhead the implementation of one-minute management.

The division drew up the vision, mission and goals within the context of the overall corporate vision. This was then discussed by the heads of the department in the Engineering Division and thereafter presented to the entire staff of that Division at a three-day workshop in Jinja. The participants showed up at the workshop with enthusiasm, excitement and anticipation, and with a sense of pride derived from the Corporation's recognition of their worth and potential to contribute to the destiny of their division.

The chief manager, Engineering Services, kicked off the workshop by presenting the proposed vision, mission and goals and by detailing what was expected of the participants during the workshop.

Thereafter, the participants broke up into groups, according to their respective department and sections.

Chaired by the head of the department, the participants in each department/ section drew up their own departmental vision, mission and goals that were consistent with the overarching Division vision, mission and goals.

Each department/section then presented its results to the division plenary session for discussion, harmonisation and adoption. Once the departmental/ section vision, mission and goals were adopted, the next step was for the individuals to go back to the drawing board to formulate their own individual visions, missions and goals in line with those of their respective departments/ sections. A team of lead champions, led by a consultant from Group To Consult, facilitated this part of the process. The formulation of individual visions, missions and goals was a laborious exercise, lasting one and a half days or 50% of the entire workshop time.

Individuals had to ensure that their goals were specific, measurable, workable, time bound and consistent with those of their respective departments and the division as a whole. The individual visions, missions and goals were then presented to the departmental sessions for discussion and adoption.

The construction of individual visions and missions and the setting of individual goals were truly amazing experiences for the entire headquarters staff of the Engineering Division. For the first time in the history of the Corporation, staff engaged in defining where they wanted to go and mapping out how and when they wanted to get there. For most of them, this was a dream come true.

This refreshing experience was reflected in some of the individual visions that came out of the workshop. For example, the declared vision of one office attendant was 'to be the best office attendant in the world'. Similarly, the vision of one senior engineer was 'to be a world renowned engineer'. This show of enthusiasm meant that the one-minute management was taking root in the Corporation with promising results.

Once the vision-, mission- and goal-setting exercise was over, the participants turned to the essential inputs that were required for individuals, as well as for the departments and the Division to accomplish their respective tasks. The logistical inputs were divided into financial requirements, which were constrained by the budget for the 2003/04 financial year. Other inputs took the form of an individual incentives mechanism through which managers created an enabling environment to motivate and facilitate individuals to realise their goals within agreed time frames.

The participants also identified the individual staff responsible for particular input requirements for purposes of monitoring and accountability. The input requirements were presented to the chief managers for consideration and approval in the shortest possible time. Astonishingly, 90% of the Engineering Division inputs were approved while the remaining 10% were out of line with the overall corporate vision, mission and strategy.

The approved visions, missions and goals and input requirements were translated into detailed individual, departmental and division work plans, which were subsequently submitted to the NWSC top management for final approval before actual implementation.

The one-minute management workshop of the Engineering Division was officially closed by the Board chairman. The motivation, commitment and diligence and hard work that the participants had shown during the workshop were very impressive. From the daily briefing I had received during the workshop, it was clear to me that all our staff had huge reserves of energy to move mountains in NWSC provided those reserves were properly tapped.

Both the chairman and I commended the participants on a job done well and expressed optimism that their efforts would, sooner rather than later, yield positive results. When we returned to Kampala from Jinja, I called some individuals to my office and gave them one-minute praise for their exemplary performance during that workshop.

Other divisions and Areas come aboard: The one-minute management experience in the Engineering Division was so encouraging that we decided to extend the Programme to all the divisions at the headquarters immediately. Messrs Odonga, Mugisha and Amayo, our most experienced and reliable lead champions, spearheaded the extension. They, together with the Group To Consult consultants, led a team of 12 key staff to guide and facilitate the preparation process in the four divisions.

Again a three-day workshop was held in Jinja, where the staff of the four divisions (Commercial and Customer Care, Finance and Accounts, Management Services, and the Internal Audit divisions) went through the rigorous and demanding process as the pioneer division had done. Like the Engineering Division, after the workshop, the four divisions produced their own detailed work plans, which were similarly approved by top management at headquarters. After stretching headquarters, the one-minute concept was combined with what should have been Stretch-out III and extended to all other Areas.

In addition to the Area Stretch targets, each individual staff member had to formulate his or her own vision, mission and (individual) one-minute goals that were aligned to the vision of the respective Areas and goals directed towards achieving the Area targets.

The one-minute manager commissioned: The climax of the one-minute management preparation process for both headquarters and Areas was the signing of the memoranda of understanding binding all the individual staff to the realisation of their one-minute goals. Each individual had to sign a MoU with his supervisor – committing himself or herself to the accomplishment of the goals agreed upon.

Before signing the memoranda, the lead champions worked day and night to ensure that all documentation was completed before the signing ceremony. All of them received one-minute praise for their tireless efforts. On the D-day, all the invited dignitaries, including the minister of state for Water and her permanent secretary, as well as the NWSC Board, management and headquarters staff, assembled at the Corporation's training centre for the signing ceremony.

The minister thanked NWSC management for the innovations: from the 100 days Programme to the one-minute manager. The guests were clearly impressed by the staff zeal and commitment to performance improvement in the Corporation. This was proof enough, if any was needed, that significant performance improvement in public utilities is feasible without recourse to 'hiring people from Mars'. What matters most is the right management strategy to achieve wonders with locally available human resources.

Building high performing teams

Unlike previous change management programmes, one-minute management was not a substitute for, or successor of, stretch-out. It was not a separate programme. In conception and in practice, one-minute management and stretch-out were merged into one programme; they were inseparable sides of the same coin, so to speak. The one-minute management concept and practices were deliberately embraced to inform and augment stretch-out – not to replace it – for obvious reasons. Individuals do not live and work in isolation. They are first and foremost social animals and as such they live and work within groups, such as families, sections, departments, divisions and organisations.

In any organisation, both individuals and groups need each other to foster teamwork, increase production, be creative, share experiences, make decisions and solve problems. After all, collective effort is a sum total of individual efforts. Therefore, while at NWSC, we introduced one-minute management to set individual goals, praise good performance and reprimand those who did not meet our expectations, we also knew that its central purpose was to refine, catalyse, beef up and sustain the stretch-out momentum.

That is why it was important to focus on individual incentives while, at the same time, to continue to build up high performing stretch-out teams. In order to inject the one-minute management principles and practices into stretch-out team building, we turned to the ideas that Kenneth Blanchard and his colleagues expound in their book, *The One Minute Manager Builds High Performing Teams* (1996).

For Blanchard and his colleagues, the idea of building high performing teams in any organisation is anchored on the conviction that 'none of us is as smart as all of us'. In other words, two heads are always better than one. However, Blanchard and his colleagues hasten to warn that if high performing teams are allowed to be too large (more than 15 or 20 people) they become unmanageable; as the English say, too many cooks spoil the broth.

Groups that are too big to manage tend to be unfocused and to negate cohesion and productivity. In contrast, small, coherent, purposeful and well-managed teams tend to enhance a sense of ownership and commitment and to foster group performance, capacity building, skills development and satisfaction, which, in turn, culminate in higher productivity and the realisation of individual, group and organisational visions, missions and goals.

This means that high performing and results oriented teams in any organisation must be properly constructed, motivated and facilitated in accordance with the one-minute management principles and practices.

According to Blanchard and his colleagues, the essential characteristics of high performing teams are purpose, empowerment, relationships and communication, flexibility, optimal performance, recognition, appreciation and morale. A common purpose or vision enables teams to set clear, challenging and SMART goals to work out strategies to achieve the team goals and to define individual contributions to group activities.

Empowerment gives individuals and teams a sense of collective power, spells out policies and practices to support team activities and enables the teams to access resources to accomplish their goals and objectives.

Building high performing teams demands honesty, openness, listening, mutual respect and recognition of different opinions. This means that in teamwork individuals treat each other with warmth, courtesy and companionship. High performing teams must be flexible enough to anticipate and to adapt to change. Properly constructed and focused teams culminate in high productivity, individual and group actualisation, group cohesion and team spirit, a sense of pride and satisfaction.

Building high performing teams starts with identifying the problem, understanding the circumstances in which it has arisen, its implications on the organisation and what needs to be done, at individual and group levels, to solve it. Once the problem is identified, group and individual roles and tasks must be defined clearly; problem-solving strategies and modalities must be worked out, just as resources must be mobilised.

A participatory approach to decision making, goal setting and task accomplishment is an essential part of building high performing teams. Blanchard and his colleagues hasten to remind us that in teamwork conflict is inevitable but it is the duty of high performing team leaders to detect it early enough and resolve it before it causes irreparable damage to the work, cohesion and morale of the team.

Blanchard and his colleagues have identified four stages of development in building high performing teams. All groups begin team building in what Blanchard and his colleagues have called the orientation stage, with a mixed sense of expectation and anxiety. Group members may not be clear about the purpose of the team and what is expected of them in general and as individuals who may not know each other well in terms of temperament and skills.

Therefore, the challenge at this stage is to create a common sense of purpose; clarify team goals and individual roles and relationships; spell out team norms, ground rules and values; and consolidate team cohesion.

Stage two of this development, which has been called the dissatisfaction stage, is characterised by power struggle, conflict, clash between hopes and realities, frustration and confusion. This turbulent stage must be handled with care so as not to stifle creativity. Therefore, all elements of discontent, all group differences and all points of contention must be discussed and resolved.

In stage three, which Blanchard and his colleagues have called the resolution stage, the team learns the roles and settles down to balance hopes and realities and to resolve disruptive animosities and misunderstandings. This stage is characterised by harmony, mutual support, trust and mutual respect, as well as by self-esteem, confidence, feedback, shared responsibility and a common team language. At this stage, the team challenge is to avoid euphoria and too much self-satisfaction and to guard against relapsing into conflict and turbulence, which would invariably undermine productivity.

In stage four of development of high performing teams, called the production stage, the teams reach the climax of enthusiasm and productivity. This stage is characterised by excitement, full participation and collaboration, mutual appreciation and recognition, clarity, confidence and strength to move mountains, and optimal performance. In this stage, the group is stretched to the limit; the individual feels good and valued; the realisation of group goals is around the corner and the sky is the limit.

Blanchard and his colleagues taught us the beauty of combining one-minute management and stretching. We realised that the success of NWSC and the realisation of its vision, mission and goals – of its hopes and dreams – would henceforth be shaped by a combination of teamwork and individual enthusiasm.

Since 2003, we have purposively and deliberately built high performing teams at the headquarters and in the Areas – in all sections, departments and divisions – based on the combined concepts of stretching and one-minute management. The results have been beyond our original expectations, namely increased public appreciation of our services, customer satisfaction and, above all, a much better corporate image.

Monitoring and evaluation

To ensure sustainability and continuous performance improvement, a mechanism was put in place to give feedback and continuously monitor and evaluate performance. Individuals were required to assess their own performance (self-management). The immediate supervisors were required to monitor the progress of achievement of each individual's one-minute goals. For purposes of the objective

The Individual Incentive Mechanism | 161

evaluation of accomplishment of tasks, quarterly evaluation of achievement of milestones within the individual's one-minute goals was carried out.

This exercise involved individuals giving proof of accomplishment of specific tasks, and for those tasks that were still ongoing, a new plan would be discussed and agreed upon between the individual and the supervisor regarding when those activities would be accomplished. This exercise also helped to update goals and tasks of each individual in accordance with the prevailing situation and challenges that the Corporation was facing. As highlighted earlier, the progress of each individual's level of competence and commitment in a particular goal and hence the development level (D1, D2, D3 and D4) was assessed on a quarterly basis. The latter assessment resulted in appropriate change in the leadership style to be applied subsequently.

In addition to individual activity monitoring, a monitoring and evaluation system for accomplishing the planned deliverables/milestones for the section/department/division was also developed. The monitoring and evaluation of activities under this category were coordinated by the secretariat, which was composed mainly of staff from the research, monitoring and evaluation, and operations departments. The secretariat provided regular reports to the top management concerning the progress of achievement of milestones and any constraints that had arisen along the way.

Towards a new corporate culture

One-minute management was a wonderful, enriching and refreshing experience in all NWSC spheres of work – in all the sections, departments, divisions and Areas. At a practical level, one-minute management beefed up, catalysed and refined stretch-out by harmonising group and individual incentive mechanisms. This was reflected in the increased water production, water distribution and service coverage, substantially reduced NRW, more efficient and effective billing, increased revenue collection and reduced arrears. But much more important was the fundamental and irreversible transformation of NWSC's corporate culture in barely a year.

Participatory goal setting: From April 2003, the commitment of workers to their duties throughout the Corporation improved dramatically. This commitment was reflected in the visions, missions and goals that have been set by individuals, sections, departments and divisions, and against which MoUs were signed in June 2003. Since then, all the NWSC members of staff, from the top to the grassroots, have been totally committed to their schedule of duties. Once one-minute management was launched, all the staff started reporting for work on time. They did not miss meetings without a good reason and completed their tasks on time. They became more candid and open in dealing with their colleagues. This significantly improved the working atmosphere across the Corporation. Participatory goal setting has become the norm in all Areas.

Situational leadership: Secondly, the introduction of one-minute management saw more staff becoming more confident of expressing their opinions and taking the initiative. People no longer waited for directives from the bosses. Bosses were no longer all-knowing, omnipotent autocratic leaders. They had become situational leaders. They explained, justified, consulted and involved the members of the staff in decision making, goal setting, activity implementation, monitoring and evaluation. The bosses were required to determine the development levels of their subordinates in order to know when to apply directive, coaching, supporting or delegating styles of leadership. By adapting a participatory approach and tapping into the inputs of all their subordinates and delegating much of the work, the managers had more time to do strategic managerial work.

After the introduction of one-minute management, I too took the backseat and found more time to think, listen, consult and do public relations for the NWSC. My desk was and is no longer overflowing with documents. In, pending and out paper trays of the old days vanished from my desk. Managers no longer resort to the old, tired excuse that action has not been taken because they are still waiting for answers or decisions from the MD. Such inaction is no longer acceptable. It deserves and, indeed, often carries a one-minute reprimand. Under one-minute management, all the members of the staff are required to get on with their jobs without hesitation or excuses. In NWSC, procrastination is no longer acceptable or tolerated.

Praise for good performance: Confidence-building measures and initiatives up and down the Corporation gradually raised staff morale and enhanced a sense of job satisfaction. By boosting morale, people not only became more competent at and motivated for doing their work, but also shared responsibilities and became more open to the culture of criticism, consultation and collective pride. NWSC indeed became a fraternal organisation in which the success of any of us was the success of all of us.

Partnering: Thanks to one-minute management, the old vices of intrigue, backbiting, gossiping, jealousy and envy were banished to the dustbin of history. Since the advent of one-minute management, NWSC has become an open book. All staff, from top to bottom, are free to candidly pinpoint operational shortfalls or problems and to suggest possible solutions. Under one-minute management, we focus more on what people do than on what they are. We are more interested in performance and productivity than in personalities. During one-minute praising and reprimanding, we distinguish the actions for which someone is to be reprimanded from individual worth and self-esteem. This activity-centred approach to management leaves no room for victimisation and complaints about bosses being unfair to their subordinates. Any complaints of this nature are raised openly and are resolved promptly. The principles are clear to everyone, and we all play by the same rules regardless of our respective positions.

Continuous staff appraisal – the ABC format: One of the benefits of one-minute management was the review and refinement of the NWSC staff appraisal system and procedures. Previously, staff appraisal in NWSC was an annual ritual with little reference to the individual's performed activities, and without transparency. Supervisors used to evaluate the performance of their subordinates and forward the evaluation forms to the human resources department.

These forms would be stacked up without being analysed and no feedback would be given to the individuals concerned. The forms were simply filed and that was the end of the matter. They were only referred to when it came to matters concerning individual discipline or promotion. For obvious reasons, this appraisal system served no useful purpose to the individual or the organisation, it was open to abuse and prone to the whims and prejudices of those conducting the appraisals.

After the introduction of one-minute management, staff appraisals became more open and individual-centred. Firstly, by determining the development level of the individual employee, assignments and tasks were synchronised with the employee's capacity and competence and any appraisal had to take this into account. Secondly, appraisals were based on the agreed and negotiated goals and targets to be accomplished within specific time frames. Thirdly, through one-minute praises and reprimands, the appraisal system became a continuous process. At any given time, individuals knew exactly how they were performing.

Most important of all, during monthly, quarterly and annual appraisals, individual employees were fully involved in the evaluation process, which was done through a three-stage process. First, the individual completed Form A in which he/she rated his/her performance measured against the set goals on a scale of 1 to 100. On their part, the supervisor completed Form B in which individual performance was rated using the same scale. Then the supervisor and the individual met to exchange notes about their respective ratings.

Where there was any rating discrepancy, it was the duty of both the supervisor and the individual being appraised to discuss the differences until they reached a consensus. Once the supervisor and the individual had reached a consensus and agreed on the individual's new development level, the supervisor would complete Form C, taking into account the ratings from Forms A and B to ensure objectivity and fairness. Finally, the individual was given feedback and rewarded if the performance was exemplary in terms of achieving the set goals. On the other hand, if the individual had not lived up to the expectations, the shortcomings were communicated and the individual was advised on what to do to improve. This new appraisal system worked very well because nothing was done behind the scenes.

Individual rewards and incentives for good performance: One-minute management also helped us to shift the incentives system from the group to the individual. In previous change management programmes, we recognised and rewarded group (Area) achievement through trophies and cash bonuses. Thanks to one-minute management it became possible to pinpoint and reward individual performance.

Individuals who achieved all the milestones set out in their individual one-minute goals earned 25% of their quarterly basic salary. This was in addition to the stretch-out incentives, where if the quarterly evaluation showed that at least 95% of stretch quarterly targets were achieved, every individual member of the staff would get an additional 15% of their quarterly basic salary. This incentive mechanism greatly improved individual performance, motivation and productivity across the Corporation.

Our gains

The Corporation's performance continued to improve, posting far better results than ever before. One-minute management once more demonstrated the merits of change management programmes and the virtues of experimenting with good ideas that have worked for others elsewhere in the world and adapting them to the local corporate environment. People have been wondering how NWSC recovered and gained robust health. The answer lies in the fact that there are good ideas out there. What one needs to do is to find the relevant ideas and apply them in the pursuit of the respective corporate or individual vision and mission. It works! We did it at NWSC! And it is applicable elsewhere.

Lessons learnt

- Focusing on group incentives is a good strategy but it is not sufficient to sustain organisational change. It should be coupled with individual incentives and recognition.
- In any organisation, nothing is more important than the human resource(s). Any organisation that fails to invest in people, risks failure. Invest in your people, recognise and appreciate their efforts and reward them accordingly.
- Set goals and praise good performance but also reprimand failure to meet expectations. In reprimands, the purpose is to correct behaviour and improve performance, not to punish or humiliate the individual. And once the reprimand is done, forget the incident, do not haunt the individual.
- Managers are partners, not bosses – they are not commanders. They should always involve individuals they work with at all stages of decision making, implementation, monitoring and evaluation.
- Always remember that 'working smarter' is more rewarding than 'working harder'.
- Choose your management style, taking into account different situations as well as the workers' backgrounds, skills and competencies. Assigning duties that are either too difficult or too easy for individuals could breed despondency, frustration, lack of confidence and resignation.
- In any organisation, individuals need to work together to build team spirit, for collective effort is strength; two heads are better than one and 'none of us is better than all of us'.
- Devise a staff appraisal system that is open, fair, objective, interactive, negotiated and mediated in accordance with agreed rating criteria. Analyse appraisals and use the results to improve individual performance.

Chapter Eight

Reshaping the Partnership with the Government

Second Government Performance Contract

As a way forward, this ministry will work closely with NWSC and other stakeholders to conclude the new Performance Contract by the end of August 2003. I would like to commend the management of NWSC for the improvements that have been registered and would urge them to consolidate and safeguard these achievements.

Letter from the permanent secretary, Ministry of Finance, Planning and Economic Development, to the permanent secretary, Ministry of Water, Lands and Environment
4 July 2003

The first Performance Contract (PC I) signed between the government and the NWSC, as discussed in earlier chapters, gave fresh impetus to our home-grown management initiatives. Our set of solutions halted financial haemorrhage in the Corporation; the hitherto pedestrian delivery of services to our customers had been revitalised and the weak work ethos, which had characterised most of our staff, had improved greatly.

Our interventions were a far cry from what could have possibly been achieved by the expatriate private sector moguls. Hitherto, there had been a lot of hype from some quarters that private sector participation was the only panacea to the Corporation's plethora of problems. What the proponents of such an argument forgot to point out was that, such a private sector's *raison d'être* is accumulation of normal and abnormal profits because they do not have social-mission objectives, such as delivery of water and sanitation services, especially to the poor. Those that claim to have social-mission objectives and many of them do, will not deliver the service for free, someone somewhere must pay for it!

When the first performance contract expired, the questions asked were whether the Corporation's work ethic and delivery of services had really changed for the better. Had our set of solutions put the Corporation on a more robust managerial, financial and economic footing? Or were we just hyping our efforts?

The Ministry of Finance, Planning and Economic Development, which together with the Ministry of Water, Lands and Environment, were in charge

© 2009 William T. Muhairwe. *Making Public Enterprises Work*, by William T. Muhairwe.
ISBN: 9781843393245. Published by IWA Publishing, London, UK.

of administering the PC, in a letter dated 4 July 2003, stated:

> ... the Corporation has registered progress and achieved most of the targets as stipulated in the performance contract.

The Performance Contract Review Committee (PCRC), which had been appointed by the government to assess the performance of the Corporation in light of the contract, in its report of 17 December 2003, noted rightly that

> ... since the signing of the PC I between the GoU and NWSC in August 2000, the Corporation has achieved a turn-around in its financial and operational situation and at the same time expanded its services to new customers at a fast rate, surpassing the target connection rates of PC I.

Such praise notwithstanding, as pointed out by the PCRC report and confirmed by the World Bank, the Corporation was still not in a position to meet its debt service obligations as envisaged in the terms of the first performance contract. For the Corporation to continue to operate in a financially sustainable manner, it was therefore necessary for the government to continue with the debt-servicing freeze. The debts, as mentioned earlier, were mainly due to the legacy of poor investment decisions. In spite of the fact that all performance and operational indicators showed that NWSC was beginning to reach efficient levels, diversion into debt repayments was unsustainable at this level and would undermine further efficiency gains.

Against this background, we agreed with the government to enter into a follow-up performance contract (PC II) to consolidate and build on efficiency gains, which we had realised in the first performance contract. This chapter highlights the key features of the PC II and demonstrates how this tool was used to reshape the partnership with the Government.

Key objectives of PC II

The purpose of the second performance contract was to increase efficiency further by consolidating and enhancing the financial and commercial sustainability of our operations and to prepare the Corporation for transition to a higher level private sector participation mode. In addition, the renewed contract sought to maintain the financial equilibrium and sustainability of the achieved financial viability, to improve operations, and to expand services. We wanted to build upon the achievement of PC I while identifying other areas in which critical improvements could be made. We also wanted to refocus on two key areas of management and operations and tariff reform to ensure sustainability.

Lastly, the contract was aimed at providing a stand-alone framework for the implementation of the reforms as a contingency to the delayed procurement of the transaction advisory services for the involvement of the private sector in the operations of the Corporation.

Box 8.1: Objectives of PC II

- Further increase efficiency by consolidating and enhancing the financial and commercial sustainability of the operations of NWSC.
- Maintain financial equilibrium and sustain financial viability to improve operations and expand services.

Distinguishing features

The significant difference between the first and the second contracts was the emphasis on sustainability of improved performance. The Corporation had moved from a point of take-off to that of buoyancy. This was analogous to renowned economist Walt Whitman Rostow's third stage of 'take off' in his growth model. The take-off stage, according to Rostow, is the interval between the old blocks and resistances to steady growth; [when hardships] are finally overcome. The second contract recognised the fact that the NWSC had come a long way in establishing systems, which had ensured improvement(s), and which had to be sustained.

The second key difference between the first and the second contracts was that the monitoring indicators had been enhanced to include ratios as opposed to absolute figures. The indicators were in line with the Millennium Development Goals (MDGs) and were in the form of clear trackable service indicators. Accordingly, the second contract stated that the NWSC would maintain financial equilibrium intended to meet the operational and maintenance costs, on the one hand, and depreciation costs, on the other.

The second performance contract continued to target increased productivity so as to sustain efficiency gains already registered. Table 8.1 below highlights some of the key performance targets the Corporation was expected to achieve by the end of the contract.

The third difference was that the second performance contract took care of some of the limitations of the first contract such as the inclusion of the MoU with the government on the payment of government bills. The second contract provided for the following:

> It is recognised that the GoU as a consumer of utility services in the past has not been settling its water bills as and when they fall due. GoU now undertakes to settle all its bills as provided for under the Memorandum

Table 8.1: PC II Performance Targets

Target	Yr 2003	Yr 2004	Yr 2005	Yr 2006
NRW (%)	39	38	37	36
New Water Connections (No./month)	960	958	1,000	1,041
Total water Connections (No.)	87,172	99,593	11,940	124,730
Proportion of Inactive Accounts to Total Accounts (%)	21	17	14	12
Metering Efficiency (%)	95	95	95	96
Staff numbers (No.)	950	980	1,010	1,040
Staff Productivity (staff/1,000 connections)	11	10	9	8
Overall collection ratio	95	112	103	103

of Understanding signed between the Ministry of Finance Planning and Economic Development and the NWSC in April 2003.

The fourth difference was that the government recognised the internal reforms that were being carried out by the NWSC. It is stated in the second performance contract that

> Specifically, the management of the NWSC will undertake to continue to implement the current performance programmes under the current arrangements and uphold the institutional reform initiatives by the current management to effect the transformation of areas into autonomous business units.

The second performance contract, however, introduced more stringent measures to monitor the use of internally generated funds for the expansion of services, especially the application of these funds to flagship investments. To that effect, an appendix entitled 'Investment Activities' provided a detailed list of investments to be undertaken over the three-year period. Accordingly, the contract stated that the network extension project, as targeted, would be financed partly by cash savings arising from the debt service restructuring and funds generated by the Corporation for the duration of the contract.

In order to accelerate the achievement of the expansion plan, NWSC was to work towards the establishment of separate special funds from internal surpluses. These would include a 'network expansion fund' to be earmarked for implementation of the three-year network expansion plan; a 'new connections fund' to cater for connection costs, especially to the urban poor and a 'maintenance fund' from residual surplus, to maintain a minimum cash surplus deemed to be adequate to handle liquidity requirements. It was also stipulated that, when implementing the above network expansion programme, due regard be taken to ensure that a maximum percentage of active connections is maintained.

Furthermore, the contract spelt out that flagship investment should be undertaken; the NWSC would have to fully or partially finance a number of these investments. These included, among others, extension of water services to Seeta and Mukono; refurbishment of the Gulu Water Supply and Sewerage System to increase production; refurbishment of the Soroti, Arua and Bushenyi water supply systems; counterpart funding for the construction of a new Gaba III Plant and the refurbishment and expansion of the Entebbe Water Supply and Sewerage System – in addition to carrying out expansion of the water distribution network in all towns serviced by NWSC.

Implementation of the second performance contract

The commencement of the second performance contract required the management to re-invigorate its strategies towards meeting the new objectives and targets. The transformation of Area Performance Contracts (APCs) into Internally Delegated Area Management Contracts (IDAMCs) was one of the key strategies adopted in the implementation of the second performance contract. The IDAMCs, which are presented in detail in Chapter 9, were formulated to further consolidate operations at the Area level. The principal objectives of these contracts were increased autonomy of Area management teams, increased team accountability and the introduction of performance-based remuneration. The management moved fast, and by December 2003, the first seven Areas were placed under the framework of the IDAMCs.

The Corporation was also required, through the Monitoring and Evaluation Department, to develop an effective tracking system to allow for the free flow of information to and from the Areas. With this tracking system, the Corporation has since been able to build a reliable, accurate and comprehensive database, not only for internal use but also for all water sector stakeholders.

The idea of performance contracts took a new approach, on the one hand between the government and NWSC, and between the Corporation and its constituent Areas, on the other. This approach was instrumental in entrenching the corporate culture of rights, responsibilities and obligations. The management revolution, through which the Corporation has gone, has transformed it into a utility role model and a reference centre for some African countries. Some of these include Tanzania and Zanzibar, Malawi, Kenya, Zambia, Rwanda and South Africa, in addition to developing countries in other parts of the world.

Reflections from PC II

- Governments should never be complacent, especially when dealing with its institutions. Praises and reprimands should be given where they are due, but most importantly there is always a need to review and redefine direction.
- The contracting parties should be challenged continuously with new and stimulating responsibilities and expectations as well as corresponding rewards and penalties.
- Never set new performance targets/expectations without carrying out a comprehensive review of the previous agreement/targets.
- Each contract or challenge should come with a clear and well-defined implementation mechanism. You should not expect to get different results if you are doing the same thing, the same way.
- Never be satisfied with the status quo, change is a constant and new challenges keep coming up. These should be reflected in the redefinition of your direction.
- Monitoring and evaluation are essential.

Chapter Nine

An Alternative Approach to Privatisation

Internally Delegated Area Management Contracts

Partnerships between public utilities and the private sector have been introduced in a large number of water sectors in developing countries for over the past two decades. While the main principle of using private business oriented management expertise is still widely accepted, various innovative approaches of introducing and shaping "private sector principles" in providing water supply services have emerged during the past couple of decades.

Susanne Mauve
GTZ technical adviser to NWSC
June 2004

In Chapter 5, I described the Area Performance Contracts (APCs) between the NWSC Head Office and the Areas as the operational arm of the PC I. First introduced in 1999, the APCs were designed to transform NWSC Areas into businesslike enterprises driven by commercial principles and geared towards enhancing customer satisfaction and commercialisation of water and sewerage services. The subsequent Stretch-out Programme (Chapter 6) and the introduction of the one-minute management concepts across the NWSC (Chapter 7) catalysed the APC through the combined use of the group and individual incentive mechanisms.

However, despite the incremental performance improvements that had been brought about by successive change management programmes since 1998, pressure for private sector participation or even outright divestiture had continued to mount in government and donor circles. Indeed, a study of the urban water sector submitted in 2001 had recommended an international lease contract covering 33 major towns in the country, most of which were under NWSC.

Even after signing the second performance contract (PC II), various stakeholders within the government still believed that privatisation or at least some form of private sector participation was the only panacea for the challenges facing the NWSC in its struggle to improve the delivery of water and sewerage services. But were there no alternatives to private sector participation? Did the use of international expertise necessarily and always guarantee the turnaround, let alone the success of public enterprises such as the NWSC?

© 2009 William T. Muhairwe. *Making Public Enterprises Work*, by William T. Muhairwe.
ISBN: 9781843393245. Published by IWA Publishing, London, UK.

What were the lessons from the two private sector experiments in Kampala that involved international firms?

Personally, I was not convinced that the NWSC should rush into an international lease contract, or concession, for that matter, without exploring and exhausting local alternatives, for three reasons. To start with, our internally initiated change management programmes, from 100 days to one-minute management, had worked or were working fine, to the astonishment of all stakeholders. If the previous change management programmes had yielded positive results beyond the expectations of the doubting Thomases, with limited international involvement, why couldn't we be given more time to continue with the internal reforms?

In this regard, our previous change management programme successes were our best allies. If we had done it before, we could surely do it again. Not even the most determined advocates of private sector participation could ignore our record. True, many stakeholders among government officials and donors were willing to listen to our alternative proposals and to give us the benefit of the doubt.

Secondly, the negative aspects of the experiments of two international water three-and-a-half-year contract with Gauff Ingenieure, a German firm, to manage water supply in Kampala, excluding water production and sewerage services. Under the Kampala Revenue Improvement Programme, the operator undertook to increase billings and revenue collection and to pass on technical expertise to NWSC staff in return for a fixed management contract. In February 2002, ONDEO, a French firm, took over the operations of water supply in Kampala, this time including sewerage services, but not water production, for two years with a provision for extension to a third year, pending the introduction of an international lease covering 33 towns. In return for its services, ONDEO received a performance based management fee to cover the operational costs as well as the salaries of its expatriate staff.

The performance of both management contracts demonstrated that international private sector participation was not the best way to go. The process of recruiting an international operator, with all the competitive bidding requirements and procedures, was too slow and too bureaucratic for our internal reform process momentum. If allowed to take its course, international private sector participation would have hampered or even derailed our internal change management initiatives. While the management contracts were expensive to operate, the results were not impressive. Ms Susanne Mauve, the GTZ technical adviser, acknowledged our disappointment, which forced us to seek an alternative approach. She said 'One of the reasons for the introduction of the IDAMC was the negative experiences encountered with two

An Alternative Approach to Privatisation | 177

international private operators in Kampala (expensive transaction costs, high salaries for not necessarily well performing expatriate staff, poor performance in some key areas like financial management and reduction of NRW).' The fat salaries of expatriate staff were not reflected in improved performance and, in many areas, the services registered marked deterioration. The international contractors were not familiar with the local environment. And it took them a long time to learn the ropes (some never cared to do so) to cope with the local situation. By now it was clear enough, as internal and external assessments of Gauff and ONDEO proved that international operators were not the best option. Another reason in favour of a home-grown alternative was that, over the years, especially since 1998, the NWSC had built a critical mass of professional managers and engineers who were capable of operating water and sewerage services as businesslike ventures without direction or supervision from the headquarters. Many of these managers had received postgraduate training in management and had been exposed to management principles and best practices.

> Ms Susanne Mauve recalls that This experienced pool of staff was working with enthusiasm and ready to take on any challenges. Individuals like the very dynamic Managing Director together with ambitious key NWSC staff who went through the experiences of dealing with the two international private operators, were keen to prove that they could perform as well as, if not better than the international private companies. (Susanne Mauve, *The implementation of IDAMC as one of the key elements of the Urban Water Sector reform in Uganda*, NWSC working document 2004.)

The local staff and management had played the role of champions in the development and implementation of Stretch-out and one-minute management. Through APC, NWSC Area managers already enjoyed substantial autonomy in the delivery of water and sewerage services. Moreover, the principle of delegating, which was enshrined in one-minute management, already underscored all NWSC operations and methods of work, both at the headquarters and in the Areas).

My question to agitators for private sector participation was why not use the local pool of human resources as building blocks to build a NWSC internal reform alternative to private sector participation? I did not have to wait long for their answer. I knew we could do it and went for it.

In the subsequent sections, I give a detailed account of proactive steps taken to strengthen commercialisation of NWSC operations through a new concept of internally delegated area management contracts (IDAMCs). By successfully implementing the IDAMCs, the NWSC has demonstrated that there are many ways of skinning a cat – there are alternative approaches to outright privatisation. You just need the right mindset.

Initiation of IDAMC concept

As already discussed, under APCs, Area managers enjoyed operational autonomy. What was needed was to give maximum autonomy to Areas by delegating a whole range of activities and by transforming Areas into quasi-private autonomous business units – in other words, to mimic private sector participation. That is how we came up with the idea of Internally Delegated Area Management Contracts (IDAMCs) as an alternative to direct private sector participation. The IDAMCs were therefore designed as successor initiatives to the APCs and conceived purposely to help operationalise the second performance contract.

We opted for IDAMCs for a number of practical and pragmatic reasons. Firstly, there were no local investors with the resources, skills and commitment to enter private operator management contracts with the NWSC as some stakeholders wished. In any case, even if such operators were readily available, the protracted negotiation processes would inevitably have created an atmosphere of suspense and possibly paralysis in the delivery of water and sewerage services in the NWSC Areas. In contrast, in-house, IDAMCs were quick to draft, sign and flexible enough to implement.

Secondly, contracting external private operators would probably have provoked resistance from people who feared that their jobs and careers would be jeopardised.

Thirdly, sorting out the maze of complex legal technicalities, entailing the identification, valuation and transfer of assets, would have been too disruptive for the good of the Corporation. And in any case, the experience in Kampala had already indicated that the private sector did not necessarily deliver the expected output.

In opting for IDAMC, the following were some of the key guiding principles. Firstly, operationally IDAMCs had to become the alternative to external private sector participation; they had to conform to, and satisfy international standards of management contracts. As will be explained later, for the case of Kampala Area, the IDAMC was closely modelled on the erstwhile ONDEO management contract.

Secondly, although the IDAMCs were internal contractual arrangements, they were designed to give the NWSC Areas the highest degree of operational autonomy and at the same time consolidate and promote local management capacity in all Areas.

Above all, the IDAMCs were aimed at operationalising the PC II. The contract concept was therefore largely informed by the stretch-out and the one-minute management principles and practices.

But how different were the IDAMCs from the good old APCs? In a way, the two forms of contract were similar. Both covered NWSC Areas. Both were anchored on the principles of delegation and autonomous self-management. Both were NWSC internal arrangements aimed at operationalising the GoU/NWSC contracts. But there were four major distinctions between the two forms of contract. Firstly, under IDAMCs there was clear separation of roles and responsibilities, with more rights and responsibilities transferred to the Areas than under APCs.

Secondly, IDAMCs entailed the creation of Area partnerships that signed the contracts with NWSC headquarters. This was different from the APC structure, where the contractual relationship was between Area managers and headquarters. Thirdly, unlike the APCs, the IDAMCs introduced an element of competition where the partnerships had to compete for the right to manage a particular Area.

Marketing IDAMC

Although IDAMCs were easier and more flexible to craft and implement than external management contracts, selling them to the various internal and external stakeholders in the delivery of water and sewerage services was no easy task.

Some of the top NWSC managers feared that the limitations associated with IDAMCs would impact negatively on their decision-making powers or their freedom to guide Area operations as they deemed fit, as they had done in the past. Other stakeholders objected to the IDAMCs on the ground that they were not very different from APC. Key decision makers in the government, especially in the Ministries of Finance, and Water, Lands and Environment, would have preferred an international lease contract or even complete divestiture.

Accordingly, we had to marshal all our powers of persuasion to lobby for and market the IDAMCs as an 'alternative way' of private sector participation. Eventually we managed to convince some of our conservative colleagues and the concerned ministries.

Preparation of the IDAMCs

The drafting of IDAMCs and the development of their implementation strategy were entrusted to a core task force drawn from the staff of the operations and monitoring and evaluation departments, under my guidance. This task force worked closely with the Areas to draft the contracts, work out the modalities for their implementation, evaluate the business plans and to coordinate and harmonise the preparation process.

The task force would also eventually monitor the progress of IDAMCs by reviewing monthly, quarterly and annual reports from the Areas and provide appropriate feedback to the Area partnerships. As was the case in the previous change management programmes, we decided to pilot the IDAMCs in seven Areas first and later draw in the rest of the Areas.

The selection of the pilot Areas was no easy task because this time every Area wanted to be the pioneer. We finally decided to use the criteria of size of the Area targeting our 'cash cows', and the Area's enthusiasm in the previous change management programmes. This is how we ended up with Jinja, Mbale, Entebbe, Masaka, Mbarara, Kasese and Fort Portal as the pilot Areas. The preparation of the first seven IDAMCs took place between October and December 2003.

As already mentioned, the IDAMCs were designed to be as similar as possible to the international management contracts. This was no easy task and it involved a lot of hard work, planning and consultation with the different stakeholders. The task force did a tremendous job; the drafting of the contract framework for the seven pilot Areas took three months.

During this period, a number of workshops and consultative meetings were held to obtain stakeholders' comments and input. The process of coming up with the final contract went through 10 different draft stages. In addition to the contribution from the IDAMC task force, and NWSC members of staff, drafting the contracts also received input from other external stakeholders. The KfW-finance transaction advisers made valuable comments on various drafts, which in the long run helped to meet international standards. The GTZ long-term technical adviser was actively involved in drafting and implementing the IDAMCs, thus giving the process 'donor respect and credibility'. Another development partner that contributed to the process was the World Bank, an agency that had taken a keen interest in NWSC's work for many years.

With the contract preparation process under way, both the Areas and NWSC headquarters management had to fulfil certain conditions. To start with, the Areas had to create 'businesslike' partnerships consisting of key staff in the Area, for example, the Area manager, Area engineers, accountant, commercial officer and personnel officer. Each Area had the freedom to choose the name and the composition of the partnership on the basis of a set of guidelines and agreed criteria.

Under the IDAMCs, the partnership was bound by the Deed of Partnership signed by all members of the partnership and it clearly detailed the duties, rights and obligations of each partner. The Area manager assumed the role of the managing/lead partner. For example, the Margarita Water Partnership, which was in charge of Kasese Area, comprised the Area manager, Area engineer, accounts officer and senior commercial assistant as the partners.

The idea of partnerships was memorable to most of our managers. Most of them freely accepted that this was the beginning of full autonomy for the Areas. Adong Margaret, the Entebbe Area personnel officer, recalls that

> The programme gave staff the opportunity to identify themselves with their Area, which was the sole reason for names such as Greater Entebbe Water Partnership or any other like Kiira Water Partnership. IDAMC brought in the personal touch and created ownership of the business.

According to Andrew Sekayizzi,

> ...it was during one of our planning meetings for the IDAMC that we agreed on the name of our partnership and that marked the birth of the KIIRA WATER PARTNERSHIP a.k.a. the PACE SETTERS.

The second requirement for the Area partnerships was drawing up a comprehensive two-year rolling business plan. These constituted the vision and mission statements, Area SWOT analyses, the annual targets, the proposed strategies or actions, the resource requirements and financial management plans. They also contained human resource plans giving details on human resources and management requirements. The business plans under the IDAMCs were operational tools to guide Areas to achieve IDAMCs targets. During the preparation process of the business plans, the Area partnerships received assistance and guidance from the IDAMC task force and a local consultant, in addition to consultative workshops. The business plans prepared by the different partnerships were weighed against those from other partnerships before a winner would be declared.

As mentioned earlier, one very salient feature in the IDAMCs was competition for the right to operate a particular Area. Competition for the market is one of the intrinsic values of such contracts. To legitimise competition, we introduced the aspect of preparation of business plans and expressions of interest by the competing managers.

Any manager interested in operating a particular Area was expected to submit an expression of interest and a business plan for that Area. The IDAMC task force was then charged with the overwhelming task of evaluating all the business plans and identifying the best one for a particular Area. This was one of the most challenging but exciting aspects of the IDAMCs. Most Area managers thought that this was a deliberate attempt to eliminate some of them, and such fears were clearly indicated in their various recollections. According to Mr Sekayizzi, the Jinja Area Manager:

> The bombshell was later to be thrown at us towards the end of the workshop. That the autonomy we had welcomed so warmly in the morning was not to come on a silver plate; all the Area Managers were to compete for the seven pilot Areas. Honestly, this announcement was received with mixed feelings among us. Those who had special attachment to their respective Areas felt threatened whereas those who had always nurtured ambition to go to "better" Areas saw this as a once in lifetime golden opportunity to realise their dreams. Each Area manager was required to put in Expression of Interest (EoI) for at most three Areas, which would then later be evaluated, and whoever would give the best bid would take that Area. For sure, this announcement caused anxiety among us, and as we moved out of the workshop hall, I couldn't imagine the effect on all of us that day; our Areas had been floated and were up for grabs. We had now become "opponents"; each unsure of the other's moves. Whether your hitherto perceived "friend" would be the very one to unseat you.

Sylvia Tumuheirwe, the Entebbe Area manager, recalls thus:

> Whoever gains power fears to lose it, and one would be behaving abnormally if he/she behaved contrary to this! This is exactly the position I found myself in at the introduction of the IDAMC. What was unique was that the Area Manager and his/her key staff were required to put up a comprehensive Business Plan and all Areas were to be subjected to a competitive bidding process. This was probably the most challenging part of the programme especially for the Area Managers (Lead Partners) and me in particular who had just been recently appointed Area Manager and was still trying to master the Area management techniques.

However, as events turned out, all the sitting Area managers, using the experience gained from the previous change management programmes and the knowledge of their respective Areas, prepared impressive and sound business plans. None of them had to go through the ignominy of losing his/her Area.

After evaluation of the business plans, negotiations were organised between the headquarters and the respective winning Area partnerships. Both parties would then agree on the IDAMC performance standards, the incentive package, the human resource and other requirements. Once they reached consensus on all these issues, the Head Office would have to second staff and to provide support services and technical advice to the Area partnerships. Both the Area partnerships and the NWSC headquarters also had to carry out inventories of all assets to be transferred to the partnerships and to establish all the liabilities for which the Head Office took full responsibility upon the signing of IDAMC.

Launching the IDAMC

Once the groundwork was completed, the IDAMCs for the pilot Areas were officially launched and signed at a colourful ceremony on 17 December 2003, with the effective start date for operation being 1 January 2004. As required by the contracts, the Areas had formed partnerships, namely; Kiira Water Partnership for Jinja, Elgon Water Partnership for Mbale, Greater Entebbe Water Partnership for Entebbe, Buddu Water Partnership for Masaka, Rwizi Water Partnership for Mbarara, Margarita Water Partnership for Kasese and Rwenzori Water Partnership for Fort Portal.

They were responsible for providing water and sewerage services in the respective Areas. The signing ceremony was witnessed by several people; the contracts were endorsed by the chairman, Uganda Public Employees Union, the NWSC Board chairman, the Corporation secretary and I. The duration of the IDAMCs was two years with a provision for renegotiating the performance targets after the end of the first year, based on the shared experience of the previous year. As mentioned earlier, the OSUL contract was to last two years starting from February 2002 with an option to be extended for another year through negotiations. The negotiations commenced in January 2004, one month prior to the expiry of the first two years. After a lot of consultations and discussions, OSUL decided to exercise the option of not extending the contract, arguing that the offer made by NWSC did not meet their expectations. This created an operational gap in Kampala and an immediate solution was needed urgently and yet no one was willing to go through the long and costly process of procuring another international private operator.

Besides, the experiment with the two international operators had failed to deliver the expected outputs, and there was no guarantee that another private operator would do better. On the contrary, our change management programmes had a track record; we had delivered, and in some cases, such as stretch-out, even beyond the expected outputs.

The IDAMCs had already kicked off in a high gear and preliminary results were already impressive. Based on this, the management, with support of the donor partners, decided to implement the IDAMC model in Kampala. The Kampala IDAMC was finally signed in June 2004 between Kampala Water Partnership as the operator and NWSC headquarters as the client.

The initial positive results that were registered in the seven pilot Areas impelled the management to extend the IDAMCs to the other Areas. By August 2004, all Areas had signed IDAMCs.

The IDAMCs for the pilot Areas were later extended to the end of June 2006 to coincide with the expiry of the PC II and also give ample time for the preparation of the next phase.

Duties, rights and obligations

The IDAMC clearly spelt out the contractual duties of the Area partnerships, on the one hand, and the NWSC Head Office, on the other. They were detailed enough; every effort was made to minimise grey areas and to plug any loopholes that could become sources of dispute once the contracts came into effect. At the same time, the IDAMCs were flexible, with in-built qualifiers in many of the provisions to allow for any eventualities.

In essence, the IDAMC entailed delegating the NWSC statutory functions and responsibilities to the Area partnerships, but in conformity with the corporate vision and mission, corporate plan and budgets, PC II, and in line with accounting instructions, regulations and procedures, as well as terms and conditions of service.

Responsibilities of Area partnerships: The contractual duties, rights and obligations of the Area partnerships were wide-ranging and overarching, covering a whole spectrum of water and sewerage services, from water production and distribution to marketing, public relations and customer care. Precisely, each Area partnership was 'to effectively discharge all the duties and obligations of a water supply and sewerage service provider'.

Specifically, the Area partnership was, among other things, obliged to produce and distribute quality water; collect, treat and dispose off sewage; rehabilitate and extend the water and sewerage network; increase customer connections; serve customers with bills and collect revenue; market water and sewerage services and ensure customer care/satisfaction.

The discharge of this duty had to be carried out diligently, efficiently, cost-effectively and professionally in accordance with 'sound management practices and appropriate technology, and in the best interest of the Corporation'.

In exercising its duty of collecting revenues, the Area partnership could also engage services of independent debt collectors, subject to headquarters' approval, and at a maximum commission of 10%, to handle debts that were otherwise difficult to recover by ordinary means. All collected revenue had to be deposited in the Corporation's accounts without fail.

The partnerships were also required to prepare and submit sound business plans for approval by headquarters. The primary purpose of the business plan was guide the Area partnerships in their contractual commitment of achieving or, better still, exceeding the agreed performance standards and targets. Such a plan had to follow the Corporation's strategic objective and planning principle and processes. The business plan had to be reviewed and updated annually to take into account any new corporate policies or performance improvement developments.

Although the NWSC headquarters reserved the right of ownership of all Area assets, the Area partnership had the right to access, use and control them in the best interest of the Corporation. In exercising this right, the Area partnership was obliged to keep the assets in good condition but had no power to dispose of any asset without express written permission from the Head Office, and in the event of receiving any such permission the proceeds would have to be remitted to the Corporation's coffers.

In the area of human resources management, the IDAMC required the Area partnership to draw and submit a human resource plan to the Head Office, in which it had, among other things, to lay down the rules, conditions and terms of employment, including job classification and description, standard employment contracts and remuneration packages. The human resources plan also had to spell out the required staff levels, methods of staff recruitment, training, promotion, disciplining and assessment in accordance with the one-minute management-based appraisal principles and procedures.

In addition, the Area partnership had certain obligations to the members of the staff seconded from the Head Office. The Managing Partner had powers to promote seconded staff under certain conditions but could not terminate their services. The Area partnership could only refer seconded staff to the headquarters on fully justified and proven grounds of misconduct, under-performance and patronage-induced insubordination.

With regard to the Area partners, the names of individuals specified in the Deed of Partnership had to be submitted to headquarters for 'non-objection'. The Head Office had to express any objection with good reasons within seven days. Failure to do so would be construed as acceptance of the partners. The fact that the Head Office has not objected to the nominated partners in any of the Areas so far indicates that the partnership teams have been constituted with due diligence and utmost care, which indicates a high sense of professionalism among our Area managers.

In return for fulfilling their contractual duties and obligations, the Area partnerships enjoyed certain rights and entitlements, some of which were reflected in the duties and obligations of the Head Office, which will be described shortly. But for the moment, it will suffice to mention those of a procedural and financial nature. To start with, the Area partnership was entitled to get feedback from the headquarters within the specified time frames. This was deliberately done to ensure that the Area partnership's pursuit of performance standards and business plan targets did not fail due to feedback delays from the Head Office.

Secondly, and more importantly, the Area partnerships were entitled to management fees to enable them to fulfil their operational duties and obligations, and to compensate them for their labour. The management fee had to be settled within a specified time after the Area partnership invoiced the Head Office. In addition, each Area partnership was entitled to a working capital advance within seven days of signing the IDAMC to enable it kick-start the fulfilment of its contractual duties and obligations.

Furthermore, Area partnerships were entitled to financial relief for expenditure deficits beyond their control and for charges not originally anticipated in the business plans. Ultimately, therefore, they enjoyed the protection and financial underwriting that would not have been available to outright private sector participation entities.

In any case, if any of the Area partnerships failed to perform its duties and obligations, the Head Office would reserve the right to resume full responsibility for the delivery of water and sewerage services in that Area.

The role of headquarters: The IDAMC also stipulated the duties, rights and obligations of the NWSC Head Office. These imperatives had to be consistent with the Corporation's vision, mission, corporate plan, PC II, budget framework and commercial objectives, as well as with its statutory and social mission responsibilities. It was the duty of the Head Office to review tariffs, fees, rates and other charges, and inform the Area partnerships accordingly. However, in exercising this right, the Head Office was duty bound to take into consideration the business plan requirements and advice of the Area partnerships.

Head office was also required to provide logistics and asset management needed for the effective and timely implementation of the Area partnership business plans. This also included bulk procurement of materials and equipment in accordance with the requirements of the Area partnership business plans and to deliver them without any undue delay, which would disrupt the smooth operations of the partnerships.

Furthermore, the Head Office was obliged to review additional Area partnership requirements regularly and expeditiously and to ensure the harmonisation of the partnerships' business plans with the Corporation's annual budgets. All in all, the Head Office had to ensure that the Area partnerships were fully supported and facilitated in terms of logistics to enable them meet their IDAMC obligations.

Another head office responsibility was to monitor and evaluate the progress of IDAMC. This process entailed data validation and performance analysis, ensuring conformity to the IDAMC standards and the provision of comparative data from other Area partnerships.

In general terms, the Head Office had to continue encouraging the spirit of competition between Areas, to protect and promote customer rights and to gauge the degree of customer satisfaction through regular customer surveys. Although the Head Office reserved the right to initiate new change management programmes in the wider interests of the Corporation, the Area partnerships had to be fully consulted to make their input in such programmes.

To maximise the partnerships' autonomy, the IDAMC explicitly prohibited NWSC headquarters from interfering in the day-to-day Area operations. For example, the Head Office had no power to issue instructions to Area partnerships regarding operational decisions or actions. And it was not allowed to influence Area partnership–customer relations and dealings. The Area partnerships were free to ignore undue interference from the headquarters in contravention of the provisions of IDAMC, and the Head Office could not withhold payments due to the partnerships or impose any other functions on the ground that its instructions had not been implemented.

Breaches, penalties and arbitration

Every good contract contains clauses regarding what happens when either party fails to live up to its duties and obligations. The IDAMCs were no exception. These contracts specified clearly what constituted breaches of contract, what penalties to impose for such breaches and how to terminate the contracts and the process of arbitration. For example, an Area partnership would have breached the IDAMC if it failed to protect and maintain the Corporation's assets and properties. A more serious breach of contract would be failure to remit bank collections on the Corporation account on a day-to-day basis. Throwing the name and image of the Corporation into disrepute through professional misconduct or negligence would also constitute a serious breach of contract. Depending on the severity of breaches of contract, the penalties would vary from warnings and reprimands to withholding the management fees, to termination of the contract.

On its part, the Head Office would have breached the contract if it failed to settle payments due to the Area partnership within the stipulated time frames. Similarly failure to provide the logistics and assets within the agreed contractual period and hence jeopardising the partnership's performance would be a breach of the contract.

The penalty for failure to settle payments due to the Area partnership was the imposition of interest charges on amounts outstanding. The penalty for failure to provide enough logistics and assets would be deemed serious enough to warrant the termination of a contract.

The IDAMC could be terminated either when it had run its course and was not renewed or under certain stipulated provisions in the contract. The Head Office could terminate the contract by giving notice to the Area partnership upon failure of the Area partnerships to maintain assets, ensure good housekeeping, observe professional standards and ethics, uphold the good name and reputation of the Corporation, meet minimum performance targets and deposit the collected revenue to the Corporation's bank account. The Head office could also terminate the contract, after 30 days notice, for purposes of initiating and implementing other performance improvement programmes. On its part, the Area partnership could terminate the contract by giving written notice to the Head Office upon failure of the Head Office to pay the management fees due to the partnership, comply with its obligation of non-interference, or provide logistics and asset management for three consecutive months.

Once the contract was terminated for whatever reasons, the assets of the Area as well as its employees, including the partners, would revert to the Corporation.

Disputes between the Head Office and the Area partnership could, first of all, be resolved amicably through mutual negotiation, and in good time before the dispute got out of hand. Beyond that, the issue would be referred to the NWSC the managing director for adjudication. If the latter's action did not satisfy either of the two parties, the dispute would be referred to the NWSC Board of Directors, whose decision would be final and binding.

Since the commencement of the IDAMCs, there have not been many cases going beyond the first stage of dispute resolution. Most have been amicably settled between the Head Office and the Area partnerships concerned.

As the managing director, I acted as an adjudicator in only a handful of cases. Not a single dispute was ever referred to the Board of Directors. This is not to say that it was smooth sailing as demonstrated by the challenges discussed later on in this chapter.

IDAMCs' management fee structure

One of the benefits that come with management contracts is the management fee which is negotiated and paid to the operating party on the basis of terms and conditions detailed in the contracts. The management fee structure under the IDAMC was performance based, unlike some other conventional management contracts. It was composed of three parts, namely the base fee, performance fee and the incentive fee. The base fee component covered all the fixed or non-controllable operating costs and part (75%) of the controllable operating costs.

This was a fixed amount, negotiated and agreed upon before the signing of contracts. On the other hand, the performance fee component covered the remaining part (25%) of the controllable operating costs of the Area and was subject to achievement of the minimum performance standards stipulated in the contract. The performance fee component also included the salaries of the partners, which meant that the Area partnerships could only receive the full operating costs, including the salaries of the partners, if all minimum performance standards were met. In a way, this was a penalty to guard against complacence and backsliding.

On the other hand, the IDAMC offered a more attractive incentive package than any of the previous change management programmes. This was in recognition of the fact that as a result of the other change management programmes, Areas had reached a level where any improvement in performance required more effort than ever before. Also, the Areas had taken on more responsibility, which had to be matched with an attractive package. The incentive fee structure allowed the partnerships to share in the excess cash-operating margin though subject to the level of achievement of the key targets.

These key targets were in the domain of cost control and income generation, reduction of NRW, increase in new and active connections and timely collection of bills. The other interesting feature of the IDAMC incentive structure was that, unlike previous programmes, we had a fixed incentive formula that cut across the board. Under the IDAMC, each Area negotiated its incentive package, depending on its offer to the headquarters and the challenges it faced in achieving its targets. Using this principle, Areas ended up with very attractive incentive packages, to the extent that partners could earn as much as 300% of their basic pay if all the targets were met.

IDAMCs' implementation

The context in which the IDAMCs were designed made their implementation process very easy and smooth. The fact that each party's role and responsibilities were clearly laid down meant that there was no room for conflict of interest and blaming one another for failure. Everyone knew what they were expected to do and the implications of failing to live up to their obligations. Areas submitted their monthly reports as required by the contract, and the headquarters took up its task of evaluating performance and giving prompt feedback.

Monitoring and evaluating the performance of the IDAMCs focused on both output and input technologies. Head office evaluated the outputs on a monthly basis using the reports submitted by the Areas. The quality of the Area reports and their content clearly showed that the future of building local managerial capacity through Area partnerships and home-grown alternatives was the right direction to take. On a quarterly basis, workshops were organised by the headquarters in accordance with the contracts; this was aimed at reviewing the progress of the Programme and addressing the constraints or obstacles that were hampering its smooth implementation.

A partnering approach was adopted to address the constraints, and this ensured a win-winsituation for both contracting parties. The workshops were also used to praise pacesetters with sets of trophies and cash awards and at the same time shame laggards. In addition, a monthly magazine called the *Water Herald* was established to provide a benchmarking platform for the partnerships where they could share their innovations, identify binding constraints and receive credit for achieving their respective targets. The introduction of the Checkers System, which is the subject of Chapter 10, in May 2005, greatly improved the momentum of IDAMCs' implementation. The system incorporated activities that promoted competition among the partnerships for successful compliance with the IDAMC objectives and requirements.

IDAMC I achievements

The performance of the Corporation in terms of the delivery of water and sewerage services continued to make commendable and sustainable progress, all thanks to the IDAMCs. The partnerships and the Corporation, as a whole, registered performance improvement on most of the key indicators, which contributed to the Corporation performing well above the PC II targets.

The Head Office, as required under the IDAMC, carried out quarterly, bi-annual and annual evaluations to establish the extent to which the Area partnerships were complying with the contractual requirements. The evaluation results showed improving performance trends in most of the Areas, which was a demonstration of the positive impact of the IDAMCs. The Performance Review Committee appointed by the permanent secretary, Ministry of Water, Lands and Environment, as was required under the PC II, examined the performance of the Corporation in respect to its obligations and targets (under the PC II) and also came to a conclusion that overall the Corporation's performance was outstanding. Table 9.1 shows the performance of Kampala and other Areas as well as the overall performance of the Corporation during the IDAMC period against the PC II targets.

As shown in Table 9.1, the overall performance of NWSC, with the exception of the collection ratio, was well above the PC II contractual targets.

As part of the business plans for the Areas, in respect to water production performance, we developed replacement plans for electromechanical production assets to ensure their efficient and reliable operation. In addition, preventive maintenance plans to extend the life of the production facilities were also enforced in all the Areas. These initiatives, together with an increase in the demand, explain the good water production performance during the IDAMCs and PC II period.

An Alternative Approach to Privatisation

Table 9.1: Performance of IDAMC I

Performance indicator	2003/2004				2004/2005				2005/2006			
	Kampala	Other Areas	Overall	PC II Target	Kampala	Other Areas	Overall	PC II Target	Kampala	Other Areas	Overall	PC II Target
Water production ('000 m³/month)	3,208	1,383	4,591	4,425	3,605	1,477	5,081	4,592	3,603	1,469	5,071	4,725
NRW (%)	45	21	38	38	41	17	34	37	36	15	30	36
New Water Connections	727	444	1,170	958	1,015	837	1,852	1,000	1,659	718	2,377	1,041
Total Water Connections	60,077	40,398	100,475	99,593	74,777	48,269	123,046	111,940	95,191	56,947	152,138	124,730
Proportion of Inactive Accounts to Total Accounts (%)	20	13	17	17	17	10	14	14	16	8	13	12
Metering Efficiency (%)	98	94	97	95	97	99	98	95	98	100	99	96
Staff Productivity (staff/1,000 connections)	7	13.5	10	10	6	10	9	9	5	8	7	8
Income (UGX million/month)	2,300	1,251	3,551	3,210	2,965	1,517	4,482	3,406	3,210	1,669	4,878	3,573
Collection Ratio (%)	99	108	101	112	88	90	89	103	92	87	90	103
Operating Surplus (UGX million/month)			1,031	929			1,120	1,038			1,168	1,119

The improvements in NRW performance were largely attributed to the decentralisation of key functions of operations to the branches (especially in Kampala), following the introduction of IDAMCs in 2004/2005 as well as the introduction of the new connection policy under which the Corporation assumed responsibility for provision of standardised materials for all connections. This helped reduce NRW arising out of use by consumers of substandard materials.

The Corporation's performance in regard to new connections was also impressive and contributed immensely to the government's efforts to meet the Millennium Development Goals. This good performance was possible due to numerous factors, among which were network extensions, especially to the peri-urban areas (a total of 668 km of water mains extensions were laid during this period), and the pro-poor water connections policy, through which customers within a radius of 50 m were offered subsidised or free connections. This greatly accelerated service coverage, especially in the urban poor areas. Due to the impressive new connections performance, there was a progressive increase in the total number of water connections and a decline in the proportion of inactive connections to the total water connections.

During the IDAMC's implementation we continued implementing the customer service policy, which contributed to the increase in the number of connections, reduction in the ratio of inactive connections to total connections and increase in revenue. A number of other customer care initiatives were also established, including establishment of a customer call centre in Kampala to improve access to customers and introduction of direct debit and over-the-counter bill payment systems with commercial banks. Staff training in customer care also helped to make the Corporation more efficient. Customer surveys, which were upheld, became important feedback tools for service delivery improvement. Through all these initiatives, the Corporation demonstrated and lived up to its slogan of 'the customer is the reason we exist'.

The bottom-up appraisal method was used to improve staff productivity in all Areas using the stretch-out and one-minute management concepts. Staff at all levels were actively involved in the day-to-day running of the systems and taking decisions concerning their respective Areas. This enhanced ownership and greatly improved staff productivity. On the other hand, as a result of the increased workload, mainly due to the expanded network and customer base, and takeover of a new operational Area, the Corporation recruited an additional 104 members of staff. This was carried out concurrently with the reduction in the number of permanent staff in preference to contract staff. By end of June 2006, 54% of the staff were on contract compared to 35% in June 2004. Notwithstanding the increase in staff numbers, we were still able to achieve the staff productivity targets of the PC II. In recognition of our impressive performance improvement in staff management, we received a number of recognition awards, which are described in Chapter 13.

Most significant of all, the Corporation progressively generated more income during the IDAMC period and performed well above the PC II targets as demonstrated in Table 9.1. The improvement in total income was attributed largely to improved billing efficiency and particularly an increase in water sales. Similarly the operating surplus before depreciation and exceptional items was well above the PC II targets. However, collection efficiency was consistently below the target throughout the PC II period. This was mainly attributed to poor payment by GoU agencies. The collection efficiency of GoU agencies dropped from 123% in 2003/2004 to 74% in 2004/2005 and 77% in 2005/2006.

From the surplus generated, we undertook a number of flagship projects, which included extension of water services to Mukono and Seeta; refurbishment of Gulu water supply and sewerage system; refurbishment of Soroti, Arua and Bushenyi water supply systems and extension of 668 km of water mains across all the Areas.

Critics proved wrong again!

All in all, we did very well and lived up to the expectation of the government and all the stakeholders. We managed to achieve and in most cases surpass most of the PC II targets. Even where we fell short of the targets, the improvement trends were once again upwards and generally satisfactory. We proved the critics wrong again and defended our argument that 'a home-grown alternative to private sector participation was another sound way of sustaining water service delivery in Uganda', as opposed to the view that PSP was the universal remedy. NWSC had enough local human resources to deliver water and sewerage services efficiently and indeed we delivered accordingly. The Performance Contract Review Committee also shared this judgement and accordingly recommended a 25% bonus to the NWSC management as was provided for in the PC II. The government also reconfirmed the achievements and accordingly entered into a third performance contract for another three years from 1 July 2006 to 30 June 2009.

The third performance contract 2006–2009

After the successful completion of PC II, the Corporation's management was once again invited by the Finance ministry to enter into fresh negotiations, leading to the signing of the third performance contract (PC III). PC III was meant to safeguard the hard-earned gains and raise the Corporation's sustainability profile to meet the increasing demand for the services and to expand into new markets. The objectives of PC III in addition to those of the first and second performance contracts are presented in Box 9.1.

Box 9.1: Objectives of 2006–2009 performance contract

In brief, the objectives of the third performance contract were to

- complete the unfinished business of PC II;
- overcome the constraints that had hindered the full realisation of the objectives of PC II;
- consolidate and build on the gains of PC II while identifying other areas where significant improvements could be achieved;
- enhance the operational and financial performance of the Corporation through innovative managerial programmes aimed at advancing the commercialisation and social mission mandate of the Corporation;
- capitalise the Corporation and enable it to operate on a firm financial footing by restructuring the long-term debt (UGX 154 billion) into equity;
- continue implementing the performance improvement programmes and uphold the institutional reform initiatives to effect the transformation of Areas into autonomous business units and
- ensure evolution of IDAMCs to DAMCs, [and] encouraging participation from outside NWSC.

The overall objective of the third performance contract was to sustain the change management momentum that had been blessed by the government through the first and second contracts. The specific objectives as presented in Box 9.1 were intended to place the Corporation on an irreversible performance path. The operational arms of the third performance contract with the government were the second phase of IDAMC referred to as IDAMC II.

The second phase of IDAMC

The IDAMC II contracts were designed to operationalise the underlying principles and explicit objectives of the third government performance contract. The overall objective of IDAMC II was to improve the performance of the Areas and more specifically to increase the operational and financial performance of the Corporation; consolidate and build on the achievements of IDAMC I; strengthen and increase managerial autonomy and at the same time shed more operating risk to Areas; and have better incentive plans and performance targets.

The implementation of the first phase of IDAMCs faced challenges that had to be addressed in the second phase of IDAMCs. These included lack of failure limits for each performance standard below which the Head Office could invoke penalties. This tended to encourage partnerships to state 'any' level of performance standards for purposes of winning contracts. In addition, the incentive formula did not distinguish between the different levels of performance for each parameter. Therefore, the incentive formula did not effectively motivate partnerships to allocate effort optimally. Other constraints included low fixed incentive ceilings that acted as perverse incentives for ambitious managers who performed beyond expectations; limitation of competition among internal staff and therefore limited managerial efficiency; managers at the centre who did not easily 'let go' some of their much cherished powers and tendencies by some regional teams to manipulate data.

Furthermore, some partnerships tended to abuse portions of delegated powers through irregularities in staff recruitment and illicit activities related to delegated investment activities. Most of these constraints/challenges were addressed in the second phase of IDAMCs. The design of incentive formulae was tailored to the specific problems of each Area. In addition, failure limits were introduced and incentive ceilings were elevated to allow more room for enhanced performance. Furthermore, during the bidding process for the second phase, employees who had earlier retired from the Corporation without record of malpractice or incompetence were allowed to compete for managerial responsibility. The design of IDMAC II limited the possibility of irregularities relating to recruitment and delegated investment activities.

IDAMC II was officially launched and signed in August 2006, with the effective date for its operation being 1 July 2006. The implementation of IDAMC II demonstrated remarkable progress towards achievement of PC III objectives as shown in Table 9.2.

It is fair to say that the IDAMC II built on the achievements of IDAMC I and further improved the performance of the Corporation as demonstrated in Table 9.2. The achievements during the second phase of IDAMC did not go unnoticed as the Corporation was honoured and recognised in different circles as revealed in Chapter 13.

Looking back and further ahead

I would like to point out that the implementation of IDAMC II was not all smooth sailing. The Corporation encountered a number of challenges and some of the decisions taken to arrest or rectify the situation, to a certain extent, contravened the IDAMC spirit. Some of the partnerships took advantage and misused the autonomy and freedom provided under the IDAMC. The consequence was the dismissal of the culprits and disbanding of three Area partnerships (Mbarara, Kasese and Fort Portal). Unfortunately, among those dismissed were some of our best and most mature managers. This, in a way, was a big setback for us, and the Head Office had to improvise and take

Table 9.2 Performance of IDAMC II

Performance indicator	2006/2007				2007/2008			
	Kampala	Other Areas	Overall	PC III Target	Kampala	Other Areas	Overall	PC III Target
Water production ('000 m³/month)	3,564	1,551	5,115	5,067	3,741	1,559	5,301	5,183
NRW (%)	39	18	33	33	39.8	18.5	33.5	31.5
New water connections	1,226	809	2,035	2,113	1,224	808	2,032	2,083
Total water connections	106,522	74,175	180,697	180,520	119,897	81,942	202,559	204,822
Proportion of inactive accounts to total accounts (%)	16	9	13	12	15	9	13	11
Metering efficiency (%)	99	99	99	98	99.6	99.7	99.6	99
Staff productivity (staff/1,000 conections)	6	7	7	8	6	8	7	7
Income (UGX million/month)	3,808	2,060	5,868	5,787	4,326	2,681	7,007	6,222
Collection ratio (%)	99	91	92	95	94	87	92	95
Operating surplus (UGX million/month)			1,531	1,540			1,493	1,941

over the management of those Areas until we found suitable replacements through a competitive bidding process.

In order to safeguard against occurrence of similar incidents in the other Areas, we took drastic decisions and transferred staff who had overstayed in particular Areas to other Areas. What I noticed was that the IDAMCs had encouraged staff overstaying in an Area. Although I had originally considered this a positive idea, I later learnt that the longer the staff stayed in an Area, the more relaxed and complacent they became. There were also cases where some of the lead partners saw the difficult task ahead and opted to retire as was the case with the lead partner of Soroti. Some Area managers reached their retirement age and no longer had the vitality to drive performance. Such cases affected the performance of the partnerships as they were not anticipated originally. Consequently, we ended up with some new managers who lacked the managerial capacities required to match the challenges of arms-length management demanded under IDAMCs. We recruited more staff within the provisions of the IDAMC to strengthen and compliment the capacity in the Areas and with the view of getting better managers for the future IDAMC.

Similarly, there is perceived interference from the Head Office managers under the guise of strategic guidance – the boss element still exists. The role of the Head Office as a performance monitor has not structurally crystallised. This function needs to be reviewed and strengthened. This is evident from a letter from Mrs Evelyn Otim, the chief manager in charge of commercial services, in 25 April 2008 in response to an accusation that she had failed to supervise commercial activities effectively. She states

> ... there is a 2 year running contract signed between head office and Greater Entebbe Partnership (GEP) ... GEP as an independent operator has been granted a high degree of autonomy in the contract to manage its day-to-day activities with no interference from head office. Clearly stated in Section 9.1 (a) any instruction from head office to the operator will require amendment of the contract... this actually leaves us in the commercial division at head office in position of a "lame duck". We have been told to supervise the Areas but the IDAMC makes it difficult for us to enforce disciplinary measures ... A free slave in confinement is still a slave, however much we are flogged to supervise the Areas, we shall be frustrated by the IDAMC... we are given responsibility without authority...

The Corporation still faces challenges associated with securing capital for 'huge' investment requirements, e.g. expansion of water treatment plants, distribution mains and large reservoirs. This challenge is exacerbated by the fact that the tariff cannot be increased beyond certain limits. However, there is still a lot of vigour to tackle these challenges. The NWSC is looking at possibilities of attracting private sector financing through stock markets. This is expected soon and NWSC managers are looking at modalities of realising it.

All in all, we are changing course and rethinking the IDAMC framework in the light of the challenges highlighted above. The Corporation intends to strengthen IDAMCs further to allow them evolve into regional private sector entities. This is being considered cautiously, cognisant of the private sector profit-maximisation tendencies. In all cases, the principles of strengthened incentive plans, allocation of more operating risks to the service providers and the outsourcing of segments of operations to private operators where it is cost-effective and efficient will continue to play a significant role.

The next chapter discusses the Checkers System, a monitoring and oversight tool that revolutionalised our monitoring approach.

Lessons learnt

- Privatisation or private sector participation, whether local or through international investors/operators, is not the only guaranteed way to improve the performance of public enterprises. So keep all options open and let policy makers decide what works regardless of the fashions of the day.
- Private sector principles and practices can be infused into public sector management and operations successfully without recourse to full divestiture or concessions.
- Introducing delegation and decentralisation in public enterprise management can enhance performance and optimise efficiency gains.
- Public enterprises can have large pools of business-minded managers whose potential can be tapped and translated into star performance without recourse to privatisation. So keep 'doors' open.
- Be bold enough to try out unorthodox ways of doing things, provided you are convinced that they have a good chance of yielding results. Be careful not to rush for ready-made solutions, which may have worked wonders elsewhere but may not work for you. Be discerning; pick the good from the 'false'.
- Be bold and change your course if you realise that you have been misled or that you need to correct previous commitments or decisions.
- Remain innovative; there are always new ways of doing things, remember, as you change course, detractors are also changing theirs. Make sure they do not beat you at your game.
- Market your new changes properly.

Swinging the pendulum to the grass root. Eng Johnson Amayo, the manager Operations, signing the KW-IDAMC on behalf of Head Office, as a commitment by Head Office to support KW and fulfil its obligations. He is cheered on by (L-R) Eng Silver Mugisha, Mr Chirstopher Kanyesigye and Eng Alex Gisagara.

Above: The managing director, Dr William Muhairwe (right) handing over a signed APC Contract to the Area manager of Masaka, Mr Andrew Sekayizzi (second left). Left is the corporate secretary, Mr David Kakuba.

Internal delegation cascades to the branches. Right: The GM, Kampala Water, Eng. Harrison Mutikanga (right) exchanging the BPC with the Branch Manager of Kitintale, Mr Jackson Turyahurira (left).

Chapter Ten

A New Approach to Performance Monitoring

The Checkers System

Watch, therefore, for you do not know when your Lord is coming. But know this that if the Master of the house had known what hour the thief would come, he would have watched and not allowed his house to be broken into. Therefore be also ready for the son of man is coming at an hour you do not expect.

Matthew 24:42–44

The most single important thing to remember about any organisation is that there are no results inside its walls ... there are only cost centres. Results exist on the outside ... and are produced for the customers and other stakeholders ... Managers must take time to evaluate how well organisations are managed and the value they deliver to the customers and other stakeholders.

P. Drucker
Management and the World's Work, Harvard Business Review
September–October 1988

Incentive-based performance management is, by and large, associated with the philosophy of non-interference. The intent is to give the operator maximum incentive for innovation and to reduce oversight costs by the performance monitor/regulator ... non-interference might lead to weaker performance than if the regulator/monitor (the principal) collaborated with the operator in a meaningful way ... Most performance monitors passively wait for quarterly (or annual) reports of performance results, thinking that they will enforce the contract in case of non-compliance to performance outputs. As a result, the customer – who is the final recipient of the services – bears the ultimate cost of poor service.

Silver Mugisha, Sanford V. Berg and Heather Skilling,
Water
Volume 21, 2004

Of all our successive change management initiatives, none was as ambitious and daring as the IDAMCs, which were designed to practice the principles of

© 2009 William T. Muhairwe. *Making Public Enterprises Work*, by William T. Muhairwe.
ISBN: 9781843393245. Published by IWA Publishing, London, UK.

delegation, decentralisation, empowerment and privatisation in the management of water and sewerage services at the Area level. Under the IDAMC, the Area partnerships, people who knew the local conditions, got the opportunity to put their ideas and priorities into practice and to effect change without looking over their shoulders – in short, to run the show without the fear of interference from the Head Office. The IDAMCs, which were a contractual extension, refinement and consolidation of APCs, were intended to build local management capacity and confidence through delegation and decentralisation; to create dynamic quasi-private and financially viable business units and to inculcate a sense of operational business self-reliance, responsibility and reliability. The ultimate goal of IDAMC was to transform NWSC into a corporate 'confederation' of autonomous, profitable and self-sustaining water and sewerage services delivery 'action centres' potentially capable of independent existence from the parent Corporation.

At the same time, the IDAMC posed a great challenge, that of exercising appropriate oversight without undermining partnership confidence or contravening the non-interference provisions of the contracts. The IDAMCs were a sensitive act of faith, a great leap in the dark, of which the direction and consequences were difficult to fathom. They were a new adventure that had not been tested to establish its operational efficacy. On the other hand, the failure of any IDAMC was likely to undermine confidence in the entire project and to throw in doubt our change management momentum. And without doubt, the management, which is responsible for the implementation of policy, would have paid the heaviest price.

In initiating change management programmes, it is easy to be optimistic, to hope for the best and to overlook or minimise any operational constraints. It is always easy to climb to the top but much more difficult to stay there and deliver the expected results on a sustainable basis. In the case of IDAMC, the Area partnerships were eager to get on with their contractual obligations. They had already gained valuable experience from the implementation of APC. We had the confidence that the Areas had the right management capacity, enthusiasm and professional commitment to achieve the IDAMC performance objectives. But the question was whether the Head Office was ready to meet its side of the bargain and whether it had the right monitoring capacity to keep the partnerships on track regarding the avoidance of any misuse of delegated responsibility.

Were the existing performance monitoring mechanisms appropriate for the operational environment of delegation and decentralisation? Were there any loopholes in our traditional monitoring approaches or methods that needed to be addressed to cope with the new performance monitoring challenges posed by the

A New Approach to Performance Monitoring | 203

IDAMC? Did we have the monitoring capacity, both in terms of human and financial resources, to monitor all the Area partnerships effectively and timely to ensure that nothing important escaped the attention of the headquarters? As the IDAMC got off the ground, it dawned on me that there were no easy or ready-made answers to the above-mentioned questions. We knew for sure that our performance monitoring tools needed strengthening to cope with the IDAMC operational challenges and demands, but we did not yet know how to do so.

This chapter presents a description of how monitoring approaches have progressively evolved in NWSC and how an alternative monitoring concept called Checkers System was adapted in NWSC operations. I point out that sometimes it is better to balance monitoring of outputs and inputs regarding the various production processes.

Traditional monitoring approaches

We had developed what was in many ways a strong performance monitoring function. We had set up a fully fledged Research, Monitoring and Evaluation Department, headed by one of the best brains in the Corporation and one of the lead champions of our change initiatives since 1998. Over the years, this Department had done a wonderful job in collecting, compiling and analysing data from the Areas and disseminating the recommendations that informed subsequent initiatives in the Corporation. Thanks to the work of this Department, the Corporation had built up an invaluable data bank that is accessible to water utilities around the world and is a treasure chest for many researchers in the water business, including postgraduate students. However, given the volume and diversity of the technical information coming from the field, the monitoring control systems designed to separate and analyse technical, financial and management information for end users were stretched to the limit, often to the detriment of speed, accuracy and reliability.

The traditional monitoring approaches were found wanting for a number of reasons. To start with, performance monitoring tended to be more passive than proactive and spontaneous. The monitors used to wait for the submission of monthly, quarterly and annual reports, which would then be analysed, and the findings and recommendations would be disseminated for action. Monitoring was left to the Research, Monitoring and Evaluation Department, which was thin on the ground and as such did not have sufficient capacity to monitor a complex service delivery process. Once in a while, the monitors made 'flying visits' to the Areas to verify information, but they did not have sufficient time to collect detailed information on all management, financial and operational aspects of the Corporation. Besides, the monitors did not have sufficient capacity, means and power to ensure that the monitoring feedback and recommended follow-up actions were implemented to sustain performance improvements.

Traditional monitoring in Areas tended to encourage defensive posturing and self-justification. In the competition for awards and bonuses, some Areas were tempted to exaggerate their performance achievements and to paint their performance in the best light possible, which did not reflect the realities on the ground. The monitors had no foolproof mechanisms to detect such data manipulation, especially in areas of technical complexity. Operators did not have specific guidelines around which to organise information, ease access and ensure verification by monitors.

Therefore, monitors not only chose what they wanted to verify but they also wasted time going through mountains of not so useful or irrelevant documents. Due to time constraints, monitors tended to rely too much on written reports, to take the information in those reports at face value and ignore or overlook crucial performance indicators.

Performance monitoring tended to focus on the capture of quantitative data relating to water production, distribution, revenue collection, NRW, arrears, etc. But even the data capture systems were not properly conceived, designed and structured to rule out the possibilities of manipulation. Monitors did not pay much attention to the qualitative aspects of service delivery, such as appearance, cleanliness, maintenance, ambience, presentation, the quality of customer service, staff attitudes and behaviour, and, of course, the image of the Corporation. Quantitative data are easier to measure and to use as performance indicators to demonstrate improvement(s). In this area, we had done very well indeed, as consistently evidenced by monthly, quarterly and annual reports.

Qualitative data are more elusive, sensitive, defy exact measurement and are prone to subjectivity and hence of doubtful reliability. And yet it goes without saying that, in business, as in life, first impressions, presentations and images can be as important as quantifiable facts and figures. How a business looks or presents itself to customers and the public or packages its goods or services can prove more striking and decisive in customer choices than their intrinsic value or quality. Therefore, it was important to pay equal attention to quantitative and qualitative performance monitoring to give a comprehensive and balanced representation of what actually happens on the ground.

Our previous experience in Kampala, under both KRIP and OSUL water and sewerage service management contracts, had shown that private operators tended to ignore the maintenance of the physical infrastructure in order to achieve agreed-upon performance targets. By concentrating on performance, operators thus tended 'to milk the cow without feeding it' leading to the deterioration of the infrastructure and Kampala Area's performance. There were already some signals that some partnerships were abusing the autonomy and not following the laid down procurement procedures, as well as the human resources management guidelines. Accordingly, we had to put our house in order urgently before it was too late to keep the IDAMCs on track. We needed to rationalise monitoring, to avoid time wasting duplication of tasks, to maximise monitoring capacities and competencies and to harmonise the day-to-day working relationship between the partnerships and the Head Office.

Apart from the usual routine monitoring duties, the monitors had to carry out periodic process-related audits to check on the technical and financial performance of the systems. As a result, monitoring tended to be selective, variable in quality and in some instances incomplete. We focused on weaknesses that could compromise data standardisation, reliability and comparability. More hands and brains were thus needed from within and without to boost the performance monitoring function, to ensure IDAMCs compliance and sustainability. The spread of the monitoring function across the Corporation's departments would release the staff of the monitoring department to carry out other functions such as research, evaluation and external services.

Another major weakness of the traditional approach to performance monitoring was that monitoring schedules were announced in advance, giving the Areas sufficient time to put on their 'Sunday best' so to speak and to sweep the day-to-day 'dirt' under the carpet well beforehand. Since the Area partnerships knew exactly when the monitors would be coming, it was possible to stage grandiose receptions and hospitality and to conduct guided tours to divert the visitors from inquisitive prying and asking awkward questions. Some of the inspectors tended to see performance monitoring trips as casual tours, to take a break from the usual routine work at Head Office and to earn some subsistence allowances. Therefore, for some of the monitors and operators alike, the monitoring exercises were not handled with the purposeful professionalism it deserved. In any case, the monitors were not, as a matter of course, armed with detailed checklists and guidelines against which the thoroughness of their monitoring activities could be measured.

Given the conceptual and operational shortcomings of traditional monitoring approaches, we had to devise new, reliable, accurate and scientific monitoring tools to keep the IDAMCs on course, to detect and pre-empt problems before it was too late to correct them, and to comply with the terms and conditions of the contracts. We needed a monitoring tool that would minimise subjectivity, address specific as well as general issues, improve accuracy and reliability and take into account Area peculiarities and variations. From time to time, I had discussed these monitoring issues with colleagues, especially Mr Silver Mugisha, the manager, Research, Monitoring and Evaluation, and considered various options and scenarios, but we had not yet translated our ideas into action(s).

Whatever we did to improve monitoring, we were anxious to avoid imposing our ideas on the partnerships. We were determined to avoid the accusations of paying lip service to partnership autonomy and non-interference from on high while, behind the scenes, pulling the strings to lord it over our partners. Partnering demanded full consultation with, and participation of, partners based on mutual respect and reciprocity. However, while it was important to uphold the principles of delegation, autonomy and empowerment, it was also imperative to maintain the flow of new ideas and initiatives from the Head Office, or any other source for that matter, to achieve our performance objectives. It was in the course of debating and agonising over what needed to be done to beef up and refine our performance monitoring strategies that I landed, by coincidence, on the checkers concept, which henceforth revolutionised our IDAMC monitoring activities and approaches.

The genesis of checkers

I have already pointed out in this book that, as NWSC chief executive since 1998, I was open to ideas and concepts (trophies, stretch-out and one-minute management) that informed our change management initiatives. In the case of checkers, the honour for its genesis goes to the late Mr Daniel Kasasa Bwete, an old friend and German national of Ugandan origin and a mine of information on business management. Daniel had been a friend since my student days in Germany and I always treasured his insights and reflections on a variety of subjects, including management. Although he was then living in retirement, he had during his working life accumulated a wealth of varied experiences and knowledge, which he was generous enough to share with friends. During the course of his working life he had worked for many German employers, including one of the most successful car hire companies in Europe.

Whenever I visited Germany, as I did in early 2005, privately or on official duties, I always made a point of contacting him to exchange notes about our good old days, private lives and professional careers. In 2005, the most pressing subject on my mind was monitoring and I could not help talking about it to whoever cared to listen. So when I met Daniel, I invariably talked about the challenges of monitoring IDAMCs, and he immediately came up with the checkers, which turned out to be a brilliant idea to address our monitoring dilemma, one which has since then turned the process of monitoring in NWSC upside down.

According to Daniel, the checkers concept was a simple and straightforward idea built around the elements of surprise and unannounced visits but based on known and agreed-upon benchmarks, being conducted by anybody, including non-employees, and timely feedback of the findings and recommendations to all stakeholders in the subject organisation. Basically, the Checkers System sought to promote preparedness, ambience, efficiency, quality, prompt responsiveness and better time management. The Checkers System was an on-the-spot activity, designed to inculcate predictability, objectivity, fairness and uniformity in the monitoring practices of an organisation. Daniel assured me that the checkers concept had worked very well for his company and he recommended it as the most appropriate answer to our IDAMC monitoring dilemma. In retrospect, I am glad I took his advice. The checkers idea was so appropriate and timely that since then, I considered him my 'checkers monitoring expert' – of course, at no cost to my employer!

In the case of Daniel's Frankfurt-based car hire company, the managers used to send a checker to any branch of the company at any time to investigate what was going on in the branch, without informing the branch officials before hand. The aim was that the checker should be 'invisible' in his work so that business could go on as usual without the fear of being watched and judged. After the inspection exercise, the checker would announce his/her presence, hold an exit conference with the branch employees to discuss the findings and recommendations, map out the way forward and submit the report to the Head Office for follow-up action, to correct any operational deficiencies that had been observed.

The first checking exercise served as a warning to branch officials to pull up their socks or face the consequences. If a second check established that the branch had not implemented the checking recommendations from the previous visit without good reason, they were liable to disciplinary action or even dismissal. This on-spot checking served to keep all branches on their toes; they would always be prepared for arrival of the 'guest' any time, without prior notification.

I listened carefully to Daniel's checkers narrative and by the time he ended his story, I knew that the checkers monitoring tool was the right thing for the NWSC. From experience, I had learnt how to tell good ideas when I heard, or read about, them. To me, the checkers concept was a godsend and an unexpected solution to the IDAMC monitoring puzzle, which had eluded us for almost a year. The challenge was to internalise the basic essentials of the Checkers System and to adapt it to our own unique operational circumstances. Whenever I landed on a good idea, such as checkers, I always shared it with my lead champions at the Head Office to explore its workability in our organisation.

On this occasion, I did not even wait to go back to Kampala. I immediately phoned one of my line managers, Mr Silver Mugisha, and briefed him about my discovery. I asked him to think through the idea, discuss it with colleagues and explore the possibilities of developing it into a monitoring tool for the IDAMCs. I also asked Mr Mugisha to talk to Daniel, to hear the checkers concept from the horse's mouth and to find out how the Checkers System worked. I also requested Daniel to jot down some notes on the checkers concept, which could be used in our subsequent discussions in Kampala.

As soon as I returned to Kampala, I called a meeting of three of my senior managers – Messrs Mugisha, Johnson Amayo and Charles Odonga – to brainstorm on whether and how to push the checkers concept forward. Mr Mugisha had already prepared a draft concept paper for discussion. During the meeting, my colleagues welcomed the checkers concept as 'an excellent idea' but they also raised pertinent questions about how it could be implemented without compromising the letter and spirit of the IDAMCs.

How could checkers concept be implemented without interfering in the partnerships' operational autonomy that was entrenched in the IDAMCs? Was the checkers concept compatible with the existing monitoring systems? If not, how would the two be reconciled and harmonised without disrupting the entire monitoring system? What were the cost implications of introducing checkers concept? Was it a good strategy to introduce so many performance drivers in quick succession without pausing to take stock of the impact of ongoing programmes and processes?

While acknowledging the validity of the above-mentioned questions, I assured my colleagues that the checkers concept was not incompatible with the ongoing change management initiatives and processes of the time. The checkers concept was not meant to be a new performance programme. It was conceived to be a new approach to performance monitoring – a new monitoring tool – whose purpose was to improve, refine and perfect the existing monitoring systems and processes rather than to disrupt or dislodge them. I did not expect the checkers to undermine the non-interference provisions of IDAMCs. Nor did I anticipate it to cost much more than the traditional monitoring systems and processes. In any case, additional checkers monitoring costs would be more than justified by improvement(s) in quality, objectivity and spontaneity.

Moreover, by involving more people from all departments, and even from outside the Corporation, the checkers would ease the heavy burden of monitoring on the Research, Monitoring and Evaluation Department, which handled monitoring single-handedly in the traditional mode. Under the Checkers System, all workers of the Corporation, even outsiders, including customers and members of the public, would be potential checkers. Therefore, in my view, the challenge was not the intrinsic value of checkers *per se* but rather the modalities of its implementation to improve and perfect monitoring performance over and above what was already being done.

Once the checkers concept was accepted as the most appropriate monitoring tool for IDAMCs, the next task was to work out detailed proposals for translating the checkers concept into action in the specific NWSC work environment, both at the Head Office and in the Area partnerships. Who would be involved in checkers monitoring and how? What would be their specific assignments? Would the monitors require training and induction? How would the checkers organise and conduct their work? What would be the frequency of checkers monitoring? How would the checkers findings and recommendations be used to correct shortcomings and to feed into subsequent performance improvements?

Mr Mugisha was tasked with drafting a checkers working document, discussing it with his monitoring colleagues in order to build requisite consensus and to report progress or hurdles in the next top management meeting. Fortunately, the preliminary groundwork went according to plan and Mr Mugisha's positive report from his department paved the way for wider buy-in consultations across the Corporation, which eventually culminated in the formal launching of the Checkers System on 18 May 2005.

Widening the buy-in planning process

With the endorsement of the Research, Monitoring and Evaluation Department and some key lead champions, we had to move as quickly as possible to implement the checkers concept to ensure the success and sustainability of IDAMCs. There was no time to waste! The buy-in planning process, involving the entire NWSC family, had to be initiated as a matter of urgency. So on 19 April 2005, I called a meeting of the top management, which was also attended by officers from the monitoring and evaluation departments of Kampala Water and the headquarters.

I kicked off the meeting by reminding colleagues that, under the IDAMC framework, the NWSC Head Office had a statutory obligation to ensure good performance by all water partnerships under the jurisdiction of NWSC across the country. With the IDAMCs being operational in all NWSC Areas across the country, the headquarters' role had shifted to an emphasis on performance monitoring, major capital investments, assets management and external services. All managers were expected to focus on these areas and no complacency or backsliding could be tolerated.

In order to boost our performance monitoring function, we had to initiate the checkers concept for monitoring the IDAMCs and we wanted all the divisions in the Corporation to buy in and internalise the checkers concept and to integrate it in their day-to-day operations. The checkers had to be all encompassing, transparent, objective and cost-effective. All divisions had to get involved in monitoring and to take responsibility (whether for success or failure) for the outcome(s) of performance monitoring activities concerning their respective divisions.

The aim of widening the checkers network was to maximise human resource utilisation, to build networking among divisions and to create a NWSC working environment without boundaries or operational barriers across the divisions. Under checkers, monitoring was to become a collective and collaborative effort in which every division, department and worker had to play his or her part. In other words, monitoring was to cease being the sole responsibility of the Research, Monitoring and Evaluation Department, which would henceforth concentrate on monitoring coordination, research and data analysis and dissemination.

Although all of us were soon to become checkers, we had to do so without infringing on the autonomy of the partnerships guaranteed by the IDAMCs. We had to be conscious of the partners' sensibilities and to resist the temptation of command and control – of the boss element – in our new approach to performance monitoring. We had to heed the following words of Mr Mugisha and his colleagues:

> ... command and control has failed as a technique for mobilising and allocating resources efficiently. Even if the operator is a local company fully aware of the local conditions, the practical situation in low-income countries is that such operators generally do not have knowledge of best practice. The gap in the operating knowledge is fundamental in low forms of commercial contracts, which by their nature do not have strong incentives for efficiency improvement.
>
> In addition to the usual performance monitoring/regulatory tool of metrics benchmarking, monitors need to carry out careful process benchmarking. The latter activity serves as a means of widening the knowledge for operating utilities, enabling local managers to implement tailor-made innovations and performance improvement initiatives. However, process benchmarking, although apparently vital for information dissemination (as evident in NWSC commercial programmes), can verge on interference in the operator's business. Monitors need to adopt a core consulting principle: "I have advised you; build on the idea or try another approach. The choice is yours. The rewards for good performance are yours as well." (Silver Mugisha etal., 2004, Water 21)

After some discussion and clarification, the NWSC senior managers quickly embraced the checkers monitoring tool as the best way to strengthen the performance management system. The managers were particularly excited about the new participatory and collective checkers approach to monitoring the IDAMCs. Under the new approach, departments and divisions were to monitor all aspects of its operational mandate. In practice, this meant that all the Head Office divisions – management services, engineering, commercial and customer services, internal audit and finance and accounts, as well as the Monitoring and Evaluation Department as the overall coordinator – would be directly involved in the IDAMC monitoring and evaluation processes.

In light of the increased decentralisation of functions from the centre to the Area water partnerships, the attributes of transparency, reciprocity and the combination of elements of surprise and unpredictability inherent in the checkers monitoring tool were timely additions to existing IDAMC monitoring and evaluation processes. Once the NWSC top management embraced the checkers approach to performance monitoring, the next challenge was to work out the modalities and practicalities of implementation that involved all the people – from top to bottom – in all NWSC departments and divisions in checkers.

In terms of the modalities and practicalities of checkers implementation, all departments were required to develop detailed criteria/checkers guides, checklists and activity forms regarding their operational mandate(s), spelling out their monitoring coverage and expectations. These checkers criteria/guides were to be anchored on concise objectives developed by the Monitoring and Evaluation Department at the Head Office. The underlying aim of these objectives was to make everybody understand and own the checkers monitoring tool. The objectives that sought to guide the formulation of the modalities of checkers implementation were to

- ensure that all Areas, including Kampala Water, achieved their performance targets in accordance with the IDAMCs and Branch Performance Contracts (BPCs);
- ensure high-quality customer services in all Areas;
- enhance 'value for money' monitoring;
- help partnerships to maximise productivity and to tackle operational constraints and challenges;
- ensure compliance by the IDAMC parties with the monitoring processes, benchmarks and outputs;
- enhance predictability, reciprocity, objectivity and effectiveness in performance monitoring at all levels and
- reduce 'perceived' interference; and enhance all-round human resource participation and utilisation.

Based on the above-mentioned objectives, the divisions proceeded to draw up the guidelines for discussion internally at departmental and division levels, and subsequently submitted the drafts to the top management and the MD for review and comments. The revised checklists were then discussed at a broader meeting at which all the departments and divisions participated and shared ideas, challenges and experiences. In brief, the checkers guide of each division consisted of the list of checkers questions, actions to be taken and documents to be consulted by the checkers and the interviews of relevant personnel during the course of unannounced monitoring activities. The checkers guides were wide-ranging, containing checklists and activity forms, and covering working methods, the frequency of spontaneous visits, the holding of exit conferences to discuss findings, the recommendations/follow-up actions and the production and submission of checkers reports to all stakeholders in the Areas and at Head Office.

The checklists were basically a reflection of the departmental and divisional mandates and schedule of duties in the delivery of water and sewerage services of the Corporation. For example, the Engineering Division checkers guide revolved around water production and distribution, technical operations and management, assets management and maintenance and other operations-related monitoring questions and issues. The checkers guides for the Internal Audit and Finance and Accounts divisions were naturally preoccupied with checkers monitoring questions and issues to do with money, focusing on compliance with financial and banking regulations.

Closely related to the two finance-related divisions is the Commercial and Customer Care Services Division (the focal point of our services) whose checkers guide focused on issues to do with billing accuracy, revenue collection, debt management, marketing, customer care and information provision, and providing services to the poor. Finally, the Management Services Division checkers guide was primarily concerned with human resource management (covering recruitment, remuneration, staff welfare, training, performance appraisal, legal services, etc.) corporate administration and facilitation and the condition of the workers and their working environment. Collectively and individually, the checkers guides covered the role and function of the Corporation and their objective was to leave no stone unturned in our quest for improving and refining performance monitoring.

In the meantime, performance monitoring did not wait for the process of developing and formally adopting the checkers monitoring tool. Once top management endorsed the checkers concept, the system started working with immediate effect from 1 May 2005. All monitoring visits henceforth conformed to the checkers monitoring tool. As a guiding principle, the *modus operandi* incorporated surprise visits and spot checks. The Finance and Accounts Division was also requested to adjust the release of cash imprest to the new checkers approach.

As a result, new and appropriate disbursement procedures and financial report formats were designed even before the checkers became officially operational. These measures were not only undertaken to underline the sense of urgency we attached to the implementation of the checkers monitoring tool in our new decentralised IDAMC environment, but also to test the waters, so to speak.

The process of implementing checkers was very elaborate and demanded not only endorsement by senior managers at the Head Office but also the input of Area water partnerships in accordance with the provisions of IDAMCs. Therefore, as soon as the checkers guides from the divisions were drafted, the Area partnerships – the action centres – became involved in the wider consultation and participatory process, which entailed discussion, comments, suggestions and feedback. To avoid the suspicion of imposing solutions from above and to respect the IDAMC's spirit of non-interference, every effort was made to bring the partnerships on board through consensus.

Fortunately, the NWSC tradition of resistance to change had long broken down and Area partnerships had acquired and internalised the culture of being willing change agents who readily accepted the inevitability of change. Accordingly, they were quick to recognise the usefulness of checkers performance monitoring, which, in any case, they accepted as a legitimate residual Head Office function. That is why the checkers monitoring tool sailed through all the water partnerships smoothly in less than two weeks.

To crown the planning and the NWSC-wide buy-in consultation process, a meeting attended by senior managers, Area partnerships and branch managers from Kampala, code-named 'Interaction with Area/Branch Managers', was held at the Corporation's training centre to sell the idea and to obtain comments about the draft checkers guidelines for each of the Corporation's divisions. Apart from reiterating the importance of the checkers concept as a new monitoring tool and determining how its internalisation and acceptance would determine the success or failure of our IDAMC change management strategy, the meeting went through the final details and strategies of moving from planning to swinging into action.

One of the immediate strategies adopted at this meeting was to pretest the pros and cons of the checkers concept before giving it a NWSC-wide application. It was also agreed that all the draft checkers should be sent to the Area partnerships for detailed scrutiny, discussion, comments and suggestions, taking into account the priorities and peculiarities of individual partnerships. Our aim was to ensure full consultation and participation by staff of all the water partnerships in order to reach the widest possible consensus and to create a universal sense of the ownership of the checkers monitoring tool.

The Monitoring and Evaluation Department spearheaded the pretesting exercise in five branches of Kampala Water, using the Engineering Division's checkers guide, checklists and activity forms. The pretesting exercise focused on the new water connection policy, the cleanliness of the working environment and whether there had been follow-up action on the previous monitoring findings and recommendations. On the whole, the results of the checkers plot exercise were quite promising, suggesting that we were moving on the right track, backing the right horse, so to speak.

However, the pretesting exercise also revealed some unacceptable operational shortcomings that needed to be rectified quickly. For example, it was disappointing to find that, in some of the Kampala Water branches it took more than 30 days for a customer to get a new connection. This was embarrassing for a Corporation that anchored its operations on 'the customer is the reason we exist' dictum. The reasons for the delayed connections entailed such excuses as seeking permission to cross roads or failure to locate customers, which were simply lame excuses.

Similarly, failure to comply with technical specifications, such as the fact that excavation depths for the water pipes were found to be a mere 6 inches instead of at least the standard 3 feet or more, smacked of professional negligence. I was disturbed that such shortcomings persisted under the noses of the branch managers who were supposed to supervise fieldwork. By means of the checkers guides, checklists and activity forms, we hoped to detect such operational deficiencies and to rectify them in order to maintain our hard-won customer-friendly reputation.

The harmonisation process

During the meeting at the NWSC training centre, we agreed to hold a subsequent workshop to review and harmonise the feedback from all Areas as well as to discuss the findings of the checkers pretesting exercise in Kampala. Our aim was to ensure full consultation and participation by the staff of all the water partnerships in order to reach the widest possible consensus and to create a universal sense of ownership of the checkers monitoring tool.

Accordingly, on 17 May 2005 over 100 participants – Area managing partners, chief managers, some Board members and selected staff across the Corporation – gathered in Jinja, our traditional planning place, for a three-day wrap-up harmonisation workshop before officially launching the Checkers System. As usual, the workshop was marked by informality (blue and white NWSC T-shirts; no suits please) and cordiality in order to create a relaxed working environment. The aim of the workshop was to review and harmonise the checkers guides and to work out the checkers methodologies, processes and forward and backward feedback mechanisms.

The checkers harmonisation 'workout' sessions were similar to those of stretch-out and one-minute change management initiatives. The working group sessions began at 8.00 a.m., worked through the day, with short meal breaks, and lasted until the early morning hours of the next day. These working sessions were guided and moderated by the champions and the chief managers. Throughout the deliberations, the emphasis was on clarifying issues that had arisen during the consultation process, to discuss the revised checkers guide drafts – spelling out the visions, goals and objectives and expected outputs – to explain the processes and practicalities of checkers performance monitoring.

The checkers concept was expected to be cost-effective, operating within tight budgets and participatory in that, in the long run, the checkers activities were to be open to all members of the community regardless of whether they were NWSC employees or not. During the workshop, the champions reiterated the importance of unannounced visits, exit conferences, feedback and follow-up actions to ensure the implementation of the checkers findings and recommendations. Despite the exhaustion of the participants at the end of the three-day workshop, there was a sense of fulfilment and anticipation of the new checkers monitoring tool swinging into action.

Although the general atmosphere of the workshop was sombre, serious and busy, there were some light moments, reflecting the participants' high and cheerful spirits, to lighten the otherwise charged deliberations. One participant compared the checkers concept to the unannounced second coming of Jesus Christ. Another participant said that the checkers concept was like a hospital 'intensive care unit', requiring staff preparedness around the clock – day and night, even at ungodly hours of the night.

One of the participants even invoked the Rotarian motto of 'Service above Self' to drive the point home about the expectations of the new approach to performance monitoring. Other participants spiced their contributions with checkers calling for 'eternal preparedness', always serving customers 'with a killer smile', and using checkers as a life guard against unexpected performance 'crashes'. No doubt, these anecdotes were amusing but they all carried a serious message about the centrality of checkers in the success of IDAMCs. They also indicated that most participants had fully internalised the essence of the new approach to performance monitoring and were mentally tuned to swing into action to add the checkers concept to our growing list of change management achievements.

Throughout the workshop, I kept a low profile, occasionally contributing my own thoughts and jokes, here and there, so as to let the participants discuss without looking over their shoulders (the boss element dies hard). The participants' enthusiasm, high spirits, commitment, teamwork, professionalism and rigorous arguments were impressive, though not surprising given the new NWSC corporate culture and organisational behaviour, leaving no doubt in my mind, that, whatever the challenges, the checkers concept was on course. That is why, during the workshop closing ceremony, I praised the participants for embracing the checkers concept, a new addition to our change management initiatives.

By embracing the system, the participants had enhanced the possibilities and opportunities of improving productivity, service delivery, customer care, and above all, providing a sound monitoring tool for the future of IDAMCs through a new approach to performance monitoring. But I also reminded the Head Office management, the Area partnerships and the entire NWSC family that the checkers concept was a great challenge requiring all of us to redouble our efforts to remain on the winning side of corporate management in Uganda.

The checkers concept is not a policy initiative requiring formal approval from the board. It is a monitoring tool designed to enhance performance monitoring in the era of IDAMCs and as such management was not duty bound to submit it to the Board for official endorsement. Nevertheless, as a matter of practice, since 1998 I had involved the Board in whatever we were doing or planning to do, not only to secure their guidance and support but also to dispel any possible suspicion or accusation of the management operating behind their backs. This approach had helped build the trust and confidence between the Board and the management, which had underscored our previous change initiatives, and there was no cause to abandon it now.

Therefore, from the outset of the checkers concept, the NWSC Board was fully briefed about the checkers concept. We also sought the guidance and input of individual board members at every stage of the planning process and, no doubt, their contribution enriched the debate and the final harmonisation process.

Among the Board members who took keen interest in the checkers were Messrs Sam Okec, the chairman, Yorokamu Katwiremu, the vice chairman and Gabriel Opio, the chairman of the Finance Committee. At the opening ceremony of the harmonisation session, the Board chairman underscored the principles of objectivity, preparedness, transparency, feedback and follow-up, which were deeply embedded in the Checkers System. He was optimistic that compliance with these principles was the best way to move the IDAMCs forward and to create viable and autonomous business units in all NWSC Areas.

At the formal checkers launching ceremony, the Board vice chairman stressed the importance of self-monitoring and warned participants that 'if you don't check yourself, you'll be checked', contrary to the IDAMC principle of

A New Approach to Performance Monitoring | 217

non-interference in the affairs of the NWSC water partnerships across the country. He advised the Area partnerships to monitor and evaluate their own activities to 'judge themselves, without waiting to be judged' from the Head Office. He challenged them to reciprocate the Board's trust and confidence through enhanced performance.

Mr Opio reminded the Area partnerships that it was their duty to safeguard and maintain the Corporation's assets under their care in accordance with the provisions of the IDAMCs. The blessing of the checkers from the Board paved the way for swinging into action; it was a big push towards the implementation of the new approach to performance monitoring.

The Checkers System at work

The checkers performance monitoring tool, which was launched on 18 May 2005, has been a success so far. Since its introduction, it has greatly improved the momentum of the Corporation's internal reform initiatives. The system has incorporated activities that promote competition among Areas for successful compliance with the IDAMCs objectives. For example, competition is in areas of cleanliness of premises, orderliness of offices, promotion of the corporate image and, operation and maintenance.

Indeed, from May 2005, all the documents and records in the Areas have been kept in accordance with the checkers guides. As a result, the checking exercises, from the completion of the Checkers activity forms, through the inspection of the premises, and the interviews of staff and customers present alike, to the holding of 'exit' conferences, have, time and again been smooth and fast operations.

Secondly, the information collected and analysed under the Checkers System has been more complete, accurate and reliable than it used to be under the traditional monitoring approaches. This marked improvement has been verified and confirmed by the Monitoring and Evaluation Department, the overall performance monitoring coordinator responsible for data analysis, documentation, storage and dissemination. Besides, the checkers concept goes beyond the collection of facts and figures about water production, distribution, revenue collection, new connections, NRW, etc., to collect qualitative information on ambience, cleanliness, image, customers and so on. It goes without saying that in an organisation like NWSC, quality is just as important as quantity. The quality of the checkers data has also been improved by the objective and scientific (verifiable) criteria embedded in the guides, checklists and activity forms. Furthermore, spontaneous, unannounced checkers performance monitoring has helped to detect deficiencies early and to address them immediately. Specifically, the Checkers System helped to unearth the transgressions in

Mbarara, Fort Portal and Kasese Areas, whose partnerships were later disbanded, as already discussed in Chapter 9. To a great extent, I blame the traditional monitoring approach that, in a way, encouraged laxity. It seemed to encourage a tendency among the partnerships that they could do anything and get away with it. With the Checkers System, we were now able to detect such shortfalls and address them immediately. On a positive note, the outcome from the incidents in the three Areas underscored the seriousness of the Checkers System and put the partnerships on their toes, knowing that the checkers could come any day, any time.

Coupled with the one-minute management concept, the checkers concept has added new value to inter-Area competition and staff appraisal systems. Under checkers evaluation criteria, quantitative and qualitative performances, as well as cost optimisation and customer satisfaction in each Area, are assigned maximum scores, ranging from 100 to 150 points, making a total score of 500 points. An Area that achieves at least 60% of the total score earns one-minute praise for good performance. Those Areas that fall short of the 60% target without good cause are liable to one-minute reprimand.

The reprimand side of one-minute management is used sparingly to avoid the possibility of panic, resentment or even backlash resistance in the Area partnerships, to the detriment of IDAMCs. Therefore, during the first checkers exercise, those Areas that fell short of expectations were given the benefit of the doubt, on the second occasion they were reminded of their contractual performance obligations, and if they remained performing below the agreed standard on three consecutive occasions, they were reprimanded. We also used checkers scores and rankings to sustain the competitive spirit in the Corporation, with the best performers winning trophies and cash bonuses in the wake of the checkers-based quarterly evaluation exercises of the IDAMCs.

Since the introduction of the Checkers System, there has been great improvement in compliance of the Areas in meeting financial regulations, following procurement laws and adopting other managerial procedures. In terms of technical capabilities, the Areas have improved compliance for preventive maintenance programmes, operational procedures and safety. Last but not the least, staff discipline has improved, individuals even strengthened their interpersonal skills so that teams could perform more effectively.

However, the Checkers System has not been without challenges. Excessive paperwork becomes a danger when reports are available to so many individuals: when an Area manager is responsible to 'everyone,' he or she might effectively be responsible to 'no one'. This pattern has been avoided through the use of Authorised Contract Representatives (ACRs) who coordinate the reporting relationships and checking activities. The ACRs are the only ones supposed to take decisions on contractual issues. The incorporation of ACR has created robust cost-containment safeguards with strict accountability mechanisms, resulting in greater value for money. In addition, by its very nature, the checkers concept requires monitors to be ready to move to any place at any time to do their work, which calls for extremely dedicated individuals who are objective and cannot be unduly influenced by Areas. The review of the checkers implementation process has in some instances indicated that there are some checkers who grade (or score) the Areas close to maximum points even when the actual situation involves weaker performance. So this component of the system still needs greater attention. We used independent external private checkers in some cases to avoid collusion between the monitors and operating team colleagues.

For all the challenges facing the implementation of the checkers concept, it is fair to say that it has revolutionised the monitoring approach and significantly contributed to the effective implementation of the IDAMC.

Lessons learnt

- The decentralisation and delegation of authority in a corporate environment require new monitoring approaches to maintain quality standards, to detect and rectify any shortcomings and to ensure the achievement of agreed performance goals and objectives.
- Share your professional challenges with willing listeners, regardless of their background(s). In the world of business, or in any human activity for that matter, no one has a monopoly of ideas. Ideas and solutions are universal – they can come from any source: books, private informal conversations, friends and relatives, the young and the old.
- Performance monitoring should not be a speciality function of a single department or group of individuals in an organisation. It should be a holistic approach, drawing monitors not only from the entire organisation but also from other stakeholders, including customers and members of the public.
- It is important to design monitoring tools and methods, which capture both quantitative and qualitative data so as to reflect a global picture of performance during the process of evaluation.

Part Three
SO FAR SO GOOD

Chapter Eleven

Performance Balance Sheet

> Until recently, the argument that a government parastatal could be profitably run would never have held any credibility in Uganda. This argument may now be revisited after NWSC released financial statements this year that showed that it had more than doubled its profit ... could this be the turning point for the view that government-run organisations can succeed alongside the often high standards of the private sector?
>
> The Monitor, *Wednesday 4 December 2002*

> ... innovative change management programmes have successfully turned around NWSC from a loss making organisation to a profit making government parastatal, which is now a benchmark for best practice both nationally and globally among water utility organisations.
>
> The Monitor, *Tuesday, 26 June 2007*

The previous chapters contain a compendium of actions and initiatives that we worked out and implemented with the Board, management, staff and other key stakeholders to improve financial and operational performance of NWSC and, in turn, provide reliable services to the customers. As I extensively enumerated before, we went through tribulations, fun and insightful experiences during the entire change management process. These events can best be described as 'magic of management' in the world of unknowns. I must say, apart from remembering the path-line that we followed, most initiatives and innovations happened with the power of spontaneity and unpredictability. Even our programme approach, which consistently paid off, could not lend itself to accurate prediction of results beforehand. Water business operating environment is fragile and requires continuous innovation and improvement: the ability to move back and forth between success and failure without being discouraged. This is how we worked and that was the overall belief that I inculcated in the management and staff. This spirit, unquestionably, inspired the reforms that we undertook, registering progress at each stage.

Very few people gave us the chance of recovering in the beginning especially given the fact that the Corporation was lined up as a candidate for privatisation. But as the Swahili proverb goes, 'If the rhythm of the drum beat changes, the dance step must adapt', after a decade of rigorous work, many stakeholders

© 2009 William T. Muhairwe. *Making Public Enterprises Work*, by William T. Muhairwe. ISBN: 9781843393245. Published by IWA Publishing, London, UK.

including my staff, government and our development partners changed heart and started looking at the Corporation as a viable venture on the right track to world-class best practice. I have to admit, though, that I could never have imagined doing half of what we have managed to do in NWSC, especially after the unpleasant scenes and the shocking encounters that greeted me on my early days at work.

In this chapter, I provide a logical break to the long list of actions and strategies implemented to improve the performance of NWSC. I have synthesised the achievements, constraints and challenges we encountered along our various production processes and came up with a consolidated balance sheet. My intention is to provide a snapshot overview of the net effects of the various sub-programmes implemented and how they contributed to the overall improved performance of the Corporation. But before I discuss the details of the summary achievements, let me recap on the strategies that have brought us this far for clarity and recollection. As I had described in detail before, prior to the 100 days Programme, we realised that the organisation's most malignant cancer was the employees' pathetic work ethic. Notably, among others, this was characterised by late-coming, shabbiness, drunkenness, gossiping, ethnic-based cliques, 'god-fatherism', insubordination, absenteeism, theft and an 'I-don't-care' attitude towards the customer. As a result the performance of the Corporation was largely inadequate. The management team through the 100 days Programme zeroed in on five core areas to turn the Corporation around. These five critical areas included water production, distribution, revenue collection, cost reduction and customer care. But above all, we were aware that the work culture had to change radically.

The importance of customer care took a centre stage during the 100 days Programme as we worked on the five focus areas. Specifically, we took stock of what constituted and what caused poor customer relations. All in all, the Corporation responded late to complaints, issued estimated bills, but above all, NWSC staff were indifferent, insensitive and rude to customers! In our efforts to enhance customer relations, we brought the print and electronic media on board. We put in place programmes which were meant to re-ignite customer confidence. This is how we introduced grace periods for those who had defaulted on payment of water bills, cancelled some contested bills and regularised illegal connections.

Our new motto 'the customer is the reason we exist' evolved out of these efforts.

The next management initiative we undertook was SEREP – from August 1 1999 to July 31 2000. The main rationale of this programme was to consolidate and enhance the gains from the 100 days Programme. SEREP's major focus was on reducing costs of production, such as the huge work force, and the high electricity bill. We also wanted to build and consolidate customer confidence in our ability to deliver the required services and thus develop their willingness to pay bills. To be able to get customers to view us in positive light, we had to increase water coverage and ensure prompt response to bursts and leaks. We had to contend with other sensitive issues to do with our customers such as: how to disconnect customers without offending their pride or sensibilities.

Other programmes in the row included the Area Performance Contracts (APCs), which were aimed at decentralising power and responsibility to the regions/branches. Within the APCs, we introduced support programmes, which incorporated 'stretch-out' of the group and eventually the individual, tailored to get staff to walk an extra mile for collective good. Eventually, we introduced the Internally Delegated Area Management Contracts (IDAMCs). The latter were aimed at creating more commitment on the part of regional/area managers, their immediate management teams and seconded staff. We gave more managerial autonomy, incentives, timely logistical support and strategic guidance necessary to create a conducive operating environment to Areas/regions.

To give a strong governance anchorage to APCs and IDAMCs, we entered into the first and second performance contracts between the government and NWSC, prior to each set of contracts, respectively. The last of the change management initiatives that was aimed at turning around the Corporation was the Checkers system. The Checkers system was introduced as a monitoring tool, which kept the headquarters and all Areas on their toes. It was based on prior agreed checking criteria and unannounced visitations by checkers. In this way, performance was moved a step higher in which designated personnel, the checkers, focused on quality service rather than on the cosmetic achievement of targets.

What has happened in NWSC points to the need for utility managers to be creative, ambitious and daring in reform implementation. Specifically, we have learned that the use of managerial tools like incentives should be adequately targeted, addressing a certain focus performance area. The positive results that we have registered have been a result of vigorous creativity, which called for proactive benchmarking to cross-fertilise managerial thought with best practice and building desire for peer excellence. We were able to achieve what we did because we customised our operational plans to an environment characterised by uncertainties of water distribution operations, e.g. illicit activities, meter vandalism and consumption without payments. Our focus was, therefore, on the alignment of strategic plans and attendant operational plans in such a way that achievement of outputs and performance trends was through a host of flexible approaches. I encouraged my management to 'walk' their jobs and provide timely corrective measures to match changing customer needs and service delivery challenges.

Snapshot of our performance

The presentation of the performance balance sheet, in both table and graphical formats, has looked at the overall structure of the water business. No business can thrive without matching production with the demand side, namely the market. In this analysis, the market is characterised by the service coverage and the constituent customer connections and I want to 'photograph' the situation in 1998 and that in 2008. I have never heard of any business that thrives on external support, without it tackling internal inefficiency predicaments. In this regard, I also look at productivity and efficiency of running business and how this has progressed throughout the change management process. For purposes of simplicity and understanding, staff productivity and non-revenue water are natural choices for this measurement. Service delayed is service denied. Lack of funds can really delay expansion of services to deserving customers. Therefore, financial considerations cannot be wished away by those who seek strategies for achieving sustainability in service provision. As such, this analysis also puts income, expenditure and surplus for investments into focus.

A snapshot at the Corporation's performance after almost a decade of transformation in comparison to the performance of 1998 (see Table 11.1 below), tells the whole story of the impact of the initiatives that were undertaken. To come this far, we encouraged the staff to know the difference between success and failure. If you are easily upset by performance shortcomings or obstacles, then managing a water business will always be a disappointment for you. In my view, the distinction is found in one's attitudes towards setbacks, handicaps, discouragements and other disappointing situations. In addition, our approach towards implementation of change management programmes clearly demonstrated that through creativity teams can come up with seemingly illogical ideas geared towards significant inroads into performance enhancement. The stretch-out approach became the clearest 'signpost' to this seemingly 'mad' approach to changing things, but which eventually paid off handsomely!

According to a Bantandu proverb, 'A forest cannot be cut with a broken axe.' Therefore, leaders need the right tools to transform an organisation. As pointed out throughout this book, in this case, the tools involved the introduction of change management programmes. As a result, the reform initiatives from 1998 to 2008 have had positive impacts, which are described in greater details in the following sub-sections:

Service coverage and customer base: The service coverage has consistently improved (Figure 11.1) from about 47% in 1998 to 72% in 2008. This upward trend is mainly due to increased customer base (Figure 11.2), which in turn

Table 11.1: Comparison of performance between 1998 and 2008

#	Performance Indicator	1998 performance	2008 performance
1.	Service Coverage (%)	47	72
2.	New water connections (No./year)	3,317	24,384
3.	Total number of water connections (No.)	50,826	202,559
4.	No. of Employees (No.)	1,850	1,423
5.	Staff per 1,000 water connections	36	7
6.	Non-Revenue Water (%)	51	33.5
7.	Income (UGX billion)	21.9	84.0
8.	Operating profit (UGX billion)	−2	3.8

Source: NWSC 1997/98 and 2007/08 Audited Accounts

resulted from stepped performance in new connections as shown in Figure 11.3. The steady improvement in new connections resulted from improved new connection procedures and response times to new applicants. Coupled with this approach, new connection fees were reduced, from an average of UGX 150,000 to about UGX 50,000 in 2001. In addition, as pointed out before and in the subsequent chapters of this book, a free new connection policy was introduced in 2004 which greatly enhanced new connection hook-ups.

Non-revenue water and water production: Non-revenue water is computed as a percentage of water that is not billed to the water delivered to the distribution system. In the case of NWSC, water production has consistently improved (Figure 11.4) over the entire reform period. The stepped increase in 2003 was due to the new towns of Bushenyi, Soroti and Arua that were gazetted to NWSC in 2002. In addition, the old water treatment plant (Gaba 1) in Kampala Water supply area had been refurbished, increasing its practical capacity from about 30,000 m^3/day to about 50,000 m^3/day. Furthermore, the steep increase in 2004–2005 was due to the introduction of the new connection policy that jacked up the customer base and hence increased the level of consumption. Given that there was still ample idle production capacity, the production had to increase correspondingly to match the consumption. During this period, the operating efficiency of NWSC was also enhanced, resulting in a more rapid improvement in water billed. These efficiency gains were obtained through increased vigilance to reduce illegal connections, reduce physical losses and improve metering efficiency. Metering efficiency (proportion of metered accounts to total accounts) has increased from 65% to 99.6% while connection efficiency (proportion of active connections to total connections) has improved from 63% to 94%.

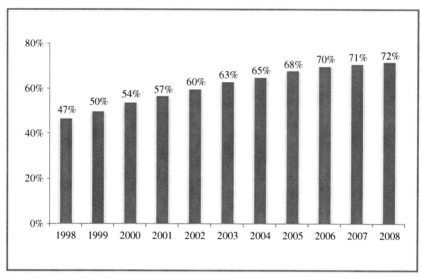

Figure 11.1: Service Coverage (%).

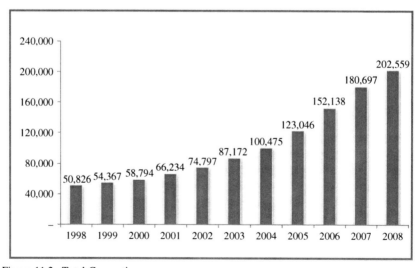

Figure 11.2: Total Connections.

Performance Balance Sheet

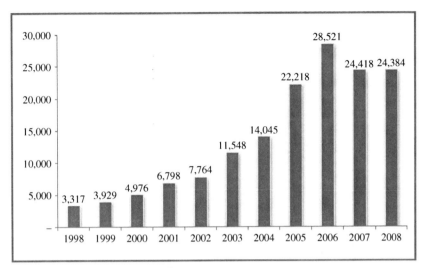

Figure 11.3: New Connections per Year.

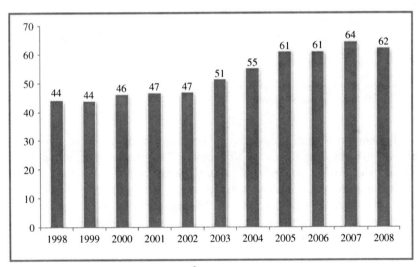

Figure 11.4: Water Production (million m³).

The ultimate achievement has been reduction of non-revenue water (NRW) from more than 50% in 1998 to less than 35% in 2008 as shown in Figure 11.5. In fact, if it were not due to high levels of NRW in Kampala Water supply area (about 38%), the NRW for all other Areas currently (2009) stands at less than 16%. Kampala still has high NRW due to the high level of illegal connections and meter tampering activities, which the Corporation is comprehensively handling through the introduction of district metering zones (hydraulic zones) which will help to identify priority problem areas. At the same time, the gazetting of small towns with high levels of NRW to NWSC has consistently disrupted our improvement efforts.

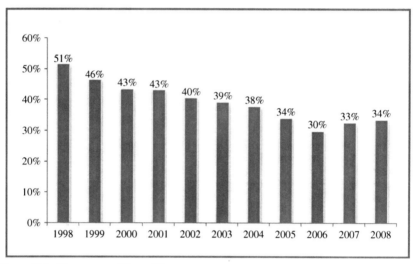

Figure 11.5: Non-revenue Water (%).

Staff productivity: Apart from gains in operating efficiency, NWSC reforms also resulted in positive shifts in staff productivity. The first five years of the reforms realised the highest productivity gains (Figure 11.6) as a result of lucrative severance/retirement packages that encouraged excess staff to resign in large numbers. Overall, more than 50% of the original 1,850 staff were laid off to match the staff strength with the business volume. This process, together with increased customer base, consistently improved staff productivity from 36 staff/1,000 connections in 1998 to about 15 staff/1,000 connections in 2002. Thereafter, productivity growth was not as fast as it was in the previous period because new, competent staff had to be recruited to match skills requirements and the growing customer base.

Financial performance: We could not afford to operate outside the balance scorecard. Financial performance can only be left out of a priority list to the organisation's own peril. In the case of NWSC, focus on financials was an obvious requirement if the Corporation was to find its way out of the bankruptcy maze that had become very apparent in 1998. With strong cost-containment measures and commercial orientation characterised by a stream of activities outlined in the book, the operating margin improved (Figure 11.7) from a loss situation during the financial years 1998–2002 to surpluses (after depreciation) beginning with 2003 onwards. This means the Corporation's capacity to finance its own operational activities, replace assets and carry out medium-sized investments improved. Needless to say, improved financial margins resulted from increased revenues (Figure 11.8) that jumped from about UGX 20 billion in 1998 to about more than UGX 84 billion in 2008. As pointed out in this book, this was not only due to rationalised tariff levels but also due to increased customer base and improved commercial efficiencies.

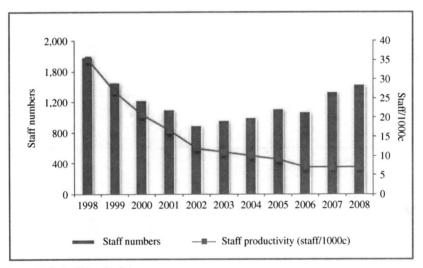

Figure 11.6: Staff Productivity.

Other service-related performance: Regarding the quality of services to our customers, the implementation of performance improvement reforms also enhanced supply reliability, with most of the customers in our Areas of operation receiving close to 24-hour water supply. Similarly, the quality of the water supplied in the system improved greatly with majority of the quality tests carried out complying with the national standards for drinking water. The Corporation has, in addition, established an excellent customer care service system with toll-free numbers which ensures that, among other things, customer complaints are responded to, on average, within 36 hours, and not two weeks as was it was in 1998.

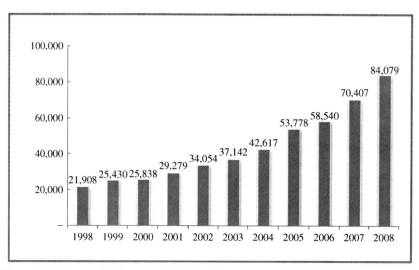

Figure 11.7: Revenue Performance in UGX million.

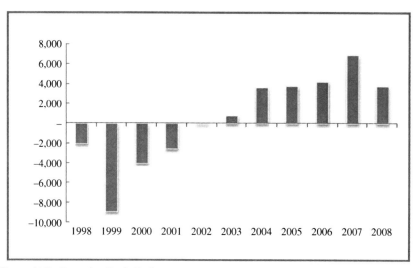

Figure 11.8: Operating Profit Performance.

The outcome of the reforms, as presented in the Audited Accounts and the Annual Reports, has also been well documented in the press and has also attracted a lot of attention within and outside the region. During a workshop held in Kampala in 2004, as a follow-up workshop to the World Summit for Sustainable Development, 2002, Dr Allen Eisendrath, the USAID senior finance adviser, described NWSC as a role model of successful reform in the water sector worldwide. Similarly, Mr Usher Sylvian, the secretary general of the African Water Association, during the Water Operators workshop held in Kampala in July 2008, recognised NWSC's role in uplifting the 'less well performing utilities' in a professional and sustainable manner.

In the same respect, the press ran a series of articles describing and acknowledging the success registered by the Corporation. Some of the headlines read as follows: 'National water body showcase of success' (*The Monitor*, Wednesday, 4 December 2002); 'Dr Muhairwe; the turnaround king' (*The Monitor*, Tuesday, 26 June 2007); 'Government companies can be profitable' (*The Weekly Observer*, 22–28 November 2007) and 'NWSC sets the pace for African water utilities' (*The Sunrise*, 4–11 July 2008). In short, through untiring efforts, we stopped the bleeding of the Corporation and inculcated a culture of change, self-belief and ownership of the business among all the staff and, above all, built a customer-focused organisation with less bureaucracy.

The challenges ahead

"Do not throw away the oars before the boat reaches the shore."

Mpongue proverb

Despite the accomplishments, NWSC still faces challenges in the area of sewerage, where the coverage is less than 10%. The rest of the sanitation needs are mainly provided through on-site septic tank facilities. The sewerage investment costs are inherently very high, and the Corporation is currently finding it hard to devote resources to such investments, given the payoffs to other uses of those funds. Therefore, achieving the Millennium Development Goals (MDGs) remains a distant goal on the part of sanitation. However, we are working closely with the government and donors to implement affordable sanitation technologies in areas of our jurisdiction.

NWSC also faces the challenge of serving poor communities where cost recovery is a daunting task. The infrastructure in such communities is very poorly planned, and extending services to such areas involves significant difficulties. Nevertheless, the organisation continues to explore cost-effective ways to carry out this task. The original approach of NWSC to serve the poor was through the use of water kiosks/communal taps. However, based on staff observations, the water vendors at these points sold water to the users at a price four to eight times that offered by NWSC. This 'middle-man' effect defeats the whole objective of the pro-poor tariff. It affects the willingness and ability to pay and restricts consumption, thereby obstructing health enhancement initiatives. In order to address this service problem, NWSC came up with a new connection policy, which aims to subsidise access and charge consumption at affordable rates. The policy also incorporates a network intensification activity in poor communities in order to reduce connections' lengths to individual households. Consequently, each household in poor communities is encouraged to connect a yard tap and pay NWSC directly. This approach is working very well so far.

At the same time, water losses in Kampala are still high, with an old and hydraulically unstructured network. This tends to complicate meaningful management of water losses in Kampala, and hence remains one of the key challenges that the Corporation is facing. Furthermore, as mentioned earlier, the Corporation still faces challenges associated with securing capital for 'huge' investment requirements, e.g. expansion of water treatment plants, distribution mains and large reservoirs. This challenge is exacerbated by the fact that the tariff cannot be increased beyond certain limits and at the same time maintain affordability of services. We have explored different options of accessing innovative financing options, namely the bond market and low-cost capital loans among others. The challenge is how to access loan funds without negatively affecting the balance sheet and hence the cash flow situation of the Corporation. It is evident that the Corporation has come a long way, but one could say that the sprint is over and the marathon has just started. The future of the Corporation lies in constant reinforcement of the reforms and not succumbing to the arrogance of success, well knowing that the best days for the Corporation are yet to come.

None of the above success and achievements could have been achieved without the support and helping hand of the different stakeholders. As the Tanzanian Nogoreme proverb says, 'If you refuse the elder's advice, you will walk the whole day', we neither wanted nor tried to walk the whole day! We were not alone in this game. We got help (both technical and financial), advice and guidance from many friends and partners, namely donors, government and other well-wishers. What follows in the next part of this book – not alone – is a change in the direction of the storyline from the preceding chapter to an account of the role of the different stakeholders in the turnaround of the Corporation.

The Corporation has put emphasis on ambience to ensure that water is treated in a clean environment. Above: Manafwa Water Treatment Plant in Mbale that was rehabilitated during the internal reform programmes.

The new and modern Gaba III Water Treatment Plant.

The Masindi Water Treatment Plant is one of the recent additions to NWSC.

236 | Making Public Enterprises Work

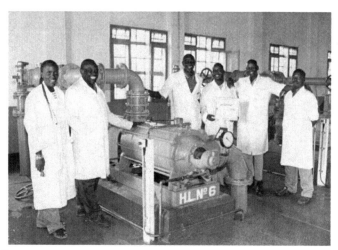

Total quality management, a sign of excellence: most of the Corporation's operations are ISO certified. Staff of Jinja Area showing off their ISO certificate.

The Corporation's improved image is reflected in its offices. Above: the newly constructed Entebbe Area office.

Performance Balance Sheet | 237

Above: the upcountry office block of Mbale Area.

The Corporation's offices have been restructured to create ambience and provide better customer care. Above: the customer care desk office of Kitintale Branch, Kampala.

Pro-poor initiatives: the Corporation made efforts to improve services to the urban poor. The president commissioning an urban poor source in Rubaga, Kampala: from left, the president, Hon. Suzan Nampijja, Hon Maria Mutagamba and Hajji Nasser Ssebaggala.

Customers at some of the water points constructed for the urban poor in Lira Area.

Performance Balance Sheet | **239**

The Corporation has contributed immensely to social programmes under its corporate social responsibility. Above: the Corporation joining hands with the church to provide safe water to the community. From left, Cardinal Emmanuel Wamala, Mr Richard Muhangi (branch manager), Dr William Muhairwe (MD) and Mr Charles Odonga (GM, Kampala).

The Corporation has been recognised for its achievements at different fora! Dr William Muhairwe (second left) receiving the UMA Best Exhibitor award from President Yoweri Museveni (second right), as Mr Abid Alam, the chairman of UMA (right), and Mr Charles Odonga look on.

The Corporation has been recognised in East Africa as a star performer. The Board chairman, Mr Ganyana Miiro (third left) receiving the award from Mr. Francis Kamulegeya, Partner PWC (second left) as the MD, Dr William Muhairwe (first right), the Union secretary, Mr Peter Welikhe (second right) and Board member Ms Florence Namayanja (left) look on.

Dr William Muhairwe (third right, front row) joining other winners of the 2007 Most Respected Companies Survey for a group photo after the award ceremony.

Performance Balance Sheet

NWSC has been recognised as Employer of the Year three times by the Federation of Uganda Employers.

NWSC was recognised as the best exhibitor, service sector, by the Uganda International Trade Fair (left). The Corporation also received an Energy Management Award in 2008 for its innovations (centre), plus the Bentley Award, USA, for innovation in water resources (left).

NWSC has been recognised with the Golden Star for Management Merit (left) from the International World Marketing Organisation, Amsterdam, Netherlands (left) and with the Vantage Award by URA as the most compliant taxpayer 2006/07 in the category for electricity, gas and water (centre). The big one: NWSC is East Africa's Most Respected Company 2007 in the public sector category.

The press ran a series of publications describing and acknowledging the success registered by NWSC. Some of the press releases are presented above.

Part Four
NOT ALONE: PARTNERS' ROLE

Chapter Twelve

Winning the Much Needed Goodwill

The Donors

"Work on their hearts and minds, and you will operate on their psychologies and weaknesses", 48 Laws of Power (Rule 43) Robert Greene. By working on the emotions and theories that these partners held dear, we all overcame the fear we had of each other and became closest allies in the fight for improved service delivery mechanisms through available and suitable local resources. Of course we knew that in this game of survival, you are surrounded by people who have absolutely no reason to help you, unless it is in their interest to do so.
The author's viewpoint on use of Greene's Rule 43

Show him your watch and he tells you what time it is. [But] the watch is yours, you bought it, you can read the time but someone wants you to pay somebody else to come and read the time for you.
S.L. Okec, NWSC Board chairman, on foreign consultants

Uganda, like any other low-income country, has been receiving support from development partners. The financing challenges that are faced by other development sectors in Uganda have also been ubiquitous in the water sector. The NWSC has not been excluded from this challenge. The change management programmes so far attempted were, therefore, conceived and implemented amidst varied financing constraints and modalities. Consequently, the role played by development partners in taming such constraints is of paramount interest in a book like this one. In the case of NWSC, the most notable development partners included the World Bank, the German government through KfW and GTZ, the European Union, and the French and Austrian governments. The involvement of these key stakeholders in the transformation of the NWSC dates back to the 1970s. The focus of the government and the donor community during that time was on the development of infrastructure for both water and sewerage services. However, throughout the 1970s and early 1980s, the infrastructure was run down due to the political upheavals of the time that affected the operating environment. Consequently, the later part of the 1980s was dedicated to infrastructure rehabilitation. With the improved infrastructure in place, the 1990s and the new millennium have been devoted to harnessing resources to invest in efficiency improvements in the provision

© 2009 William T. Muhairwe. *Making Public Enterprises Work*, by William T. Muhairwe.
ISBN: 9781843393245. Published by IWA Publishing, London, UK.

of both water and sewerage services. It goes without saying that a number of challenges were experienced by the NWSC management during this phase of development. This chapter outlines the multiplicity of challenges relating to why and how the relationship with donors has gradually shifted from a frosty relationship in 1998, through *a wait and see phase*, to one of *mutual confidence, partnership* and *harmonious cooperation*. It was indeed a daunting task to gain the donors' goodwill.

At the lowest ebb

The change of NWSC management in 1998 happened at a time when the relationship between the Corporation and its principal donors was at its lowest ebb – in the doldrums. During the initiation stages of my appointment, the 'smell' of discontent and disgust from development partners was a common phenomenon. I knew that we could not carry out our plans effectively without partnering with donors, especially where planned activities and strategies, unavoidably, required budget support. According to my managers, donors' discontent was premised on a number of factors. In the following sub-sections, I highlight a few very important ones for clarity:

Poor commercial and organisational behaviour: After carrying out a number of investments in terms of both infrastructure development and organisational development, donors and indeed all other stakeholders expected 'stepped' performance in service delivery methods and approaches. Unfortunately they were disappointed! As has been pointed out already in Chapter 1, this was clearly expressed in one of the World Bank *aide memoire* of March 1998 to the government of Uganda, where it was remarked that

> "over the last 10 years, the GoU in partnership with the World Bank and other Donors have made significant investments (over USD 100 million) in the Urban Water and Sewerage sector. These investments have contributed immensely in rehabilitating the existing infrastructure under the NWSC management. Unfortunately, these investments have not been matched with the necessary efficient commercial and financial management capacity that can ensure the delivery of sustainable services in the medium to long term".

The World Bank, from the above assessment, did not see demonstrable value from funds injected in the various infrastructural schemes within the NWSC areas of operation. The commercial orientation and organisational behaviour of NWSC management and staff fell short of what was required to optimally reap from the investments undertaken. According to one of the World Bank task managers, it was a case of *'the surgery was successful but the patient died'* – infrastructure was built but the company became more bankrupt after the investment than before. This trend was later to be addressed in the ensuing change management programmes, starting with 100 days Programme and SEREP as has been described already in detail in Chapters 2 and 3, respectively.

Management of the Second Water Supply Project: The abysmal situation was compounded by frustration about the way a number of transactions in the Second Water Supply Project were handled and managed during the formation and implementation phases. Notably, the World Bank was concerned about the way the Project was managed. The Project ended up with significant cost and time overruns, which did not only taint the image of the Corporation but also tainted the World Bank itself. The main questions in the minds of many stakeholders included whose fault was it? Or was it a fault anyway? Were these overruns due to increased scope and therefore increased value to be paid for? These and many other questions notwithstanding, all the transactions were initiated by NWSC project managers and obtained no objections from the World Bank task managers. So, who of these should have faced the music? Certainly, both parties shared equal responsibility for the whole process. Consequently, it later became obvious that all parties involved had to sit down and resolve the issue because the contractors and consultants who had carried out the works on behalf of the client were knocking on the doors impatiently for payment. Although, as elaborated in Chapter 13, the cost overruns of about USD 10 million were later resolved through a joint effort of NWSC management, government and the World Bank, the whole issue left a dent in the NWSC and the World Bank relationship.

Never-ending institutional problems: As if the cost overruns issue was not enough, the financial situation of NWSC was also appalling. There were no sufficient funds to cover vital operational inputs like chemicals, spare parts, and power and staff salaries. There were other institutional problems that caused me sleepless nights and also concerned the donors. The Corporation was plagued with illicit activities in the form of corrupt staff, illegal connections, adulteration of revenues in the field and mysterious expenditures that greatly hampered cost-containment strategies as elaborated in Chapters 2 and 3. The situation was pathetic! *Although my predecessor had already taken tough operational and strategic actions (e.g. introduction of KRIP) against the malpractices that were pulling the Corporation down, he had not yet accomplished this compelling task.* This would otherwise have given me a good beginning. As a new entrant into the Corporation, therefore, I had not only to continue and sustain my predecessor's initiatives but also to take on a multitude of other challenges that the Corporation was facing, including the not-so-good donor attitudes. I needed many prayers to manage this unprecedented situation. I was a real prisoner of too many important things to handle and found prioritising a daunting task. However, I was optimistic, gauging from the initial results of the first change management programmes (100 days and SEREP), that fortune would sooner or later smile our way. For the World Bank the situation was a recipe for umbrage. Apparently, the World Bank was convinced that under public ownership and management, the NWSC would not recover to improve service delivery, letalone become commercially viable and sustainable. *The only solution to the Corporation's woes was privatisation and the sooner the better!*

Perceived inefficiency and incompetence: Among our development partners, not only the World Bank was concerned. The German government-funded projects through KfW had stalled and the German agency, which was accusing NWSC staff of inefficiency, incompetence and any number of sins, was threatening to withhold or even withdraw its financial support from the Corporation. For KfW, NWSC was also heading for bankruptcy and it was no good sinking good German taxpayers' money after bad. They were ready to withdraw financial support if convincing remedial plans were not instituted to turn around the declining trend. By all standards, the situation in the NWSC was easier to talk about than to handle. *I am sure that even for the donors it was simpler to get annoyed about it than giving right advice and solutions.* Something had to be done. I knew that I had to do this insurmountable job to prove my worth and save the Corporation. The Corporation could not afford such widespread donor divorces. I could not afford to go down the memory lane in this manner!

NWSC managers' perceptions about donors: My uphill managerial challenges were not only with donors. I had to grapple with the perceptions of NWSC senior managers. According to most of them, the donors were *presumptuous, overbearing, intrusive* and *self-serving*. To some extent, these managers had a point. I remember one of them wondering about the double standards exercised by one of the development partner's staff in the cost overruns issue. My manager was concerned that the problems should have been handled as a collective responsibility since each of the parties, including the development partner's staff team, had played a role in the outcome of the project. This perceived attitude created a feeling among the staff and management that the donors were only interested in privatisation and nothing else. As a result, the donors were seen as proponents of privatisation, funding conditionalities and loss of jobs. They were viewed as a necessary evil, but overzealous in the way they exaggerated the Corporation's poor performance to justify the utility's immediate privatisation.

My own qualms: This charged atmosphere of mutual suspicion and recrimination was not helped by my own misgivings about the role of donor agencies, especially the World Bank, in the developing world. As a student in Germany, I used to see the World Bank as the financial agent of powerful northern countries, which had been contrived at the end of the Second World War to exploit the peoples of the Third World. At the Uganda Investment Authority, where I had been the deputy chief executive, I worked closely with several World Bank officials. My experience with them was not pleasant. Bank officials whom I met at UIA were *pompous, patronising* and *condescending*. They acted as if they 'knew it all'. They came to lecture and give conditions rather than to learn, listen and share experiences with us. Meetings with these officials were characterised by a combination of threats, censure and reprimand. It seemed to me that they were more eager to catch us doing something wrong than doing it right so that they could teach us a lesson.

When I left UIA, I felt a sense of relief, hoping that I had escaped the arm-twisting techniques of these fellows. However, when I joined the NWSC, there they were again! This taught me a lesson – *to learn to live with reality*. Donors are part and parcel of development activities in most national sectors and no one could, in any practical sense, hope to do without them! At least, not in the water business. I, therefore, had no choice but to adopt a positive attitude towards donors, viewing them as partners in development and not foes.

Inculcating a positive attitude towards donors

The performance turnaround of the NWSC in the 1990s necessitated a multitude of partners. As has been pointed out already, it was not constructive to spend a lot of time talking ill of such partners. The job at hand was to revamp the Corporation's performance, and anybody who could help was welcome. Consequently, we could no longer look at donors as scapegoats for our troubles. Our approach in our change management programmes was, therefore, clear – we had to work with the donors to help the NWSC grow into a self-sustaining institution. I knew that despite the misgivings of both the donors and NWSC management, there was a meeting point towards which we could work in order to foster a meaningful partnership. I was determined to ensure that this task was accomplished. I adopted a positive approach and told my staff and management that donors were not monsters who were out to bleed the poor Third World countries to death. I made reference to one of the Poverty Reduction Support Credit (PRSC) review meetings that I had attended in Kampala. The PRSC was at that time the guiding policy framework between the World Bank and the government of Uganda.

The meeting was co-chaired by the permanent secretary/secretary to the Treasury (PS/ST) and World Bank PRSC review team leader, Ms Satu Kakunen. As usual, the World Bank team included high-ranking and well-qualified specialists, who had the confidence to contribute during the meeting. On the government side, apart from the PS/ST and his deputy, the rest were low-ranking officials from various ministries. The permanent secretaries of those ministries had directed their subordinates to attend. Consequently, these ministry officials could not match their World Bank counterparts in discussing important issues before the meeting. There were many instances during the meeting when the PS/ST and Ms Kakunen found it hard to reach agreement in areas where the relevant permanent secretaries could have assisted if they had been present. In such circumstances, it is not fair to complain about impossible conditionalties when your own people are not proactive enough to defend or put up an alternative position acceptable to the negotiating parties.

Using the PRSC case, among many others, I sensitised my management and staff about the need to prepare well to negotiate for what would best benefit our organisation. I, over and over again, pointed out that we needed competent representatives who could make a meaningful, informed and focused contribution during negotiations. You need people who can say yes or no when it is necessary; people to discuss issues from an informed position. In most cases, donor policies are misunderstood; their objective is not to kill viable ideas or programmes, but to enhance them.

Indeed, as we implemented our change management programmes, the donors began to warm to us. Their *'prove worthy, we shall support you'* policy was beneficial to the Corporation. We have learned that more often than not donors mean well – they are people with reputations to protect and careers to advance. They do not want to be associated with failure.

Irrefutably, donors have their own interests and agendas to promote. Who does not? Donors do not force poor countries to accept assistance. In my view, they come at our invitation to help us solve problems that we cannot handle on our own. Furthermore, donor agencies are accountable to the governments and taxpayers of the rich countries, which provide the financial contribution. That is why they impose conditions to ensure that the financial assistance given is properly utilised in accordance to their home country expectations and regulations. In any case, the conditionalities, if well negotiated, could lead to win-winsituations for all parties concerned.

Take for instance the donor conditionalities which the NWSC was required to fulfil in order to qualify for more assistance. The Corporation was required to become commercially viable and financially sustainable to increase productivity, reduce costs, improve billing and revenue collection, reduce non-revenue water, etc. Now, who stands to benefit from the fulfilment of these conditionalities? Obviously, it is the water consumer. Who would take the credit for improved performance and service delivery? The answer is the Corporation's management. After all, it is the duty of the managers to improve efficiency, productivity and cost-effectiveness. That is what they are employed to do. Of course, the donors would also take credit for backing the right horse, but that is of secondary rather than primary importance. The truth of the matter is that *donor conditionalities are simple reflections of the challenges that the beneficiaries of donor assistance face rather than unexpected hurdles to make life impossible for those implementing donor-funded projects.* Donors would be foolish to give a blank cheque to aid recipients.

So why have donor conditionalities become so resented and controversial in recipient countries and organisations? In my view, it is because of incompetence, disorganisation, lack of commitment and political will on the side of the beneficiary organisations/countries. Most conditionalities are linked to clear outputs related to improved performance. As a result, when the implementing parties fail to deliver on agreed mandates, they find it more convenient to blame donors for the failures and shortcomings. The other fact is that recipients of donor assistance do not do their homework to counter any unrealistic or onerous conditionalities that donors seek to impose. They do not negotiate hard enough. They take donor conditions at face value without bothering to offer alternative proposals or initiatives or to bargain for better deals. *Therefore, donation recipients need to put their own houses in order so as to exploit the opportunities offered by donor assistance.*

Courting donors

The management change in the NWSC transpired against a background of prior significant donor participation in the Corporation's infrastructure and managerial development initiatives. Some of the donor agencies, like GTZ, concentrated on providing technical assistance (providing manpower to give advisory support) while others like the World Bank, KfW and the Austrian government gave financial support (providing financing for infrastructure strengthening and/or development) to the Corporation through the government. Given the frosty relationship between the Corporation and the donors in 1998, courting donors was one of my immediate challenges. It was an uphill, but eventually rewarding, task. The following sub-sections describe the processes and challenges we went through to bring donors to NWSC change management cause.

Realigning GTZ technical assistance: Prior to 1998, GTZ was deeply involved in NWSC's performance improvement initiatives – providing technical assistance, mostly in the form of personnel and vehicles. They had their own perspective of solving NWSC's performance problems. GTZ was left to recruit their own personnel and deploy them in the NWSC. The types of vehicles and equipment were prearranged and sometimes such facilities were incompatible with the local conditions in NWSC. This in turn resulted in high and irrational upkeep/maintenance costs. The GTZ assistance activities had peaked between 1993 and 1998. During this period, there was significant involvement in financial management, procurement, billing, project management, stores and water quality management experts, among others, courtesy of GTZ technical assistance. Unfortunately, however, this was the period when the Corporation faced significant financial and commercial managerial challenges. This trend clearly questioned the *modus operandi* of GTZ technical assistance. After due consultation with my management, I resolved that we had to change the strategic approach to future cooperation. We had to discuss with GTZ a shift in emphasis from personnel because the NWSC had built sufficient local expertise over time. In my view, what was urgently required was the beefing up of the Corporation's information technology (IT) capacity through computerisation and networking of NWSC operations countrywide. Since the existing technical assistance project was coming to an end and German/Ugandan bilateral negotiations for a new aid package were around the corner, my priority was to shift emphasis of GTZ technical assistance from *getting what they thought we needed to what we knew we wanted and was important to us*. It had to be a mixture of financial and technical assistance rather than technical assistance only. What they wanted to give us had to suit our demands rather than following a predominantly supply-driven approach of *'you have to take it or leave it'*.

The opportunity to properly realign GTZ assistance came during the bilateral negotiations that took place in Bonn in 2001. At this meeting, the NWSC team was convinced that computerisation and communication were the most important input missing from our operations. But the computerisation, as an activity, was outside GTZ's technical assistance portfolio. It was regarded as more of financial assistance, which was, arguably, under the mandate of KfW. This meant that GTZ could only meet our computerisation requirements if it accessed some of the money intended for financial cooperation, which was under the mandate of KfW. So we had to argue and haggle to persuade the German government to reallocate and earmark some of the money to GTZ for the computerisation and networking of NWSC operations.

Fortunately for us, we had Mr Thomas Albert, a clearheaded, objective and transparent civil servant as the head of the German negotiating team. At that time, he was head of the East African desk in the German Ministry of Development Cooperation (BMZ). His management of the meeting was excellent.

He clearly knew how to handle confrontational issues in a balanced manner without hurting anybody involved. We were also privileged to have an equally clearheaded, objective and experienced government negotiator in Mr Chris Kassami who co-chaired the bilateral negotiations with Albert. *Both of them made sure that throughout the discussions, reason rather than emotion prevailed.*

Mr Albert ruled that GTZ could still get its allocated share of the funds and discuss with the recipient institution how to utilise such funds. He was indeed very supportive of NWSC's point of view. Through his good management of the meeting, combined with the efforts and skills of Mr Kassami, we got all that we wanted. He prevailed over GTZ to reprioritise funds to meet NWSC demands. This guidance subsequently gave a green light to the German team to negotiate a new computerisation and networking project with us.

The implementation of this project put in place a number of management information systems (software infrastructure), namely Lotus Notes, higher versions of the billing system (CUSTIMA), accounting packages such as SCALA, website and the associated hardware IT infrastructure. The hardware IT infrastructure included, among others, purchase and installation of computers; a wide area network (WAN) connecting headquarters (Kampala) to all NWSC Areas; refurbishment of IT offices; and purchase and installation of generators to ensure continuous computer operations. In addition, NWSC staff were trained to man and service the computer network.

The computerisation and networking project was a great achievement which elated both the donor and the recipient. Indeed the successful implementation of this process was among the factors that contributed to the restoration of mutual confidence and respect between the Corporation and the German government. Both GTZ and NWSC won, namely an intervention that could not be financed easily through internally generated funds and yet was critical for operations and service delivery was implemented successfully, which was the desire of both parties.

The successful implementation of the computerisation process in NWSC using donor grants was a good learning experience. First, we acquired additional skills in realigning donor preferences and conditions to client demands and concluded that *transparent* and *objective discussions are key considerations.* Second, we underscored the significance of the role played by supervisors of field consultants in providing strategic guidance to project implementation. In this particular intervention, for example, had it not been for the excellent oversight of Mr Hansjörg Zitter (then Uganda's GTZ Country Director) and Mr Dettmering (head of East African Division at GTZ headquarters in Germany), the computerisation project could not have realised the full impact it has had on NWSC operations.

Restoring the confidence of KfW: Restoring KfW confidence in the water sector was more problematic than the rebirth of the NWSC–GTZ partnership. Before 1998, the German agency had funded all its water projects in Uganda, including those in Fort Portal and Kasese, through the Directorate of Water Development (DWD). However, after 1998, the agency preferred to work directly with the NWSC and insisted that the Kabale Project, which was then in the pipeline, be managed by the NWSC *based on the need to ensure continuity between infrastructure development and downstream operations.*

That is partly why the town of Kabale was added to the list of NWSC-managed Areas. However, despite this apparent goodwill towards NWSC, KfW had enormous reservations about doing business with the Corporation, an organisation that all donors concurred was on the verge of bankruptcy. We had to act quickly to restore KfW's confidence, and indeed that of the German Development Corporation. As a result, at the beginning of 1999 I decided to set up a team comprising prominent members of the Board, headed by the Board Chairman, Mr Sam Okec, to court the KfW leadership and convince them that a new chapter in the life of NWSC had begun and that the time had come for the two organisations to become serious partners in development. On the delegation we had Dr Shire, a member of the Board, who, like me, had studied and worked in Germany and knew the German language very well. In addition, Shire had used his long stay in Germany to make friends.

Given this prearranged composition of the delegation, I was confident that with our proficiency in German and knowledge of German society, Shire and I would prevail upon the KfW leaders to continue and hopefully increase financial assistance to the NWSC. To our great surprise and disappointment, we received a cool reception, to say the least.

No sooner had we arrived than Dr Dittmar, the KfW Director of African Department, and Dr Claudia Radecke, his deputy, dropped their bombshell. 'Sorry, immediate KfW assistance is out of the question.' 'Why?' we wanted to know. They explained but by looking on their faces, I could see a lot of hesitation to continue working with a partner that had not, properly, handled previous investments. In actual fact, I began to wonder whether our trip was not a waste of time. The Germans are famous for being blunt.

During the subsequent meetings, which took place in two working groups, there were obvious fears related to corruption, incompetence and outright bad faith on the part of NWSC management team. To make matters worse, a consultant from a German company, which had been working on one of the water projects in Uganda, who had been invited to the meetings, claimed that Ugandan officials were notorious for lacking transparency and fair dealing. He, therefore, insisted that they could not be trusted partners. Our protests that we were a new broom with a clean record fell on deaf ears. The negotiations would have collapsed there and then had it not been for the timely and welcome intervention of the owners of Gauff engineers, Helmut Gauff Senior and Gerhard Gauff – the son – who had co-sponsored our trip.

In the end, the KfW leadership reluctantly agreed to renew the working relationship with NWSC on condition that the Corporation undertook drastic reforms to become a well-governed, viable and financially sustainable utility. That is why every subsequent agreement with KfW incorporated the conditionality, *'funds will be released according to the progress of urban water reforms'*. One of KfW's suggestions, at this stage, was that the management of water and sewerage services in Uganda be let out to an international company and that NWSC either be converted into an asset-holding company or be dissolved. The Gauff engineers' intervention saved the day but it took much longer to restore full confidence and partnership between NWSC and KfW.

However, the less than successful trip to Germany had its own silver lining. It opened our eyes to what needed to be done to make the NWSC a respectable and viable corporation worthy of support. It became clear that donors were willing to lend a helping hand provided we put our house in order. The ball was, therefore, in our court. I told my Board and management that to attract donor funding we had to work hard, be accountable, transparent and give constant feedback to the donors.

For me, it was, therefore, imperative to redouble our efforts to ensure the success of the 100 days Programme and subsequent change management initiatives. It was a question of do or die for the Corporation. I was convinced that progress and success were the only ways through which we would cultivate donor confidence and win donor funding for large capital development projects. Thanks to that eye-opening trip, our actions have since served the NWSC far better than our words or negotiating skills ever could. Dr Dittmar and Dr Radecke became our very good allies and friends having witnessed the evolution of NWSC's performance turnaround since that time.

Cultivating World Bank support: Among the struggles to court donors, none was a harder nut to crack than the World Bank. Before 1998, the Bank was the NWSC's main financial supporter, but the Corporation had not fully tapped the benefits of the huge investments sunk in infrastructure development activities. According to the World Bank, the Corporation's performance was appalling. This became clear to me during the first meeting with the Bank mission, led by Mr Alain Locussol, in early 1999. He was a plain speaking engineer with an analytical mind. While I was still a novice in the language of the water industry, Mr Locussol was a water expert with a mastery of the jargon of water production and distribution such as *non-revenue water, staff productivity ratios, billing and revenue-collection efficiency ratios, sewage effluent,* etc. In summary, he combined the attributes of an astute engineer backed by a very good understanding of the financial operations in the water sector.

Indeed, I sensed that Mr Locussol had grave doubts about my suitability for the NWSC's Chief Executive Officer's job. He seemed to think that given my professional background, I did not have much knowledge of the water business and that I would therefore not be up to the task of leading the Corporation. But since he was not my appointing authority, he was stuck with me for better or for worse.

Mr Locussol and his team were in a quagmire. I could read it from their faces and actions that, according to them, NWSC was terminally sick. It was inefficient, mismanaged and ridden with sleaze. It was performing below the not-so-good African standards, and below minimum international standards. The 'take it or leave it' conditionalities were slowly becoming a reality. Non-revenue water must be reduced. Staff productivity as measured by optimal staff utilisation must improve, arrears must be reduced, revenue collection must increase, and so on. In short, the Corporation had to become commercially viable or else face the consequences. I had no illusions about the daunting task to put NWSC back on track but when I pleaded for time and patience, Mr Locussol and his colleagues just ignored me. However, I had the will to push on! I knew that their impressions and actions were justified, although I still believed that they needed to give a chance to the new management to improve things. I was determined to court them into this direction since the government had taken the same direction anyway. The government had given the Corporation another chance by appointing a new board and MD hoping for better performance. Why couldn't the donors also give us a chance? I had no doubt that we could come up with a formula to work with these vital partners. *I had to do it through action, not words.* My management and staff were ready in this regard.

I remember one time, my Chief Engineer in charge of Planning and Capital Development, Mr Alex Gisagara requesting Mr Locussol's group that 'give us a chance...don't judge us unfairly for mistakes committed in the past'. This statement assured me that I had a group of committed 'foot soldiers' ready to tackle the enormous challenges ahead.

My World Bank courting efforts were fouled by accusations in the media that the new MD was wasting and misusing the Corporation's resources for his own benefit, at a time when the Corporation's finances and services were in desperate straits. According to the press (The Monitor, 15 March 1999), the new MD had not only bought a brand new four-wheel inter-cooler vehicle and two mobile phones but he had also organised an expensive trip to Germany at the expense of the Corporation. The reporter who filed this story had not bothered to cross-check the news tip. The vehicle was, for instance, budgeted for and its procurement process had started before I joined the Corporation. In addition, the trip had been organised and co-sponsored by organisations in Britain and Germany, hoping to do business with the Corporation, especially regarding the possibilities of foreign direct investment (FDI) in the water sector in Uganda. Nevertheless, whenever such mud is thrown at public servants, especially in a developing country like Uganda, some of it tends to stick because it captures the public imagination, regardless of whether it is premised on the truth or not.

Indeed, when the above-mentioned accusations appeared in the *The Monitor*, one of the World Bank officials gleefully drew my attention to the report. This report seemed to confirm the worst fears of Mr Locussol and his colleagues about the excesses of self-serving managers of public enterprises, including myself.

To an outsider, the NWSC's performance problems could easily be described as 'irredeemable'. This was unfortunate and unbelievable for me, in particular! I had been employed to manage the Corporation, not to liquidate it. At least I deserved to be given an opportunity. The World Bank was in the equation with the usual standard textbook solution for poor performing or failing public enterprises, 'privatise them'. Indeed, at that time a consulting firm, Consult 4 Pty Ltd had been commissioned to explore ways and means of restructuring the urban water sector. Consult 4 was a consortium of several South African firms, coordinated by Sunshine Projects, a local consultancy firm based in Kampala.

This consortium was commissioned to study the urban water sub-sector and to make appropriate recommendations about the future management of urban water and sewerage services in Uganda. The sooner NWSC was privatised, the better! Everybody, mostly at the World Bank, eagerly awaited Consult 4's report. From the way things were unfolding, it appeared that this was going to be a foregone recommendation in favour of the privatisation of water and sewerage services in Uganda.

For Mr Locussol and his colleagues, the Consult 4 team was taking too long to come up with what was an obvious solution – privatisation or at least private sector participation (PSP). Actually, PSP in NWSC was a government of Uganda Policy for state-owned enterprises (SOEs). Therefore, according to Mr Locussol and his colleagues, the Consult 4 consultants were taking too long to complete the work. To skirt this delay, the Bank carried out its own internal due diligence sector work to prepare itself for supporting the government of Uganda reform Agenda. A parallel assessment of what PSP model might fit the asset situation at NWSC was carried out to inform the Consult 4 process. The objective of the parallel assessment was to provide a shadow roadmap and framework for Consult 4's findings and recommendations.

The report of the parallel assessment recommended a number of private sector participation options for the water sector in Uganda. This parallel assignment caused significant discomfort to Consult 4 consultants who were seen by the Bank as slow and over-consulting stakeholders. Amidst this confusion, I told my staff that we had to push on. I made it clear that performance required no 'colour', it only required *commitment and focused hard work*. I made sure that in every stakeholder meeting about the privatisation options, the NWSC delegates proclaimed the positive performance outcomes from our change management processes.

My intention was to broaden the debate about what works and what does not. I urged my staff to find out what was happening in other African utilities where similar interventions (PSP options) had been practiced and in turn inform the discussions accordingly. We relentlessly worked for improved service delivery to our customers and expanding services to those without. *I quietly avoided the trial and error options, which were being proposed to us by people who were not on the ground!*

In the meantime, the World Bank was determined to step up the pressure until the NWSC was reformed. Every visit and every meeting were an occasion to remind us that drastic reforms were the answer to the financial, technical and management challenges that the Corporation was facing. I took the Bank's pressure positively and told my staff and management that what the Bank wanted was performance. *I urged them to be vigilant, hardworking and committed to continuous innovation.* This was the only way to keep our enthusiastic stakeholders hopeful.

Were initial efforts convincing? The World Bank's view

Judging from the World Bank's initial misgivings, it was not surprising, therefore, that they were sceptical about the first of the NWSC change management programmes – the 100 days Programme. They saw it as a propaganda gimmick contrived to lull the government, donors and the public at large into thinking that the NWSC Board and management were committed to turning the Corporation around. During the evaluation of the 100 days in Jinja, we invited dignitaries, including ministers, senior civil servants, local notables and of course Mr Locussol to witness the progress that had been made to lay the foundation for turning around the Corporation without recourse to immediate privatisation.

Although Mr Locussol accepted our invitation and came to the function in Jinja, he was not impressed by the 100 days achievements and left in a hurry (without saying a word), even before the guest of honour arrived to grace the occasion. In his subsequent *aide memoir*, Mr Locussol alluded to the perception that the 100 days Programme did not measure up to the first step on the journey to turning the Corporation around.

On the face of it, the World Bank's attitude, pressure and conditionalities were irritating and humiliating. The Bank officials were not giving us a chance to prove ourselves! Mr Locussol and his colleagues saw no future for the Corporation as a public utility. The writing was on the wall. On our part, we thought their judgement was too harsh.

We deserved the benefit of the doubt. We too appreciated the sorry state the Corporation was in and were committed to bringing about a change for the better. But we had only been in office a few months! We needed time to think through the problems, find appropriate solutions and put our house in order. The Bank conditionalities could not be implemented overnight. Instead of appreciating our predicament, the Bank's officials were quick and eager to shoot down our first change management initiative. In retrospect, I am glad that the World Bank imposed stringent conditions on us, which forced us to concentrate our minds and to rely on our own resources.

In any case, most of the conditionalities were reasonable enough and in the long-term interest of the Corporation with or without privatisation. I, therefore, used the *spectre* of privatisation hanging over the Corporation to push NWSC staff to work harder to deliver the results – to justify our existence. I incessantly warned the entire NWSC staff that unless we all pulled up our socks and worked harder, the privatisation axe would soon sever our existence. *In this sense, I used the shadow of privatisation to spur staff productivity.*

Eventual payoff: positive attitudes start evolving

The positive trends as a result of enormous efforts to improve the Corporation's performance started capturing the attention of people we initially believed to be against us. The talking and interaction were becoming more enabling and friendlier. In fact, by the time Mr Locussol and his team visited Uganda for the first appraisal since the end of the 100 days Programme, there was cause for cautious optimism. Mr Locussol was conspicuously smiling and radiating clear signals of appreciation. Clearly, as a result of efficiency improvements, there was a thaw in NWSC–World Bank relationship.

After implementing a series of short-term improvement programmes and sharing performance results, the World Bank colleagues began to take our change management programmes seriously. Specifically, with the advent of the first NWSC/GoU Performance Contract and Area Performance Contracts, there was a clear change in the attitude of Mr Locussol who, during our discussions, passionately guided his colleagues to support our innovative insights. For once I saw an objective team leader. No wonder everybody in my management team saw him as a determined lead specialist who *hated failure and liked working with progressive managers*. He only had to be a little patient!

Mr Locussol was not the only happy man. His colleagues, Messrs Fook Chuan, Solomon Alemu and Ato Brown, were equally encouraging. Mr Fook was a brilliant financial analyst from Malaysia, who was interested in the financial performance of NWSC. On the other hand, Mr Brown, from Ghana, was a young professional with all-round strong analytical capabilities. Brown and Alemu, from Ethiopia, were always wondering why their fellow Africans could not put their house in order when they seemed to have all the qualities needed to do the job. They hated people telling stories of how things needed to be done rather than actually doing them. Mr Brown, particularly, always insisted that people should do what they are saying instead of merely talking. These World Bank officials were initially not amused with the seemingly 'lazy and insensitive Africans'. They believed that the job could be done but were wondering why it was not! As already pointed out, one of the perceived erratic actions was the MD's extravagancy that featured in the newspapers, which was cynically brought to my attention by Ato Brown – in this respect he was really behaving like an infuriated concerned 'brother'. In essence, both men looked very upset whenever Mr Locussol fiercely condemned the NWSC. They believed in the objectivity of Mr Locussol's fury but also knew that NWSC managers were capable of doing much better.

However, as NWSC achievements started unfolding, I could see Brown and Alemu heave a sigh of relief whenever Mr Locussol acknowledged what NWSC had done. It was time for Mr Brown, in particular, to celebrate. He had, in the beginning, told Mr Locussol to give us a chance but the latter had had his reservations. Mr Brown's instinct could no longer be ignored. The results were there for all to see!

As successive change management programmes and their results evolved, the World Bank team shifted from being critics to advocates of our cause. They began to relax the pressure on the government to speed up the privatisation process in order to give us a chance to turn the Corporation around through internal reforms. According to Mr Locussol and his colleagues, the Corporation was beginning to be a worthy partner. It was these hitherto sceptical officials that began to advise the government to grant us a debt moratorium and to pay its water bills promptly. For us, it was a clear message that if you do not perform, you should not expect comfort from the World Bank. Mr Locussol's orientation was, and rightly so, 'you either perform or you will not be supported'. Indeed, this is why he spearheaded the proposal urging government to enter into a performance contract (2000–2003) with NWSC. This performance improvement approach as outlined in Chapter 4 was the first of its kind in the history of public enterprises in Uganda. This welcome and positive change of tone demonstrated the World Bank's willingness to listen to local voices. The Bank was flexible enough to not only consider local initiatives but also review its own conditionalities insituations where actions spoke louder than words. *I used the positive change in attitude to push for more performance improvements in the Corporation.*

Donors become real-life partners in service delivery

Once donor confidence was restored, the door for partnership between the Corporation and its benefactors in the delivery of water and sewerage services was wide open. For example, KfW became one of the Corporation's best partners in water and sewerage services delivery. It started funding various capital projects in Kampala, Entebbe and Kabale without any reservations. They also agreed to support our efforts to secure more funding from the European Union's Africa Water Facility Grants. The era of 'we shall wait and see' was long gone. It had been replaced by one of mutual confidence, respect and cooperation. They were no longer watching with a binoculars and a whip in hand but rather they now adopted a 'working together' approach.

The NWSC relationship with the World Bank had shifted from pressure through conditionalities to proactive cooperation. It supported NWSC change management programmes from APCs to IDAMCs. It adopted the 'do-it-yourself' approach to NWSC capacity-building efforts and was against hiring consultants where it felt we could do the job. For example, in 1999/2000, the Bank insisted that we draw up the first NWSC financial model. In its view, there was no need or money to pay an external consultant as we were capable of doing it ourselves. By drawing up the financial model on our own, with Mr Fook's technical assistance wherever it was required, we were able to make our own mistakes and learn from them. 'Doing it ourselves' was the more effective and enduring way of learning. As the Chinese say, 'Tell me and I will remember for a day; but let me do it and I will remember forever.' On the World Bank's insistence, not only did we produce a financial model that measured up to international standards but also saved money. We built capacity and experience in-house and the effects have rolled over and strengthened ways of doing things in the NWSC.

Thanks to the Bank's 'do-it-yourself' capacity-building strategy, NWSC staff have progressively acquired excellent professional expertise to the extent that they have been hired as consultants to render service to water utilities in neighbouring countries. To this end, we have even set up an *External Services Department* to render specialised services in sub-Saharan Africa and beyond. A number of professionals commended the Corporation's outward-looking policy. For instance, at a Financing for Water Sector Infrastructure Seminar held in Kampala, in September 2004, Dr Allen Eisenhardt, a senior finance infrastructure advisor at the United States Agency for International Development (USAID), Washington, DC, remarked that 'NWSC had used best practices that are used in some of the world's successful water programmes and reform organisations' (*Daily Monitor* of 23 September 2004). He asserted that 'there is no other corporation in Africa that covers operation costs, makes profits and creates a fund for expansion'.

The process of in-house capacity building in NWSC and the results achieved so far inspired me to review the usefulness of what has been called 'external consultants' in developing countries. In some instances, they know less than the local people. Although they may have specialist knowledge, some of these 'external consultants' take too long to adjust to local conditions. Therefore, external technical assistance (TA) is meaningful if it is targeted and designed to complement local capacities.

Besides, such consultants are, in most cases, very expensive and eventually do not give full value for money. Most of them are either average or mediocre in their fields in their countries of origin or are young novices who are sent to the developing world to gain experience as part of their career development. They also tend to be slaves to standard textbook knowledge and theories. In addition, they provide standard remedies in standard formats; they preach what is already known or what they have been told by their employers to do.

These experts cost money and time. At NWSC, we made it clear that they should come when and where it is cost-effective and absolutely necessary to do so. Of course, not all consultants were found wanting. Some were very knowledgeable and had varied but useful experience. Some of them were open-minded, willing to listen and learn – to give and take. And some of them quickly adjusted to local conditions and settled in very well with local colleagues.

At NWSC, I met a number of consultants who I respect highly and who rendered invaluable technical assistance and advice without throwing their weight around. The outstanding examples of this type of consultants accredited to NWSC were Susanne Mauve and Jürgen Bickert. Ms Mauve, a young German from GTZ/KfW, served NWSC with distinction while Mr Bickert provided valuable short-term input at various times.

Some seemingly insurmountable challenges were faced in the initial stages of assimilating Ms Mauve into NWSC's working environment.. I must say, we misjudged her and she also misjudged us. At her tender age, I mistook her for another external consultant who had just finished her studies and was simply being foisted on us as part of her internship/training. Actually, this initial perception, I must say, greatly influenced my approach and indeed that of my senior managers towards her. As if she had also 'read it in glass' it seems she perceived us as a bunch of people who did not understand her mission very well and simply hated anything called 'consultant'. These two divergent opinions created a lot of start-up problems between us and Ms Mauve. It was after talking to her and her bosses that a number of issues of mutual coexistence were resolved and we achieved convergence between our way of conducting business and her way of executing the assigned mission.

In fact, with time, we realised that we had misunderstood each other. On the one hand, we noticed that Ms Mauve was committed, resourceful and hard working. She did not mind sitting anywhere 'so long as she had enough space for her computer'. And in spite of being a foreign technical expert, she did not demand special privileges from her hosts. She was willing to mix freely with colleagues and to learn from them. She fully participated in the conceptualisation and implementation of some of our change management programmes, in particular IDAMCs. Interestingly, Ms Mauve always complied with the Corporation's regulations and practices though she had no contractual obligation to do so. Mauve must have noticed that we did not hate consultants, *per se,* but wanted people with innovative ideas who would bring on board new methods and strategies that could create a difference for NWSC's internal reforms. We encouraged mutual learning and respect. It is, therefore, not surprising that at the time of leaving the NWSC, Ms Mauve humbly admitted that she had learnt more than she had taught. In my view this was actually partly untrue. We had instead learnt a lot from each other. Indeed, when she left NWSC, her colleagues missed her friendship and valuable contributions to performance improvement programmes.

Mr Bickert was another consultant that the staff at NWSC enjoyed working with. His ideas were so inspiring that whenever he fixed an appointment to discuss his findings, I looked forward to a worthwhile meeting. He was a German technical consultant who had been hired to monitor the use of KfW funds and to ensure that OSUL (as has been explained already, an International private operator) attained the agreed contract performance targets without interference from NWSC management. My initial perception of Mr Bickert's involvement was not favourable at all! I actually thought we had gotten another good-for-nothing man to churn out conventional consultant reports and earn money.

I am sure he read it from my face that I did not approve of him outright. However, Mr Bickert, experienced as he was, took this impression in a positive sense and worked around the clock to prove me wrong. Like Ms Mauve, he used a participatory approach, discussing every issue and offering useful 'food for thought' with all concerned parties. He never, at any time, took a rigid position. He knew his information asymmetry problem. He did not believe in conventional consultants' textbook copy-and-paste but rather in what works and what the people on the ground believed in. He was more of a facilitator and bridge between the donor, KfW, and the client (NWSC) in ensuring that *what it took to perform better was fulfilled.*

NWSC donor-funded projects since 1998

When I joined the NWSC, I found three categories of donor-funded projects and decided to build on them. The first category involved *long-term capital development projects (hardware investments)* which entailed the rehabilitation, construction and expansion of new water and sewerage facilities. By 1998, projects that had been completed with funding from the World Bank, the European Union and the Austrian government included the refurbishment and expansion of the Mbarara and Masaka water and sewerage systems, rehabilitation of the water intakes in Tororo and Mbale, the rehabilitation of sewerage networks in Kampala, the construction of the Gaba II Water Works, and rehabilitation of the Namasuba network. All these projects were within the ambit of the 'Second Water Supply Project', one of the largest projects ever implemented within the Corporation.

The challenge after the expansion of the production capacities was to improve and expand water distribution and to improve the delivery of water and sewerage services to the satisfaction of the customers.

Since 1998, donors have continued to support NWSC capital development projects through grants rather than through loans. The German government, through their development agency, KfW, funded waterworks in Fort Portal, Kasese, Kabale, Entebbe and Kampala (Gaba III Water Works). These capital development projects were subject to hard bargaining and negotiating processes between the donors and NWSC to ensure that the intended beneficiaries got value for money. The KfW has continuously monitored the German government-funded projects, pinpointed the shortfalls and bottlenecks, and suggested appropriate and timely remedies. The Germans have zero tolerance for poor performance and, therefore, the mere fact that they continued to support our projects all those years was no small recognition of the ongoing efforts to transform the NWSC into a commercially viable and sustainable enterprise.

The second category of donor-funded projects revolved around *capacity building (software investments)*. The World Bank took the lead in the NWSC staff training and development programmes. The Bank together with support from the Dutch, German, British, Swedish and Danish governments helped NWSC staff to participate in programmes ranging from short courses, seminars and conferences to masters and PhD degrees in water engineering and institutional development.

These courses exposed the beneficiary staff to new ideas, experiences and international best practice. This training in turn helped to shape the design and implementation of NWSC change management programmes since 1998. As a result, the NWSC has been able to create a critical mass of professionals whose contribution to the transformation of the Corporation goes without saying.

In carrying out short-term courses, we emphasised interaction and discussion so that our employees' ideas could be heard and enhanced. I was aware that some trainers were fond of using the *'listen and I teach you'* approach. At the outset, this was not the right capacity-building strategy for NWSC since staff already had ideas and only lacked the right triggers to roll their ideas into action. I insisted that all refresher courses for my staff had to be carried out in a free environment and be discussion based.

One example where I concretised this approach was during a management and institutional development course organised by Ms Mauve at the NWSC training centre in 2004 for the Board and senior managers of NWSC. The course was conducted by Stone and Webster, an international consulting company. At first, some resource persons had planned to use the *'I will teach you'* approach but I intervened and insisted that we should have more discussions than one-way teaching. At the end of the course, my insistence paid off; one of the resource persons actually confided in me that the approach assisted her to learn more than she had prepared and actually delivered.

The third category of donor-funded projects in NWSC was related to the use of *teams of consultants* to study, review and analyse the prevailing situation and to recommend solutions to problems that the Corporation was facing. *This was an indirect approach to design strategies that were a direct replica of donor ambitions and values.* In most cases, donors have preconceived notions about what they want to do, but they do not know how to go about it without offending the recipient's sensibilities.

So the use of consultants is a standard and convenient way of legitimising what donors intend to do. Consultants are supposed to render informed expert advice based on the analytical study of the prevailing situation. But, in practice, reports from consultants, more often than not, contain what donors want to hear and sooner or later such reports are overtaken by events. It is hardly surprising that such reports are left to gather dust soon after their submission without a follow-up action. Needless to say, several of the reports from donor-funded consultants regarding NWSC or the urban water sector in general suffered the same fate because NWSC management could not see significant advantages compared to their own home-grown performance improvement initiatives.

However, NWSC managers have gained considerable experience while dealing with teams of consultants in a number of ways. First, we observed that some consultants come with their masters' instructions and interests and want them implemented in beneficiary institutions without any alteration/modifications. This calls for an open mind, keen observation and more informed decision making on the part of local managers. A good example was a young woman professional (name withheld) from a donor agency in Uganda who had instructions from her bosses to go to NWSC and inform them about the donor's plans to engage an external consultant to help implement a certain structural reform policy in NWSC. I talked, at length, with this consultant and eventually we agreed to disagree about the strategic approach to use! While I insisted that such interventions ought to be implemented by those who understood the technologies and business processes on the ground better, the consultant insisted that her bosses could not detract from the original reform strategy agreed upon more than three years before. Our firm stand helped in this case. We took less of her input and continued working with our own reform initiatives. She backed off and the agency understood our point.

Second, the Consult 4 (mentioned earlier) process also gave us some useful insights. We noticed that emphasis on balancing conflicting stakeholder interests was good but was ultimately a waste of time and resources. In this particular assignment, which was supposed to design an optimal institutional management strategy, the desire to balance competing interests of NWSC (who would pick up the bill), the World Bank (who would give a no-objection to payment) and other stakeholders (who could morally rubbish the report) resulted into a 'redundant' lease option which had been overtaken by events on the ground. More cost-effective and efficient management options had, instead, started emerging from the ongoing innovative NWSC programmes. In hindsight, the Consult 4 study gave us more time to evolve our own programmes.

Third, we also had an interesting experience in engaging the second PSP contract – ONDEO Services Uganda Limited (OSUL) in Kampala. The transactions were conducted amidst conflicting minds on the side of NWSC Board, management and other key outside stakeholders. Some people thought that the first PSP contract (KRIP) had not performed according to the expectations and there was no need to try another PSP again. On the other hand, other people argued that PSP could perform in Kampala if past mistakes in procurement, for example, were avoided. I remained neutral in these arguments since I had not participated in the past transactions and thus did not have sufficient information to make objective judgements. I therefore opted for objectivity and leaving 'doors' open for any good ideas. In any case, I was not worried because the two options – with and without PSP – could be applied at the same time, for comparison purposes, due to the vast geographical scope of NWSC operations that stood at 15 towns at that time. Consequently, with the help of financial assistance from donors, we procured ONDEO Services Uganda Ltd (OSUL) using a team of NWSC managers and headed by an independent consultant, Mr Tony Richards, a retired British-born South African. Ultimately, a PSP contract was signed in 2002. The OSUL contract was one of the few management contracts that was designed to be paid by the tariff and not directly funded by donors. The KfW retained Tony Richards and Bickert (another German consultant) to support the contract management process, together with our own NWSC staff. In addition, PricewaterhouseCoopers (Uganda) was retained as the overall project auditors and analysts under KfW funding.

Actually, our initial contestations were not about transaction and monitoring costs associated with the consultants since these were not being met by NWSC. However, our real concern was how long the Corporation was going to retain foreign consultants to oversee another foreign contractor! My question was: If a foreign contractor needs a foreign consultant to be regulated, how costly would this be and was this sustainable in the long run? Although I continued to be haunted by the OSUL contract, I told my staff that every business management venture runs with residual unresolved issues and as such there was no cause for alarm. What we needed to do was to focus on maximisation of benefits accruing from the PSP contract. We had to abstain from crying over spilled milk!

All the above notwithstanding, *the introduction of private sector management in Kampala brought some relief to NWSC management.* I told my staff and management that we had to give the new OSUL management all the support they required to operate well. We secured goodwill from our development partners and government because we had not deviated from their overall GoU water sector development strategy, which dictated PSP involvement at all costs.

This conducive environment allowed us to concentrate our brains, energies and resources on the other 14 NWSC operational Areas. I told my colleagues that in management, this was a good strategy – trying out, at the same time, different options and assessing what works best, given the circumstances. We, therefore, had time to continue the implementation of our change management programmes on one side and give our maximum support to OSUL, on the other. I hoped that, subject to good performance, the OSUL management contract would automatically be renewed for another two years to allow us concretise the most suitable performance improvement strategy. This was a win-win situation for NWSC.

However, the facts on the ground soon proved to the contrary. *Operationally, with three teams of consultants and one NWSC authorised representative, the contract proved too expensive.* In addition, OSUL'S actual performance fell short of the agreed targets. The other NWSC Areas, where we had already introduced IDAMCs to strengthen operations, performed relatively better than

Kampala. The IDAMCs were, in many respects, similar in principle, intent and character to any conventional management contract. It was, therefore, easier to switch Kampala operations over to the IDAMCs operational mode, with minor adjustments in the contractual framework. Indeed, after OSUL chose not to extend the contract, the NWSC Board and management resolved to put Kampala operations under IDAMCs framework. It worked.

Trivialities can be detrimental to citizens

Throughout our change management programmes, we tried to avoid misunderstandings that could derail service delivery to citizens. However, situations arose during our change management processes that turned out to be beyond our control. Our reform efforts involved continuously taking over smaller towns as gazetted by the minister of Water from time to time. In fact, the NWSC towns which stood at 11 in 1998 grew steadily to 22 in 2008, primarily the biggest towns of Uganda. In all the towns we took over, we made sure that both operational and financial performance improved to the benefit of service delivery to citizens. However, there is one incident when two towns were gazetted by the minister of Water to NWSC and the process had to be reversed through the intervention by one of the donors. The reason for the reversal was given as the fact that the then-ongoing reform efforts supported by that donor had potentially superior advantages over what NWSC would bring to those two towns. Our efforts to explain that we had nothing to do with the gazetting fell on deaf ears. Other donors also intervened and tried to convince their colleagues, pointing out that this was a government of Uganda issue, but to no avail. The donor did not bother to cross-check what the civic leaders in those towns were in favour of.

Consequently, the respective gazettes in the two towns were reversed and the status quo continued. Until today, the local leaders and citizens in those two towns continue to call upon NWSC to take over those towns, yearning for improved services. Much as the two towns were eventually put under the operational responsibility of local private operators (LPOs), the initiative has not helped to improve services to the fullest satisfaction of respective citizens. This misunderstanding goes a long way to underscore the importance of harmonising donor interests with those of local citizens. Conditionalities and egoistic tendencies do not help, but rather understanding of the local situation and coming close to what works best under certain conditions help. In the case of the two towns, the donor demanded that further financial assistance would be pegged to reversal of the gazetting processes but service in those towns has not improved compared to towns that were gazetted to NWSC at that time. The donor has since ceased support to these towns!

Donors continue to talk about NWSC's reforms

The sustained impetus in the development of successive change management programmes, specifically after launching the IDAMCs, convinced the donors and made them NWSC's best allies in the improvement and development of water and sewerage services. Because of the successes associated with this management model, the Corporation started hosting enthusiastic visitors and celebrities from donor countries who were interested in NWSC's performance models and programmes. They have been engrossed in seeing whether some of the internal reform approaches can be replicated in other utilities operating in similar environments.

One such enthusiastic visitor was Dr Uschi Eid, at that time a member of the United Nations Secretary General Advisory Council on Water and Sanitation and also the parliamentary secretary in BMZ of the German Ministry for Development Cooperation. Dr Uschi came to Uganda specifically to witness the revolution in the delivery of water and sewerage services that was being implemented in NWSC. Her visit was followed by that of a group of eight German parliamentarians of the Select Committee on Development Aid. The team was headed by Dr Addicks, and according to *New Vision* of 4 June 2008 they concluded that NWSC was on the right path to improve service delivery, especially to the urban poor. Their visits had been preceded by that of Bill Gates Snr and his wife – the parents of one of the world's richest men (Bill Gates), representing the Bill and Melinda Gates Foundation. The visit by the Gates family was also aimed at witnessing how NWSC was succeeding in serving poor communities, a lesson they wanted emulated elsewhere through the Bill and Melinda Gates Foundation. All these visits have boosted our External Services Department and the image of the Corporation.

As a result, donors have sponsored senior NWSC managers to visit other utility providers in Africa and elsewhere in the developing world to render technical assistance and advice under the external services unit. On one such trip to India, a World Bank official confided in me that the NWSC, whose change management achievements he described as 'a management miracle in Africa', put NWSC on the right track to full-fledged commercialisation.

The NWSC's significant progress in improving performance has earned it considerable donor confidence. Donors have become the champions of the NWSC cause. For example, they persuaded the government to incorporate NWSC's performance indicators – NRW, staff productivity, financial ratios, etc. – in the Poverty Reduction Support Credit system. Grants and credits under this system were made conditional to continued improvements in the delivery of water and sewerage services. In doing all this, donors wanted government to do everything possible to support the financial re-engineering process of NWSC. Specifically, the World Bank and the German government insisted that the NWSC have a clean balance sheet, and for this reason they called on the government of Uganda to write off the old loans in order to enable the Corporation achieve long-term financial sustainability. In addition, they also insisted on staff restructuring so that the proceeds could be used for further retrenchment activities. Further, they encouraged simplification of the tariff structure to enhance its pro-poor focus. To improve financial sustainability, the World Bank also insisted that there be a clear operating framework under which the government could promptly pay all accrued water bills. Consequently, a memorandum of agreement was signed between NWSC and government stipulating that in future all government departments would pay their water bills in full and in advance. *These efforts and interventions were the most prominent sign of the hard-won confidence and partnership between the NWSC and the donors.*

It counts a lot to be appreciated

Once we won donor confidence and converted them from cynics to NWSC champions, the future prospects of the Corporation's collaboration with its donor partners became quite promising. We moved a long way from the days when donors doubted the likelihood of the 100 days Programme winning their hearts and minds, making them our erstwhile allies.

The NWSC change management programmes proved beyond reasonable doubt that it was possible to turn poorly performing utilities into commercially viable and sustainable enterprises with or without privatisation. The general attitude that public companies cannot perform is plainly wrong. What matters is not the nature of ownership (be it public, private or hybrid), but improvement of management systems, staff commitment to work, strong customer orientation and stakeholder support.

In retrospect, we learnt that donors appreciate success when they see it. According to one of the development partners' staff, donors in fact love the free ride with a successful utility. They are flexible enough to change their minds and accept sound recipient-generated alternatives and initiatives. What donors are entitled to demand – for he who pays the piper calls the tune – is information sharing, transparency and results-oriented management.

Donors are also entitled to insist on project-specific conditions to ensure that their assistance is put to good use. In this respect, the NWSC experience has proved that donor conditionalities are open to negotiation and that they are not set in stone. At the same time, donors are also sensitive people. For them and, indeed, any other person/organisation, it counts a lot to appreciate a helping hand extended, in any form. Some of them love to be in the limelight. Every donor official that the NWSC has worked with since 1998 has had his/her own sensibilities and idiosyncrasies. These must be recognised and accommodated to cultivate and sustain recipient–donor collaboration.

A living example is when researchers from one of the donor institutions who compiled a case study on NWSC's capacity development brief. The article appraised the performance of the Corporation, outlining internal managerial efforts that had been undertaken to improve performance. However, a representative from a different donor country strongly attacked the authors of the article, complaining about the lack of acknowledgement of her country in NWSC's performance improvement efforts. When I read through the article, I found that it was purely technical, drawing managerial lessons from a positive case study. It had nothing to do with who contributed what and in what form.

Even the parent donor agency of the researchers was not acknowledged. It seems the strong discomfort was about which institution the authors were working for, rather than about the content of the article! The donor agency complained about the author's acknowledgement omission on the ground that 'if we hadn't financed/advised NWSC throughout all these years, you would not have had a story to report on'. That donor was, of course, right, but other donors were not acknowledged in the article either. However, this example shows that donors can be sensitive, so it is important to always bear in mind their sensibilities to prevent them taking offence or even causing diplomatic incidents.

The NWSC was on the right track. The donors had done it. They helped, advised and guided, to say the least. The Corporation will continue to require donor support (financially, technically and morally) for its long-term capital development projects in order to improve its services to the rapidly growing urban population. It is imperative for the NWSC to continue jealously guarding the hard-won donor confidence, support and goodwill; to nurse it and consolidate it for tomorrow.

We need to remember that donor confidence can only be sustained through hard work, innovation, enhanced performance, openness and accountability, the drive towards financial viability, professionalism, strong negotiation skills and results-oriented management.

Lessons learnt

- Donors are open to reason and persuasion provided the force of arguments is matched by results-oriented actions. They are more interested in performance than in setting conditionalities *per se*. Do not expect a blank cheque from them; they would be foolhardy to give their money without imposing conditions on how it should be spent. The challenge is to convert conditionalities into opportunities to yield handsome dividends.
- Donors' attitudes and policy positions are not set in stone. They will back the winning horse when they see one. So develop your negotiation skills, do your homework and present sound, viable and sustainable policy alternatives and programmes; go out of your way to cultivate donor support by deeds rather than by words, and in the end you will win the argument.
- Once convinced of your case, donors will provide wholehearted support in formulating new change management programmes and strategies. They will also sustain your change management momentum to consolidate gains and register further performance improvements for the good of shareholders and customers alike.
- Sustain donor confidence through transparency, accountability, efficiency and cost-effective, demonstrable results.
- Do not expect magic answers from donors. They may give a helping hand but they are not a substitute for internal capacity to initiate and sustain change in an organisation.
- Insist on a 'do-it-yourself' philosophy, which is one of the promising ways of sustainably benefiting from donor support.

Winning the Much Needed Goodwill

Donors partner with NWSC to share its experience: Dr William Muhairwe joining a panel of experts in Berlin, Germany, to discuss global challenges in providing water for all. From right, Sire Paul Lever, Dr William Muhairwe, Dr Uschi Eid, Messrs Michel Camdessus and Simon Trace.

Donors confidence in NWSC restored: The German delegation led by Dr Uschi Eid during their visit to NWSC (L-R): Dr A. Shire, H.E. Drs Muehlen, Uschi Eid, Mr Seubert, Dr William Muhairwe and Mr Sam Okec at the NWSC Head Office.

Attracting international attention: Dr William Muhiarwe (first left) guiding Mr and Mrs Bill Gates Senior of the Bill and Melinda Gates Foundation during their visit to NWSC.

Chapter Thirteen

In the Shadow of Privatisation

The Government

I have just received the great news that the National Water and Sewerage Corporation has received an international award for excellent management. I therefore hasten to offer my warmest congratulations and felicitations on this great achievement for all the workers, the Chairman, yourself, and all the Board. This is undoubtedly in great measure due to your inspiration, drive and vision. Africa has generally been known as a death yard for innumerable ventures, and where they have not atrophied, they have been reduced to shambles. Therefore the National Water and Sewerage Corporation serves as a beacon of hope for all of us... This must not however be time to relax in your laurels, but must serve as an impetus to greater heights of achievement.

Hon. Henry Kajura
Third deputy prime minister

This is a truly remarkable achievement and reminder to the corporate world that Uganda too can have world-class organisations.

Hon. Dr Nsaba Buturo
Minister of State for Information and Broadcasting

The author received these letters when he received the Golden Star of Management Merit Award.

Until the era of privatisation in the 1990s, public utility companies – water, electricity, telephones, etc. – were state owned. In theory, these enterprises, including NWSC, were driven by commercial and social-mission objectives. While they were expected to be profitable, making money for the government, they were at the same time duty bound to deliver affordable services to the public, especially the poor. However, in practice, more often than not, commercial viability and social-mission objectives tend to be contradictory because rendering gratuitous or subsidised social services is inconsistent with the quest for profitability.

Apart from inherent commercial and social mission contradictions, public enterprises have in the past been paralysed by political patronage, cronyism, government interference and government failure to pay for the services rendered. Board of directors and top managers would be appointed on the

© 2009 William T. Muhairwe. *Making Public Enterprises Work*, by William T. Muhairwe. ISBN: 9781843393245. Published by IWA Publishing, London, UK.

grounds of political support or connections rather than merit. Ministers and senior servants in parent ministries would direct public utility managers, on the telephone or through chits, to employ or promote certain 'recommended' individuals regardless of whether they qualified.

As a result, before the era of privatisation, the face of most public utility companies in Africa was 'ugly', characterised by over-staffing, inefficiency, loss making and poor value for money service delivery. Poor performance and insubordination among the 'untouchables' who enjoyed political patronage and protection were universal norms in these utilities.

In Africa, the role of the government in the performance of public enterprises has been indispensable for a number of reasons. To start with, as the sole shareholder, the government is responsible for the capitalisation and determination of policies and priorities in such utility enterprises. It is the governments that appoint boards of directors and top managers. Secondly, since they are the biggest consumers of public utility products such as water and electricity, it is imperative that they pay their bills on time, if the financial health of these enterprises is to remain buoyant.

However, governments' failure to settle such bills and on time has been one of the most endemic challenges facing public utility companies in Africa. Furthermore, as the supreme authority in the land, the government has overall responsibility for the economic and social welfare of its people, especially disadvantaged groups, and as such, it cannot let go of essential and sensitive services such as water, whether in public or private hands. That is why, for example, the Ugandan government has persistently pushed for the social-mission objectives of NWSC even when they negate the drive towards commercial viability. As such, water tariffs can never be reviewed without government consent and approval.

This chapter describes how, despite the ever-present shadow of privatisation, the government of Uganda has, in collaboration with the donors, facilitated and supported the NWSC change management programmes, from the 100 Days Programme to the IDAMCs, since 1998. This chapter details the contributions of individual government officials, especially in the Ministry of Finance, Planning and Economic Development, who, despite their initial scepticism about the wisdom or chances of success of NWSC, 'home-grown' reforms have been converted from critics into friends and allies. It also shows how the government has gradually shifted its position from the privatisation panacea to give the NWSC's internal reform momentum a chance. The fact that the NWSC has, through its change management programmes, become a 'world-class organisation' success story with 'world-class standards of management' and 'a performance-oriented culture' is a glowing tribute to the government's policy pragmatism, patience and encouragement since 1998. In short, the chapter manifests the actions, challenges and approaches that eventually helped to deepen NWSC actions and strategies into the mind of government.

In the eye of the storm

The beginning was not easy! The corporation I joined to lead in 1998 was in the eye of the storm that not only threatened to grind its services to a halt but to end its very existence as a public utility. The corporation was in the midst of an unprecedented debt crisis with various creditors literally knocking at, or worse still, closing the doors. As mentioned in chapter two, on my second day of reporting to office, I was stunned to find the NWSC headquarter offices under lock and key by URA. As if the closure of our offices was not bad enough, the Uganda Electricity Board was threatening to cut off power from Kampala Water and Sewerage Works and other upcountry stations unless the electricity bills worth UGX 2.8 billion were settled immediately. In addition, under the World Bank-funded Second Water Supply Project, cost overruns worth USD 10 million had been accumulated and the creditors were menacingly threatening to take us to court. Many other creditors, including the Uganda Posts and Telecommunications, were also knocking at our doors, insisting that their bills must be settled at once.

The creditors' demands and actions were certainly the worst crisis in my management career so far. Why were they demanding their 'pound of flesh' barely after I had assumed office? What had they been waiting for? Were these premeditated acts of sabotage? Did they want my head and, if so, what on earth had I done wrong to warrant their wrath? Didn't they know that we could not pay because our money, too, was tied up in government arrears? Indeed, my initial euphoria as the new MD was quickly being dampened.

In any case, how on earth did the URA expect us to pay their taxes when our offices were locked? When we did not have access to our accounts and bankbooks, files and safes? The outrageous actions by creditors angered and frustrated me. In my mind, the URA and other creditors were high-handed and insensitive. They were not giving us a chance to address their demands! But, however unreasonable and inconsiderate they were, it was no good crying over spilt milk. We had to look for a solution to the debt crisis and prevent the entire NWSC edifice from collapsing around us.

As a first step, I immediately got in touch with the Commissioner for Inland Revenue for help. She did not budge. She had instructions from above to close the office. Sorry, she bluntly told me, she could not help. She advised me to pay up or else appeal to her boss. Next, I appealed to Mr Elly Rwakakooko, then the URA commissioner general, to authorise the opening of our office, at least temporarily, to allow us time to address the issue of paying the taxes. He was very sympathetic. He promised to take up the matter with the secretary to the Treasury at that time, in an effort to resolve the problem. Mr Rwakakooko kept his word. That afternoon, the URA opened our offices. It was a great and welcome relief but no solution to our debt problems. But at least we had breathing space. We could now access our accounts and pursue different scenarios, to ensure that the creditors, including the URA, got off our backs.

In the meantime, I had contacted my line ministry for solace and advice. I requested the Ministry of Water, Lands and Environment to exert pressure on the Ministry of Finance, Planning and Economic Development to prevail on URA to negotiate a realistic payment schedule of all outstanding taxes without disrupting NWSC operations or starving the Corporation of financial resources to sustain delivery of water and sewerage services. Our line ministry promised to take up the issue with the Ministry of Finance at the ministerial and administrative levels. I was also advised to contact Ministry of Finance officials with specific proposals about how the tax impasse could be amicably resolved.

Accordingly, I summoned my Director of Finance and sought his advice. He informed me that although the financial situation looked terrible, it was not hopeless because many debtors, including government and its public enterprises, owed the Corporation lots of money. He pointed out that the government alone owed the Corporation more than UGX 11 billion, some of which included the infamous Departed Asian Properties Custodian Board bills worth UGX 2.8 billion, which had been outstanding for many years. If these arrears could be settled at once, the Corporation would comfortably pay off URA taxes and all other creditors, and remain with some cash to spare. This turned out to be the Corporation's lifeline to resolving the debt crisis. I therefore decided to press the government to meet its obligations, as a water consumer, in order to enable us meet our own.

Armed with this precious information, I sought an appointment with Mr Emmanuel Tumusiime-Mutebile, then the permanent secretary and secretary to the Treasury, to explain our predicament and seek his assistance. I briefed him about the URA closure of our offices, which was a source of embarrassment to me personally as I had barely settled into my new job. I cited the billions owed to the NWSC by the government and pointed out that this debt was more than enough to settle all our outstanding bills.

Meanwhile, Hon. Henry Kajura, my line minister, had taken up the issue with his colleague, Hon. Gerald Ssendaula, then the minister of Finance. Accordingly, in a memorandum signed thereafter, it was resolved that the government would immediately settle the Departed Asians Properties Custodian Board debt, and budget for all other outstanding bills. It was also resolved that pending the government's clearance of outstanding arrears, neither URA nor any other public entity would require NWSC to settle its bills.

In retrospect, the debt crisis was a blessing in disguise. By giving me an opportunity to address the crisis as soon as I took over the NWSC management thus keeping it financially afloat, I had passed my first major test on the job. I had shown the will and the stamina to take the bull by the horns before finding my feet on the ground; I had persuaded the URA and the government to come to our rescue at the time of need. The crisis paved the way for the gradual clearance of the creditors' bills, including those of the former Uganda Electricity Board (UEB) and the Uganda Posts and Telecommunications Corporation (UPTC), in addition to the terminal benefits of staff members laid off as part of the restructuring process within the NWSC.

The debt crisis marked the beginning of a proper working relationship with the URA and eventually led to NWSC becoming one of the best taxpayers; ranked among the largest 15 in the country. Specifically, on 9 December 2007, while the URA was hosting a Taxpayer Appreciation Day, NWSC was given the URA Vantage Award for exemplary tax compliance in the category of Electricity, Gas and Water Sector. Therefore, in a way, the debt crisis was an induction process that initiated me into the inner workings of the government and taught me how to make a winning case for the Corporation and to cultivate government confidence and support. This useful experience served as a launching pad for what has turned out to be a fruitful relationship between the NWSC and its sole shareholder.

The issue of cost overruns was, however, much more difficult to handle than the local creditors. At USD 10 million, the cost overruns were more than our combined local indebtedness – and involved many more stakeholders. Apart from the Second Water Supply Project creditors who were demanding their money from the NWSC as the implementing agency, there were the World Bank and the government of Uganda, which had negotiated and guaranteed the project credit.

It was not easy to establish how the cost overruns had been accumulated. The corporation did not have the money to clear the cost overruns, and neither the World Bank nor the government was willing to pick up the bill! With the creditors impatiently demanding 'their pound of flesh' and threatening to go to court, the cost overruns were building into an incalculable dilemma. The office of the Inspector General of Government (IGG) and security agencies had already got wind of the cost overruns, and was busy investigating whether there had been something fishy in the procurement process of the Second Water Supply Project, and the awarding of the contracts.

Once again we had to turn to the government for help. We held several meetings with officials in the Ministry of Finance and eventually persuaded the government to come to our rescue. The government realised that paying the overruns on our own would bankrupt the NWSC and culminate in the shutdown of its operations.

For the government and, indeed, the customers in the urban areas under NWSC, the closure of a utility that delivered water and sewerage services was too ghastly to contemplate and therefore unacceptable because 'water is life'. But since the government itself did not have a contingency budget provision to settle the cost overruns, the challenge was to source money from elsewhere. And the only option was to persuade the World Bank either to pick up the bill or to consent to the re-allocation of funds from another project.

Fortunately, the World Bank was concurrently funding the Small Towns Water and Sanitation Project (Part B), a project that had adequate unspent funds. In the end, the government and the World Bank agreed to amend the Development Credit Agreement for the Small Towns Water and Sanitation Project in order to reallocate USD 10 million from it and settle the Second Water Supply Project cost overruns. This government intervention saved the NWSC from bankruptcy and gave us an opportunity to concentrate on change management programmes without worrying about troublesome creditors or debilitating debt payments. By coming to our rescue to resolve the cost overruns issue, the government had proved that it was a friend in need – not once but twice.

In the shadow of privatisation

In 1998, the privatisation of the Corporation, as indeed of other public utilities, looked like a foregone conclusion. Under the Uganda Structural Facility Policy Framework Paper 1998–2001, which sought to build a self-sustaining and private sector-driven modern industrial economy, the government was committed to the divestiture and privatisation of all public enterprises, including the utility companies. Under the Public Enterprises Privatisation and Divestiture (PERD) Statute, the NWSC was listed as one of the public enterprises earmarked for privatisation.

The corporation was listed in the Class B category in which its assets were to remain vested in government because the water and sewerage services were considered too sensitive to be completely divested. By allowing and encouraging private sector participation in the delivery of water and sewerage services, the government hoped to phase out the drain on government resources, attract private investment, improve efficiency and catalyse the recovery of NWSC.

The government in its policy framework spelt out the steps that would culminate in the privatisation of NWSC. These steps included, among others, reviewing the water sector, exploring the possibilities and practicalities of private sector participation, allowing the NWSC to make tariff changes to cover marginal costs and targeting short-term government funding to stimulate the recovery of the Corporation, pending privatisation. In the meantime, the operations of the Corporation were to be closely monitored by the Privatisation Unit of the Ministry of Finance in accordance with the provisions of the PERD Statute.

In addition, the Ministry of Finance set up the Utility Reform Unit (URU) to oversee and monitor the utility reform process in all public utilities, including NWSC, in preparation for privatisation or private sector participation. These policy initiatives meant that the NWSC had henceforth to serve two government masters – the Ministry of Water, Lands and Environment, which was in charge of the water sector, and the Finance ministry, which was responsible for overseeing the process of privatisation.

Although the government was committed to the privatisation of public utility enterprises, including NWSC, it was not sure how to translate this policy into reality. In 1997, the government, through the Finance ministry, directed the Water ministry to prepare a framework for the reform of the water sector, and to pave way for private sector participation. When I joined the NWSC in November 1998, I found a copy of this directive on my desk, and the Water ministry was grappling with the most appropriate approach to carry out the reforms. Although the reform of the water four sub sectors, i.e. urban, rural, water for production and water resources, there seemed to be a sense of urgency to tackle the urban water sub-sector first.

As a result, a series of tripartite meetings were held between officials of the Ministries of Finance and Water and management of NWSC. This culminated in the engagement of a consulting firm Consult 4 Pty Ltd of South Africa to study the water sector and to make appropriate recommendations regarding the modalities of privatising the delivery of water and sewerage services in urban centres of Uganda. The cost of the study, worth USD 1 million, was borne by NWSC from one of its World Bank-funded projects.

Personally, I was sceptical about the government reform strategy; I had reservations about the necessity of a study at a cost of one million dollars to address the problems of the water sector, which were already common knowledge. Although I accepted that in its privatisation strategy, the government was acting in good faith, and out of frustration, because of the poor performance of public utility enterprises, I was also convinced that the option of internal corporate reforms had not been given adequate attention.

Our SWOT analysis had shown that the challenges facing NWSC were by no means insurmountable, and we had already initiated the 100 days Programme to address some of them. And as has been consistently pointed out in this book, the Corporation had enough local skills to initiate and sustain the process of internal reform. What was needed was focused management and government support to turn the Corporation around and catalyse its recovery. Since the shortcomings of the water sector were already well known, spending a whopping one million dollars on a study was not an effective way of utilising money at a time when the Corporation was hard pressed to make ends meet.

However, being new on the job, I did not want to appear presumptuous and obstructive. I did not want to be accused of frustrating government policy of privatisation or standing in the way of *perceived prospective investors*. Accordingly, I reluctantly went along with the study; we gave them our full cooperation. If the study made groundbreaking discoveries and helped us improve the delivery of water and sewerage services, well and good. After all, the stakeholders were driven by one common purpose, namely, to turn around and build a viable and sustainable water industry.

In the meantime, we were determined to pursue our strategy of internal reforms, pending the outcome of the study. Fortunately, by the time Consult 4 submitted its report, our change management programmes were beginning to yield positive results. This was a good start for us; the government was beginning to see some fundamental changes. As a matter of fact, the government started paying serious attention to our alternative internal reform strategy. Thus, although the shadow of privatisation hanging over NWSC had not completely receded, it was no longer as dark and menacing as it had been in 1998. The government's initial change in attitude gave us breathing space to continue our internal reform momentum – and eventually won the government over to our change management strategy. This was confirmed through a Cabinet Decision, which directed that the implementation of the water reforms should utilise, 'as much as possible, the current sector capacity and institutions and ... taking into account the private sector participation market situation, the cost *implications to the consumers* and the ongoing reforms in NWSC'. The Government was in effect underscoring the momentum of the NWSC internal reforms.

Actions start speaking louder than words

Since the government was already committed to the privatisation policy, we were conscious of the fact that selling an alternative internal change management strategy would be no walkover. Nevertheless, we were determined to try to persuade the government to give us the benefit of the doubt, regardless of the difficulties along the way. To do this, we focused on actions rather than words. To start with, in the corporate history of post-colonial Uganda, no management had ever committed itself to turning around the fortunes of a corporation within 100 days.

So our audacious 100 days Programme deserved to be given a chance. In order to sell our change management strategy, we resolved to keep government officials fully informed about whatever we were doing and to involve them at every stage. For this reason, we invited Hon. Henry Kajura, then minister of Water, to launch the 100 days Programme. He eloquently endorsed it as an expression of an 'innovative spirit' and promised that his ministry would mobilise the government to our change management efforts. Mr Bill Kabanda, then the permanent secretary of the same ministry, was equally supportive. This pat on the back gave us the confidence to continue our internal efforts.

At the closure of the 100 days Programme, we invited Hon. Othieno Akiika, then the minister of state for Water. He reiterated the government's commitment to the success of the NWSC change management programmes and praised the devotion and foresight of the Board and management in conceiving and successfully implementing the Programme.

Hon. Akika said he looked forward to the day when the NWSC would become a model of successful 'home-grown' reform to other public utility companies in Uganda and elsewhere in Africa. Similar support for our change management initiatives came from the Ministry of Finance, which, as already mentioned, was overseeing the process of divestiture and privatisation of public utilities. The challenge was to sustain this hard-won government confidence by proving that our change management strategy was equal to, if not better than, the privatisation alternative. And, indeed, the facts on the ground for the last 10 years have proved that we rose to the occasion and won.

In the following sub-sections, I highlight a number of actions and policies that we undertook to underpin the change management programmes. The key policy makers and/or technocrats that played a key supportive role are mentioned in all sections.

Tariff review: When I took over the mantle of leadership in the NWSC in 1998, I found that a consultant had been engaged to review the existing water tariff with the aim of increasing revenue and cultivating a customer culture of 'willingness to pay'. In his report, the consultant made a number of recommendations, the most prominent of which was to increase the water tariff by 60% on the grounds that such increment would enable the Corporation cover the running costs and to cope with the burden of debt servicing. Of course, this was welcome news to the NWSC, and the government was willing to support the tariff increment after weighing the implications of such a sensitive course of action.

Unfortunately for us, the consultant's tariff recommendations leaked to the press, and provoked stiff opposition in parliament, before the government and the NWSC could reach a consensus on the appropriate course of action. Parliament took the stand that the water tariff was already too high for most consumers and, that therefore, there was no justification for the increment. The fact that the increment was necessary to meet the running costs, to improve and extend water and sewerage services and to service the long-term debt that the government had passed on to the Corporation at the commercial lending rate of plus 3% fell on deaf ears.

The government was reluctant to force an unpopular measure through parliament and risk possible fallout. Accordingly, the proposed water tariff had to be shelved in order to avoid antagonising parliament and the public. In public utility enterprises such as NWSC, management cannot always have a free hand; it must be prepared for compromises or even setbacks and live to fight another day.

Despite the objections of parliament, the circumstances surrounding the tariff could not be brushed away without serious financial and operational consequences for the Corporation. As a compromise therefore, and in the interest of achieving a win-winsituation, the Government allowed for minor changes at this stage, which included the elimination of the minimum charge, the reduction of the connection and reconnection fees and the introduction of a service charge which was aimed at defraying the costs associated with the reduction in the connection charges.

These changes notwithstanding, it was noted that the existing water tariff had been fixed in 1994 and, eight years down the road, it had never been reviewed. Our calculations showed that the real value of the tariff had been eroded by 45% owing to inflation (local and external), the UGX exchange rate depreciation and the cost of electricity. As a result, the gap between the value of the tariff and the cost of water production and distribution inputs was steadily widening.

There was, therefore, a compelling case for reviewing the tariff in order to enable the Corporation make ends meet. In 2002, with the backing of the World Bank, we presented our case to government, pointing out that for the Corporation to cover all its costs, including debt servicing, the tariff increment would be 135%. But since such drastic tariff increment was not palatable either to the government or the water consumers, a full cost-recovery tariff was out of the question. In any case, since the government had already provided a debt relief through a debt freeze to the Corporation, we could make a lower and politically acceptable tariff increment.

To avoid public outcry against the cost of water, which would have negated the very principle of the government's objective of ensuring 100% safe water coverage, we proposed to the government *to index the tariff annually against further erosion.* The advantage of tariff indexing was that small annual increments would be less dramatic and more acceptable to the public more than a once-for-all cost-recovery tariff increase.

The Water ministry, in collaboration with that of Finance, accepted the merits of this cautious incremental approach and, accordingly, in April 2002, submitted a memorandum to the cabinet proposing annual water tariff indexing. This proposal was approved and, on 12 April 2002, Hon. Dr Ruhakana Rugunda, then the Water minister, issued a tariff instrument amending the general water rates to allow for annual indexing. *This was another landmark support from the government.*

The new connection policy: Borrowing a leaf from modern tenets of marketing and customer care, in 2004, the Corporation adopted a new water connection policy in which access to water services was eased. The modern tenets simply opines that, *why should you prevent a customer from entering your shop if you want to sell your product?* Prior to the implementation of the simplified connection policy, new customers were required to pay a connection fee and to supply all materials such as pipes and fittings between the water supply mains and the meter. This policy was flawed not only for the fact that it hindered access for potential customers, especially the poor, but it also made it difficult for the Corporation to detect and reduce the incidence of water losses between the supply mains and the customer's meter. Therefore as a matter of policy, in 2004, the Corporation decided to assume responsibility, including meeting the financial costs of trenching and providing connection materials to all new customers within a radius of 50 metres from the NWSC service mains. The cost of connection beyond 50 metres was charged to the customer but the actual materials supply and connection was undertaken by the NWSC. The policy resulted into the standardisation of materials used for connection and strengthened quality control.

The policy received overwhelming support from the government and a number of key stakeholders. For example, Maj. Gen. Kahinda Otafiire, the then Water minister, warmly approved it as 'pro-people'. Accordingly, the new policy was published in the *Uganda Gazette* on 23 July 2004. As a result, the number of new connections immediately shot up from an average of 1,200 to an average of 2,000 per month, suggesting that all along there had been a huge suppressed demand for water. Since then, government officials in other ministries have, at various fora, praised the Corporation's efforts in extending water services to more customers in areas under its jurisdiction.

Government subsidies: Given the fact that the NWSC has social mission obligations as well as commercial ones, it has, since its inception, required and deserved government help in one form or another. When I joined the Corporation, it was not receiving direct subsidies from the government. Instead, it was receiving government-guaranteed loans, which over the years had culminated in a huge debt burden. Fortunately, all stakeholders, including the government, recognised that the Corporation could not be turned into a viable commercial entity, while at the same time service its debt and carry out its social mission obligations.

It was therefore agreed that the Corporation would concentrate its hard-earned internal resources on the pursuit of its commercial viability goals. Debt-servicing payments were suspended and, henceforth, the Corporation received grants rather than loans. In addition, within the framework of the NWSC/ government of Uganda performance contract, the government assumed responsibility for funding the Corporation's social mission activities.

The NWSC has benefited from government support through targeted subsidies for the social mission obligations and for the rehabilitation and expansion of water services in the new towns as pointed out in earlier chapters. It should, however, be noted that government subsidies were purely for investment purposes and not for operations and maintenance. The latter have always been borne by the Corporation. Besides this, the government has provided counterpart funding for donor-funded projects such as Gaba III and Entebbe Water and Sanitation Project.

In these projects, the government counterpart funding covered certain categories of expenditure, including taxes and a 5% contribution to capital works. All in all, since 2003, the government counterpart funding to donor-funded projects amounted to UGX 10 billion. In addition, government allocated funds to the Corporation to provide water services to the urban poor in accordance with its social mission obligations.

Making NWSC a procurement entity: The recognition of NWSC as an autonomous procurement entity has been another positive government measure, which has speeded up the process of purchasing inputs for operations. When I joined the NWSC, all government procurements, including those of public utility companies, were centralised in the Central Tender Board. This meant that the process of procurement was bogged down in bureaucratic red tape, resulting in undue delays in the purchase of essential materials and supplies, the stalling of projects, poor maintenance and repairs and, above all, poor performance in service delivery.

To overcome these procurement deficiencies, the government decentralised the procurement function to the contracts committees of line ministries in 1999. Although this was intended to speed up the process of procurement, it was unfortunately akin to falling from the frying pan into the fire. The procurement delays got worse rather than better, which, given our sense of urgency in implementing the change management programmes, was unacceptable.

We, therefore, appealed to the government for help. And on 31 January 2002, the Corporation was granted an autonomous procurement status. Since then, the NWSC has been operating as a separate procurement entity, which has greatly reduced bureaucracy and improved the buying and delivery of inputs.

Intervention in the fish industry saga: Although the role of government in the recovery of the NWSC since 1998 has been generally supportive and positive, it has at times been fraught with difficulties and acrimony. For example, in 2001 we discovered that one of our big customers, involved in the processing of fish for export, had tampered with the water meter system, resulting in underbilling for a long time. This discovery triggered a bilateral dispute that ended up involving the Uganda Manufacturers' Association (UMA), the Fish Processing Association and, ultimately, precipitating government intervention.

When we discovered that the company in question had tampered with the water connection, we deliberately stayed legal action against it, and instead raised a demand note of UGX 31 million arising out of assessed water consumption for six months when the premises were illegally connected. This outraged the company, which demanded that not only should the bill be written off, but also that it should be permitted to install its own water borehole because our water tariff was too high and unaffordable. We stuck to our guns, reminding the company that it had a legal obligation to settle its bill. We also pointed out that installing a private water source was contrary to the urban water sector policy of providing safe water.

The UMA and the fish processors took up the case of the company, accusing us of *frustrating* investors and exports by rendering expensive services. Under the NRM government, any real or imagined frustration of investors is a serious accusation that was tantamount to a criminal offence because investors are considered not only as the goose that lays the golden eggs but, more importantly, as the *lifeblood* of Uganda's quest for industrialisation and modernisation.

Not surprisingly, therefore, the Office of the prime minister intervened in the dispute, not only demanding an explanation from the Corporation but also insisting that a quick solution to the water problems be found so that the operations of the fish processing company are not threatened with imminent interruption. Fortunately, we had all the facts to prove that the NWSC tariff was competitive in eastern Africa; that on average, the cost of water accounted for a mere two percent of the total costs of production for most industries in Uganda.

Indeed, the industries were consuming subsidised water. If the Corporation were to charge a full cost-recovery tariff, the consumers, including the company in question, would have to pay two and a half times much money. Luckily, Prof. Mondo Kagonyera, then the minister of General Duties in the premier's office, who was asked to handle the issue, had known the history of the problem associated with the fish industry. The EU guidelines were very clear. They did not allow any water other than treated potable water to be used in fish factories, which were exporting to Europe. He put up a very spirited fight in defence of the NWSC policy of not allowing untreated water from boreholes on the premises of the fish factories. Given these indisputable facts, the fish processors had no valid ground for complaining and, accordingly, the matter was laid to rest. Likewise, the issue of illegal connection charges was concluded and it was resolved that the company pays in full.

Rethinking the tariff policy: To add to the tumult of the case with the fish processors, the UMA, as if in a planned attack, also expressed their unhappiness with the differential structure of the water tariff system in which larger consumers paid more to subsidise the urban poor on the socialist principle of affordability. The Association argued that this system was inequitable. They demanded that, as large consumers, industries should enjoy substantial discounts in their water bills in order to keep down the overall costs of production and to maintain a competitive edge for their exports to foreign markets. The manufacturers threatened to look for alternative sources of water; and, indeed, many of them applied to the Directorate of Water Development in the Ministry of Water to sink their own boreholes.

At this point, it should be noted that from the outset, the initiators of the NWSC tariff had this *socialistic and conservationist idea* of the bigger consumers, presumably richer, paying more and thus subsidising the poor. It was also meant to *conserve water resources,* possibly due to the undeveloped water systems at that time, and to discourage consumers from wasting water.

For once, the government sided with the complainant, which was, however, also a blessing in disguise for NWSC. President Museveni personally intervened and directed the NWSC to review the tariff and address the manufacturers' concerns. His concern was that the rich in Uganda are already overburdened. He argued that they pay taxes, provide employment and contribute directly and indirectly to the country's economic growth. His question, therefore, was *what is the rationale of the big water consumers subsidising the poor?* The job of subsidising the poor lay with the government, and not with the large manufacturing concerns such as fish processors or Coca Cola. His intervention averted a crisis. As already pointed out, there were already several applications with the Directorate of Water Development and in the Water ministry for borehole use! This would have dealt a blow to our revenues.

Since industrial and commercial consumers, with substantial and economic clout, account for 33% of NWSC's revenues, it was imperative that we acted fast to pre-empt the quest for alternative sources of water, which would have eroded our revenue base. Partly owing to the President's directive but also out of self-interest, we decided to rebalance the tariff by exempting large industrial and commercial consumers from the annual water tariff indexing, with a view to reducing cross-subsidies to domestic consumers and eventually arriving at a fair tariff for all parties, in future.

In addition, the NWSC, on its part, pledged to maintain the quality of water and to render free technical assistance to industries in regard to rational and cost-effective ways of using water. By so doing, we restored the large customers' willingness to pay and in tandem complied with the government policy of supporting investment initiatives.

Debt restructuring: Government support culminated in the restructuring of the NWSC long-term debt into equity in the year 2008. The process of restructuring the debt was a protracted one, in which we had to use and justify our operational and financial predicament in light of the overburdening debt. As I have mentioned in this book, empowerment of staff and enhancing staff capacities is one of the pillars of a strong organisation. This is because, through the empowerment of staff, an organisation can articulate and present its case and aspirations for the future to different stakeholders, including the government. The debt restructuring process was indeed a manifestation of such an incidence.

Briefly, the genesis of the debt restructuring stretched back to 1999 with the formulation of the first performance contract with the government. During the years 2000 to 2006, the government agreed to freeze the NWSC debts, which implied that the NWSC was not obliged to make any debt service payments over that period. This notwithstanding, we knew that the debt continued accruing on our balance sheet over the years of the freeze. At that time, we presented four basic reasons for the freeze, which in brief included the fact that the NWSC tariff did not involve full cost recovery, secondly, by the nature of water projects, many of the projects undertaken with loans were supply driven with a strong social element, and therefore their revenue streams were not able to repay the loans, thirdly, the gazetting of new towns to the Corporation by the government imposed a financial burden on the Corporation, and finally the NWSC had an investment plan which it was partly financing from its own internal resources.

In 2006, at the end of the second performance contract, the government was satisfied with the performance of the Corporation, and also recognised the need for the NWSC to expand its services further to the fast-growing urban centres. However, due to financial constraints, the government urged the NWSC to seek alternative sources of funds to finance its critical projects. The government's reason for this was that the NWSC was a strong company that generated positive operating cash flows, and therefore was in a position to go to the market and leverage its cash flow to accelerate growth. One of the critical projects we had to finance at that time was the extension of the water intakes in Kampala and Jinja which was necessitated due to the significant drop in the lake levels.

It was at this point that we pointed out that we still had the debt commitment on our balance sheet which jeopardised our ability to access short-term loan financing from the financial markets. Secondly, we also pointed out that as a Corporation, we had to build a history of profitability and creditworthiness before being able to access market finance.

It was with this renewed spirit and understanding with the government that parliament passed a resolution on the 14 February 2008 converting the NWSC's long-term debt amounting to UGX 154 billion into equity. In return, the NWSC agreed to carry out an investment programme from its own internal resources totalling UGX 168 billion over the next five years. It was agreed that part of the funding would be used for the extension of raw water intakes in Kampala and Jinja.

We acknowledge the support provided to us by the Ministries of Finance and that of Water and Environment in pursuing the debt restructuring process. As the saying goes, the devil is in the detail, and this process certainly took a lot of patience, and back and forth discussions. In particular, we thank our minister of state for Water, Hon. Ms Namuyangu, and the minister of state for Privatisation, Hon. Chekamondu, who tirelessly defended the NWSC during the plenary session of Parliament when the proposal was presented. We also acknowledge the support of the different members of parliament who stood by our cause, especially those on the Committee of the Economy and Natural Resources. It was indeed a real breakthrough for all of us and a great relief to the Corporation.

Friends and allies

However supportive it is, the government is invariably *a bureaucratic maze of red tape*, procedures and regulations, negotiation and consultation, and reconciliation of competing inter-departmental and inter-ministerial interests. In such a maze, the process of decision making is painfully slow and frustrating, with exasperating correspondence, arguments and clarifications moving back and forth for months or even years.

Without patience, perseverance and, above all, friends and allies, good programmes or causes can be lost in the *bureaucratic maze*, seriously impacting on the quality of service delivery. Fortunately, since 1998 the NWSC has enjoyed the support of many friends and allies in various government ministries and departments; they have passionately supported and understood our cause through the maze. Sometimes, they have gone out of their way to facilitate the smooth and successful implementation of the change management programmes. Let me highlight their contributions in the following sub-sections.

On the political scene: Since 1998 NWSC has been lucky enough to fall under hardworking line ministers. Hon. Henry Muganwa Kajura, the first among them, was a quiet, thoughtful and resolute politician with a distinguished record in public service. He was instrumental in supporting our initial change management programmes and keeping at bay the axe of privatisation. Even after he left the ministry, he remained a friend of the Corporation and championed its causes in cabinet whenever it was necessary. His congratulatory message quoted at the beginning of this chapter bears testimony to his enduring friendship.

Hon. Dr Ruhakana Rugunda, an amiable and long-serving minister in the NRM government, was equally supportive. I vividly remember his robust defence of the Corporation against allegations relating to lack of transparency in awarding a particular contract by a certain ambassador accredited to Uganda. This incident happened as we were shortlisting companies for a project in the Kampala Water and Sewerage Service Area in which we received a total of 21 expressions of interest. However, we required only five eligible companies for our shortlist, who would be asked to submit detailed bids.

Because of the sensitivity of the project, we recruited a foreign consultant to assist us in carrying out the exercise in a transparent and fair manner. The consultant worked out the shortlisting criteria and the relevant stakeholders, including the donor countries involved, approved it. The shortlisting process then went ahead with no hitches.

However, one small company, which had also submitted an expression of interest, was eliminated by the criteria. Thinking that it could use back door and patronage donor threats, this company cried foul to its embassy hoping that the ambassador would use his *political and diplomatic influence and solid friendship* with the minister to convince him (Dr Rugunda) to overturn the results and include his company on the shortlist. To the ambassador's surprise, on the day the minister was to meet him, the minister also invited me to attend the special meeting. I went with my team and the foreign consultant who was the team leader of the shortlisting exercise. After an elaborate presentation, Dr Rugunda told the ambassador that there was no way one could overturn the decision of a transparent process to cater for the selfish interests of one company. That put an end to the matter!

Maj. Gen. Otafiire, who replaced Dr Rugunda, gives the impression of being a tough and abrasive person. However, beneath the hard and forbidding exterior is a sensitive and considerate person who supported the Corporation through thick and thin. In support of the NWSC, he always questioned the rationale of privatisation when there was already a winning team in place. For example, in one of his speeches at the launch of the ISO certification of NWSC towns in Jinja, he stated thus:

> In the past, I had only heard of private companies, especially those associated with Uganda Manufacturers Association, as being ISO certified. Today you are breaking the monopoly of the private companies to become one of the first public parastatals to be ISO certified. I wish to thank you for disproving theories about the performance of public companies.

Hon. Akiika, Hon. Maria Mutagamba and Hon. Jeniffer Namuyangu Byakatonda, successive ministers of state for water, have also been wonderful champions of the NWSC cause. Hon. Mutagamba, who during her time as AMCOW president, and considered the mother of the water sector on the African scene, has been a wonderful and relentless supporter of NWSC programmes not only in Uganda but also in international fora. In her interview with one of the weekly papers, on her assumption of the AMCOW chair, she expressed continued support to the NWSC with the following remarks:

> ...because of this Office [sic] "AMCOW", the international community will also have a lot of focus on Uganda. Uganda will get a lot of exposure. We are opening doors for Ugandans to market their skills and expertise. The National Water and Sewerage Corporation has come up with an indigenous form of privatisation, and we have started helping other African countries to see how our system works, and whether they can adopt it. I am proud of National Water and Sewerage Corporation. (the *Sunrise* newspaper 19–20 November 2004)

Other ministers of state in the line ministry were equally supportive. The then minister of state for Lands, Mr Baguma Isoke, was soft spoken and never forgot to put in a word of encouragement and support for the NWSC change management programmes whenever he had a chance to talk about the Corporation, and the water sector in general. At our gathering to mark 10 years of Gaba II operations, he informed the public, 'Government is very proud of the NWSC for breaking the myth that state parastatals are not profitable.'

He continued thus:

> The Ministry is aware of the good performance of the NWSC and I have come to pass on the congratulatory message. I want to commend Dr Muhairwe, the Managing Director, for compounding the stretch-out programme and urge him to teach others how best they can maximise profits while minimising costs. (*Daily Monitor,* 16 December 2002)

Lt. Gen. Jeje Odongo, who was the minister of state for Environment at that time, was also always supportive of the NWSC. He was one minister who would, at the shortest notice, officiate at any function of the Corporation even though he was primarily in charge of environment affairs. He was an eloquent speaker whose words always rang true. At the launch of the Mukono water supply project, he pointed out that NWSC was a government company that delivers. In his speech, before President Museveni and local dignitaries in Mukono District, he stated:

> As you are aware, your Excellency, the reform of utilities has been high on the Government's agenda in order to improve their performance. However, as the reforms progress, we have taken note of the internal reforms in the NWSC and are currently re-evaluating the approach to the reforms.

Hon. Namuyangu's support was remarkable in many ways. For example, in 2007–2008, she worked very hard, under the guidance of her senior minister, Hon. Mutagamba and the permanent secretary, Mr David Obong, to make sure that the cabinet and, subsequently, parliament approved a proposal to convert NWSC debt into equity. This move significantly enhanced the Corporation's capacity to access finance from the capital markets for investment. All these pointed to a ministry that wanted the Corporation to grow into a mature, independent and self-reliant institution.

Technocrats lend a hand: While ministers are useful in giving general political support and guidance, the actual donkeywork is usually done by civil servants. For example, the PS of our line ministry and his team offered invaluable support to NWSC in general and to me in particular. Despite his heavy work schedule, Mr Kabanda was always willing to spare time for us. Combining softness and frankness, he has been my inspiration behind the scenes. His weekly spiritual meetings with me, and the Christian fellowships with a few of my colleagues, always inspired me to do more for the Corporation. He gave me support that many parastatal managers would envy, always guiding, encouraging and urging us to go on even when our spirits were low. As already pointed out, his successor, David Ebong, continued supporting NWSC activities unreservedly.

In the hierarchy of the said ministry, we also got immense support from Ms Luzira, the undersecretary, who at one time was an NWSC Board member. This 'iron lady' was always objective and reliable. She was a technocrat who always stood by her word. Ms Luzira was actually at the forefront of all supportive MWLE policies concerning the water sector. She was later chosen as the chairperson of the Performance Contract Review Committee, which was established to monitor the Performance Contract between the government and the NWSC.

She consistently backed NWSC decisions and corporate autonomy. She was receptive to our change management programmes, and indeed witnessed some of our traditional get-togethers as the NWSC water family. She also supported the NWSC's quest to be granted the status of a procurement entity, being the chairperson of the ministry's contracts committee. In her personal assessment of the NWSC under the performance contract, she remarked that the Corporation was doing well and that *the internal reforms needed to be encouraged and treated with care.*

Although the NWSC functionally falls under the Water ministry, owing to the projected privatisation, the Corporation dealt more closely with the Finance ministry. Accordingly, many officials in the Finance ministry, from the PS to desk officers, some of whom we initially mistook for bureaucratic adversaries, turned out to be reliable and indispensable allies in our change management programmes over the years.

Mr Emmanuel Tumusiime-Mutebile, then the permanent secretary and secretary to the Treasury, who later became the governor of the Bank of Uganda, was one of the most powerful brains in the quest for enhanced NWSC efficiency, productivity and profitability in the wake of SEREP. This astute, clever and clearheaded public servant, free of intrigue and manipulation, understood the efficiency gains that NWSC had made with the 100 days Programme and SEREP.

To ensure that the reform momentum was sustained and enhanced, Mr Tumusiime-Mutebile was the driving force behind the conclusion of the first performance contract binding the government and the NWSC to the achievement of specific objectives and targets. He also backed the argument in favour of government inheritance of old NWSC debts to make the Corporation attractive for privatisation, pointing out that no private investor would want to pick up such a burden. He also played a vital role in persuading government ministries and departments to settle their water bills on time, which was in part intended to prepare for the anticipated strict bill payment regime that would accompany privatisation.

Of course, in fulfilling his mandate the PS/ST was supported by his lieutenants in the names of Mr Chris Kassami, the deputy secretary to the Treasury, who later became a full PS/ST, and Mrs Mary Muduuli, then the Director Budget. Mr Kassami, soft-spoken as he is, was the man behind the scenes regarding all finance policy matters. He exhibited a positive attitude towards the NWSC in various fora. He led the GoU delegation to Germany to negotiate a new financial assistance package for Uganda in which NWSC would be one of the beneficiaries, and made sure we got all the money we required for the expansion of our water supply and sewerage infrastructure and information management systems. Mr Kassami never failed to impress it upon the donors, in several fora, that the NWSC was doing well and needed more support, and was not dismantling. The respected Kassami always carried the day.

Mrs Muduuli, who later joined the World Bank, also provided invaluable support. Her strict and astute character said it all. She talked tough and her word was undisputed at all fora. She always ensured that sufficient funds were provided as counterpart contributions for all NWSC projects. She was also at the centre of the negotiations with World Bank officials which culminated in the signing of the memorandum of understanding between the NWSC and the Government, to pay water bills. In collaboration with Mr Emmanuel Nyirinkindi and Mr David Ssebabi (whose other roles and support are described later), she worked round the clock to ensure that the memorandum was duly signed and honoured. This decision compelled the government *not only to pay its bills on time, but also in advance, which is a rare practice worldwide.* She insisted that all ministries budget adequate funds to cater for utility bills, and that any defaulting government department would be 'disconnected' from our water supply.

Other significant support came from the outspoken Mr Keith Muhakanizi, the knowledge and data bank of the Finance ministry. At that time, he was the director Economic Affairs, a department in charge of economic monitoring and provision of advisory services to the minister. He later became deputy secretary to the Treasury. Despite his radical views on the performance of parastatals and the need to sell them off as quickly as possible, Mr Muhakanizi was privy to our internal reforms and wholeheartedly supported us in our quest to improve service delivery.

He, being a cattle farmer, argued along the lines of those who believed in *fattening the cow first before selling it*. He never failed, however, to add that the owner of the cow could change his mind if it provided the milk that was required and also produced offspring that could make the owner richer. Regarding government economic policy papers, the PEAP and the Medium Term Budget Framework, for example, he never failed to advise that the responsibility for investment in the water sector should remain under government auspices *as long as the water tariffs were not based on full cost recovery.*

On the other hand, he pointed out that raising the tariffs would make water unaffordable to the majority of the people, and thus defeat the government objective of poverty alleviation. Mr Muhakanizi was also sympathetic to our high and unsustainable debt burden. That is why he supported us in our plea to restructure the debt into equity.

Other key players in the Finance ministry included Mr Patrick Ocailap, the commissioner Aid Liaison, who was later elevated to the post of director Budget. Then charged with ensuring that all bilateral financing was channelled to the various sectors, he played a major role in ensuring that the EU funded both phases I and II of the Gaba I Rehabilitation Project, and that German funding to various NWSC water projects kept flowing. Even when a few snags arose, he was there with his counterpart, Mr Obella, to resolve the issues.

Last but not the least was Ms Margaret Kobusinge, the water sector liaison officer in the Finance ministry. She was resolute and steadfast in articulating the NWSC case, against many odds, for donor and government funding of capital investments. An astute, no-nonsense officer, she always put her foot down in defence of NWSC in the bitter struggle for the allocation of limited government resources for the urban water sector.

Support from the Privatisation Unit: The Privatisation Unit of the Ministry of Finance, which is in charge of parastatals, also rendered us immeasurable support in our performance improvement initiatives. First, Prof. Peter Kasenene, then the minister of state for Finance in charge of Privatisation, boosted our initiatives through encouragement and moral buttress. One remarkable gesture of support to our change management programmes involved a note to me, thanking my management and alluding to the fact that NWSC had become a signpost operating in a clearly focused management team operating on a principle of performance-oriented culture. In a letter to the managing director of NWSC, Prof. Kasenene wrote

> I would like to congratulate you on this achievement and to commend you and your management team for the receipt of this award of international recognition. Your achievement is a positive demonstration of what level of success a clearly focused management team operating on a principle of performance oriented culture can achieve in a Government enterprise.

For us in NWSC, this gave us added zeal for further managerial innovation and creativity. Other senior officers included Mr Emmanuel Nyirinkindi and Mr David Ssebabi, of the Utility Reform Unit in the Ministry of Finance, who proved to be invaluable NWSC friends and allies.

Initially, I had misgivings about the necessity of the Utility Reform Unit (URU) in addition to the Privatisation Unit. I felt that the Finance ministry was imposing too many bureaucratic masters, as if we could not do our work without close supervision by people whose knowledge of the water industry was limited.

Initially, our relationship with URU, in general, and Mr Nyirinkindi, in particular, was shrouded in mutual mistrust and suspicion.

As head of URU, Mr Nyirinkindi was concerned that the NWSC management was making rash and impulsive innovations, spending too much money on monthly programme evaluation meetings and seeking publicity through the press. He was in favour of a more cautious, measured approach to the reform process, which would enjoy government support and confidence, and carry along all stakeholders. This was during our 100 days/SEREP days. On my part, I thought that he was up to no good. I initially perceived his meddlesome and obstructive approach as detrimental to our change management initiatives, and I often wondered whether he did not have better things to do.

In retrospect, Mr Nyirinkindi was right to be concerned about the speed at which we were moving and about some of our priorities. *To many people, I seemed too ambitious and in too much of a hurry to take stock of what we were doing.* In our rush to do things, we were not paying sufficient attention to the concerns and interests of other NWSC stakeholders, including the government and the donors. The grandiose plans I had had when I joined the NWSC, for example, to add five floors to the current headquarters building and to buy a staff bus, at a time when the Corporation was in the financial doldrums, were insensitive and outlandish, to say the least.

Being a professional accountant, Mr Nyirinkindi had a firm grasp of the state of our accounts and totally agreed with the World Bank's inspired financial model as the best way of turning our corporate financial fortunes around. It was against this background that he proposed to the Water ministry, the appointment of Mr Ssebabi as the URU ex-officio member of the NWSC Board to closely monitor our change management activities and, *hopefully, to contain the impulsive instincts of the MD.* This appointment turned out to be a blessing in disguise.

Mr Ssebabi was not only an ex-officio member of the Board but also a member of its finance committee. He later on became a full member of the Board, working specifically as the chairman of the Audit Committee of the Board. Initially, we treated this young man with suspicion because we assumed that he had been planted on the Board by Mr Nyirinkindi to observe and report on what we were doing about enhancing privatisation. We thought he had been planted to catch us doing something wrong rather than doing something right, in promoting private sector participation. How wrong we were! He soon proved to be a reliable ally and an asset. One of his old school friends later told me that he had been one of the cleverest students in their class at school and that he was always straightforward, honest and very objective. He had a mastery of figures, spoke his mind without fear or favour and always presented sound, thoughtful arguments that often carried the day. He advised the Corporation to set its priorities right, from a purely commercial and objective angle.

Mr Ssebabi became a dependable counsellor, supporter and facilitator. Through his diplomacy, tact and familiarity with the inner workings of the Finance ministry, we almost always got what we wanted from URU.

I got to know Mr Nyirinkindi and the functioning of URU much better through Mr Ssebabi. I soon realised that my fears about the URU were misconceived. *Sometimes, we rush to judge people without giving them a chance to explain what they are up to.* This often breeds unnecessary suspicion and confrontation. I had not bothered to find out whether or how the aims of URU were similar to, or different from, our own. I had not invited Mr Nyirinkindi to share his thoughts and intentions or to explain our change management programmes.

To my pleasant surprise, when I started working with Mr Nyirinkindi, I discovered that our thoughts were similar and that, indeed, we were moving in the same direction. Both of us wanted to reform NWSC into a commercially viable utility. We all agreed that even if the Corporation fell under the axe of privatisation in due course, it was our duty to ensure that it went as a viable going concern.

Like Mr Muhakanizi, Mr Nyirinkindi also believed in fattening the cow before selling it at a premium. These beliefs and principles became one of the guiding lights of our change management programmes. As a result, Mr Nyirinkindi has not only become a personal friend but is also one of the most ardent champions of our internal reforms within and outside the government. He does not stop saying it and is not shy of openly supporting NWSC, since it is doing a good job. I am sure that the following quotation from his senior minister, Gerald Ssendaula, in a letter to me after receiving the Golden Star for Management Merit, attests to his conviction:

> On behalf of Government and on my own behalf, I congratulate you personally; the Board, Management and all staff for winning this precious award and for all the efforts made in turning round the Corporation and making it attain such international recognition. I urge you to maintain the high professional performance record attained thus far and aim even higher in delivery of quality services in the water sector.

In concluding this chapter, it is worth noting that dealing with the government officials has proved to me that they are not insensitive, self-serving bureaucrats of the popular imagination, determined to frustrate the performance of public utilities. Like donors, they also love to back the winning horse and to be associated with success. Once they realised that our change management programmes were on the right track, they supported them in every way possible and the quest for privatisation took the back seat.

Their support for NWSC since 1998 goes to show that privatisation is not always a foregone conclusion in a private sector-driven economy. There is still plenty of room for a mixed economy in a poor country such as Uganda. The public and private sectors can play complementary roles in the quest for economic modernisation.

Government now understands that what is required is a strong, committed and performance-oriented workforce and that there are many alternatives to achieving improved performance. *The challenge facing public utility corporations such as NWSC is to prove their worth through performance that matches, or even exceeds, that of the private sector – if they do, they will deserve and win government support to survive and flourish.*

Lessons learnt

- As sole shareholders of public utilities, governments are the ultimate source of authority and what they say or do invariably determines the fate of those utilities. Accept government policies and strategies and work around them without being a subservient rubberstamp or yes-man. Render frank, honest and sound advice; present SMART alternative options and the governments will support you.
- Governments are made up of pragmatic, flexible and experienced officials who are open to forceful argument and persuasion. They will accept and support initiatives that work. For these officials, no policy or strategy is sacrosanct; they will go for the winning horse. Deliver the desired results, improve performance and turn your organisation around and you will automatically win government confidence and support for change management initiatives.
- Governments are bureaucratic machines characterised by red tape procedures, regulations, negotiations and reconciliation of competing demands. Master how to find your way around the corridors of power; do not take things for granted. Identify and cultivate friends and allies influential enough to help you push through the interests and programmes of your organisation. Once you have earned government's trust, respect and confidence, your programmes will invariably sail through the maze of bureaucracy.

NWSC's success has come from a combined effort. The government, Board, staff, customers, suppliers and development partners all worked together. President of Uganda, Yoweri Museveni (with hat), the minister in charge of water Hon. Maria Mutagamba (fifth right), Dr William Muhairwe (fourth left) together with the NWSC Board, management and representative of the donor community at the commissioning of Gaba III Water Treatment Plant.

The minister in charge of water, Hon. Maria Mutagamba, joining the current Board that has ensured that the Corporation stays on course. From left, Hon. Gabriel Opio, Dr Elizabeth Madraa, Mr Ganyana Miiro (Board chairman), Hon. Maria Mutagamba, Eng. Sorti Bomukama and Eng. Dr Y.B. Katweremu. Back from left, Messrs James Ssegane, David Sebabi, Stephen Kabuye, Dr William Muhairwe (MD) and Mr David M. Kakuba (corporate secretary).

The minister of state for Water Hon. Jennifer Namiyangu breaking the ground for the Mukono Water Supply Project that was 100% funded from internally generated funds.

Chapter Fourteen

Public Partnership Reconstructed

The People's Voice

I wish to bring to your attention how grateful I am to a young man named George who works at your main reception. I have been without water for three weeks; [I] went through the proper procedure at Kitintale [a Kampala suburb] to report the problem. I went twice and our Office Logistic[s] Officer went three times. I then found that our complaint was not logged. I decided to visit your headquarters and lost my temper at George. I apologise for that. He sorted out my problem and kept me informed throughout. He is a very polite and efficient young man and I hope he has forgiven me for my outburst.

Letter from a customer to the NWSC MD,
May 2005

I appreciate your greatly and tremendously improved services ... the customer care you have continued to give us. This commitment to your work has been reflected in timely delivery of bills and reminders to settle bills so as to have continued uninterrupted smooth services. You have a wonderful team ... This may sound flattering but this remains a fact that your services have greatly improved and these are the kind of services that people used to enjoy many years back and it is what we have been longing for.

Letter from a customer to NWSC Ntinda Branch Manager,
May 2005

The change management programmes that the NWSC initiated and implemented since 1998 were driven not only by the quest for sustainable commercial viability and profitability but also by the determination to deliver efficient, cost-effective and affordable water and sewerage services, to achieve and sustain customer satisfaction. Indeed, given our conviction that *'the customer is king'* and *'the reason we exist'*, one of the missions of NWSC has been 'to be a customer-oriented organisation providing excellent water and sewerage services in a cost-effective manner'. That is why, as we discussed in Chapter 3, SEREP, the second of our change management programmes, focused on the needs, expectations and aspirations of our customers. Since then, with generous donor and government support, all NWSC programmes and activities have revolved

© 2009 William T. Muhairwe. *Making Public Enterprises Work*, by William T. Muhairwe.
ISBN: 9781843393245. Published by IWA Publishing, London, UK.

around customer happiness, contentment and satisfaction with our services. This chapter describes the evolving partnership between the NWSC, the public (especially the customers), the print and electronic media as a result of successive change management initiatives since 1998.

Negative customer perceptions and their effects

In 1998 the relationship between the NWSC and its customers, and indeed the public at large, could only be described as one of mutual distrust and hostility, if not outright mutual contempt and disgust. The customers were bitter about incessant water supply interruptions, water disconnection without notice, poor responses to leaks and bursts, inflated or inaccurate bills, belated billing and bills delivery, poor customer care, lack of professionalism, dishonesty and corruption, lack of proper field staff identification and, above all, sheer staff rudeness, condescension, indifference and insensitivity. It was, for example, next to impossible for our customers to get new connections without bribing our staff. For most customers, the NWSC was a good-for-nothing, self-serving public utility that did not deserve to exist. On their part, the NWSC staff used to habitually heap blame on customers for illegal connections, meter defilement, vandalism and theft, bribery and corruption, harassment of field staff and refusal to pay bills. This polluted atmosphere was certainly not conducive to the delivery of water services to the satisfaction of customers and, accordingly, one of the first challenges was to repair and improve NWSC–customer relations.

In rare cases where a relationship between individual Corporation staff and customers existed, it was solely for the purpose of personal gain at the expense of the Corporation. For example, it was common practice for customers to collude with our field staff to carry out illicit activities, leading to loss of revenue for the Corporation and unofficial compensation to the staff. This practice increased cases of illegal connections, meter bypasses, damaged meters, data manipulation, etc., which increased NRW to over 60%. These practices resulted in low collection efficiencies because most customers resorted to dealing directly with our staff. The Corporation became the major victim in all these activities. Consequently, uncollected bills accumulated to an alarming level of over UGX 30 billion, which represented a debt age of approximately 14 months.

As a customer, the government could not be exonerated from blame. At that time, it viewed NWSC as part of its extended system and therefore did not see any urgency in paying for the services it consumed. The government would pay lip service to the importance of settling utility arrears and would not take any resolute actions. The problems started with inefficiencies in reconciliation of water bills, which was attributed to NWSC staff who were not proactive to engaging the government. Even when payments were actually made, some

NWSC staff colluded with ministry officials to manipulate the payments at the expense of the Corporation. The result was the accumulation of huge debts by ministries and departments. Unquestionably, such debts were not acceptable and no business can survive if it is run in this manner. I was concerned about these debts because some very critical and large customers, such as the breweries, were complaining about huge water bills and contemplating alternative sources of water such as rain harvesting, sinking boreholes and localised water treatment facilities. One of my first challenges was therefore to revamp and improve NWSC–customer relations.

Measures to change customer perceptions

Accordingly, I initiated measures aimed at restoring customer confidence and satisfaction. I was aware that in order to effectively address customer perceptions and attitudes towards a public utility such as NWSC, it was important to take into account the nature and capacities of the customers, what they want and whether they know their rights and obligations. Customers or clients of businesses are seldom a uniform, undifferentiated mass. They have different needs, interests, idiosyncrasies and dispositions. While the majority of customers tend to be law abiding and honour their obligations, there is always a minority – the black sheep – bent on accessing services free of charge, by hook or crook. In extreme cases, some customers even commit crimes of vandalism and theft to get goods or services without any consideration or at minimum, deflated cost. In this regard, NWSC customers are no different from customers of other utilities in Uganda. In the following sub-sections, I describe some of the measures and considerations we employed to shift customer perceptions about NWSC conduct of business.

Assessing NWSC's market composition: You cannot effectively address customer concerns unless you know who the NWSC customers are. Therefore, as one of the first steps, we had to know how many customers were in which category. NWSC customers can be divided into the following categories. There are the water consumers who cannot afford private connections and, therefore, draw water from the Corporation-owned but privately operated public standpipes, and pay for water on demand. In 1998, this category made up a tiny minority (7%) of our total customers. Secondly, there are the domestic consumers with in-house or yard tap connections. These consumers constituted 40% of the total water users but their contribution to the Corporation's revenue did not match their numerical strength. Such customers have individual or group water meters and standing accounts with the Corporation. Delivering bills to this category of our customers and ensuring timely settlement of their bills are a daunting task. Then there are the large commercial, industrial and institutional consumers that consume huge quantities of water on a daily basis. At that time, commercial and industrial water consumers made up 15 and 9%, respectively, of NWSC customers. Usually, these categories pay for both water and sewerage services. Institutional consumers, including schools, universities, religious establishments (for example, churches and mosques), hospitals and government departments and ministries accounted for about 29% of all NWSC accounts.

Each of these categories of NWSC customers is unique and serving them individually and collectively poses challenges of varying degrees of complexity and intensity. For example, on the face of it, public standpipe consumers, who pay for water on demand, pose the least of our operational challenges. However, in practice, failure by private water standpipe operators to settle their bills ties up the Corporation's revenue, sometimes culminating in disconnection and bad debts. This not only reduces water consumption and revenue – once disconnected, it is next to impossible to collect the arrears – but also, more importantly, it undermines the Corporation's social-mission objective of delivering safe, clean and affordable water to the urban poor. The challenge here is to identify and contract responsible and reliable private public standpipe operators and to ensure that daily collections are remitted to the Corporation to avoid diversion to other uses.

Domestic water consumers are the most challenging of NWSC customers and the most difficult to serve and satisfy. They are as diverse as they are dispersed all over the NWSC operational areas. It is costly and time consuming to read water meters and to deliver water bills. In some cases, NWSC field workers find it difficult to access water meters either because home gates are locked or the customers are not cooperative and are even hostile. To make matters worse, some unscrupulous or crime-minded customers resort to illegal acts such as meter vandalism, illegal connections and intimidation or even physical assaults. These illegal activities are difficult to detect letalone to eliminate, and they partly explain why non-revenue water remains one of our most intractable challenges. Some of the customers even seem to think that water is God's gift that should be used free of charge. Therefore, serving domestic water consumers requires tact, patience, endurance and ingenuity on the part of the NWSC staff.

In theory, commercial, industrial and institutional customers should be the least of NWSC's challenges. They are educated enough to know that water production and distribution costs money. As business people, commercial and industrial customers know that there are no free services. That is why water consumption is included in the costs of production and the budgets of commercial and industrial enterprises. In any case, unlike domestic consumers, commercial and institutional customers are concentrated in relatively few and clearly defined places, easy to access and more permanent, reliable and predictable. However, they tend to accumulate huge arrears, imposing serious constraints on the Corporation's revenue projections and collection. In order to minimise costs and maximise profits, some unscrupulous commercial and industrial customers are not immune to the social vices of illegal connections and corruption. They also have powerful allies who can pull strings in the corridors of power and have, on some occasions, made life uncomfortable for the Corporation.

Building customer-oriented staff: Accordingly, the strategies and tactics to revamp the NWSC-customer partnership had to take into account the peculiarities, needs, sensitivities and operational complexity of each category of customers. Since charity begins at home, it was imperative to put our own house in order before striving to transform customer attitudes and perceptions towards the Corporation. As detailed in Chapter 3, SEREP, the starting point, aimed to change the work ethic and customer care attitudes of NWSC members of staff over whom we had direct control in the hope that customers would sooner rather than later warm up to our responsiveness, friendliness and care. We therefore went out of our way to sensitise the staff to recognise the fact that the customer was the reason for our existence and that our jobs and livelihoods depended on customer satisfaction and willingness to pay. We launched training programmes to sensitise staff about the merits and modalities of customer care. We gave our field staff identity cards and insisted on politeness in customer service at all times, even when the customers appeared to behave provocatively. We made it clear that our staff were customers' servants – not masters. Members of staff were, therefore, duty bound to respond promptly to customer queries, complaints and reports, with a smile, attention or show of concern and without any laxity or reservations, and to take quick and appropriate follow-up action.

Developing customer-oriented policies: As described in Chapter 3, we initiated and implemented a new customer care policy to address specific customer concerns and aspirations. To recap briefly, we launched a concerted publicity campaign through the print and electronic media to sensitise customers about their rights and responsibilities. We emphasised that customers deserved and were entitled to prompt and efficient services, and they had the right to demand information regarding all aspects of water production, distribution, pricing and accessibility. We granted a general amnesty to illegal water consumers who voluntarily reported themselves to the Corporation to regularise their connections. In this respect, we stressed that no one would be prosecuted. This was widely publicised in the media. The message was clear. For example, in a press release that appeared in *The Monitor* of 9 August 1999, we stated that

> The Corporation is giving a chance to this [illegal consumers] category of consumers, a 30 days offer with effect from 9 August to 9 September 1999 to report to any of our offices and regularise your connections by payment of the new connection fees only... after the expiry, all persons caught will be prosecuted accordingly. Management wishes to appeal to all consumers in this category to take the offer while it lasts ...

Furthermore, we invited customers with suppressed accounts or huge arrears to negotiate payment in affordable instalments. We offered to reconnect water defaulters without any penalty and write off erroneous and bad debts. Most of these customers had been without supply for a long time, some for more than three years. Testimony to this is an article that appeared in *The New Vision* of 25 May 1999, entitled 'NWSC overshoots April cash target'. In this article, the Board chairman justified the write-offs of bad debts as follows:

> As part of our promotional campaign, the NWSC Board has decided to write off all the sewerage bills. This is an incentive to get them [customers] back to supply and access to clean water. Water is life, people should not be put at risk because of other reasons.

In addition, we visited our customers on a regular basis to find out their concerns and problems and we invited them to attend and participate in our corporate functions. We also opened the channels of communication by setting up customer centres and units throughout the Corporation to receive and resolve complaints. All these initiatives were geared towards building customer confidence and creating an atmosphere of cordiality.

Since 2004, the NWSC has also earned ISO certification for all its operating Areas, including Kampala Water Supply Area, for quality management and maintenance of high and consistent water standards, leading to improved customer service. The Jinja and Tororo Areas championed these prestigious awards.

Creating a Customer Care Division and downstream sections: In order to enhance customer orientation, the NWSC was restructured to incorporate the customer care management function. Among other structural changes, we had to create an independent division that would focus on customer care issues, ranging from billing and debt management, to marketing our services. In tandem with these structural changes, we strengthened commercial and customer care operations in Areas in a number of ways. First, we had to tackle the issue of incompetent staff. In this regard, more competent people were identified and posted to the Areas as commercial and billing officers. Most of these personnel were university graduates who could ably handle an increased scope of duties. This was a great improvement on the way commercial issues in Areas had been handled by arrears inspectors in the past. Arrears inspectors, whose role was mainly restricted to revenue collection, did not have explicit customer care role definitions.

There was also systematic strengthening of the meter-reading portfolio in the Corporation, with diploma graduates replacing semiliterate meter readers, in a phased manner. This activity greatly improved the accuracy and quality of customer water bills. Previously, most meters in the field were hardly read because some of the meter-reading sheets were completed under trees, without the meter reader accessing customer metres.

Second, we carried out a tailor-made training programme targeting all staff who interfaced directly with customers. We equipped them with the necessary skills for effective handling of customers. The staff involved in this training exercise included commercial officers, meter readers, front desk staff, cashiers and plumbers, among others. This training was a very valuable activity because it equipped the staff with a fresh outlook, and they in turn started treating customers in a friendly and caring manner.

As described in Chapter 3, we put in place customer care sections in all our Areas and in each Area posted specific customer care officers. The customer care offices were equipped with new furniture and customer care registers were opened to register all customer complaints and the actions taken. The ambience of customer care offices was also emphasised, and this included cleanliness, smartness, tidiness and general attractiveness. In order to change the organisational behaviour of staff, we introduced the slogan: 'The customer is the reason we exist'. This slogan, together with 'Customer is King', was posted at every NWSC office and on all available notice boards. Everybody – from the MD to the lowest cadre staff – was called upon to become a customer care officer. The emphasis here was on our staff understanding that in everything they do the customer had to come first. We posted most of these key initiatives in the media. For example, in a press release that appeared in *The Crusader* of 18 February 1999, we wrote

> NWSC has strengthened its Customer Care Units in all its operational towns with the aim of attending to customer complaints in a more timely and efficient manner. Customers can now report or present their queries/complaints through any of the following telephone numbers ... NWSC appeals to the general public to report all cases of water pipe

bursts/leakages, sewage overflows, abuse or misuse of fire hydrants, illegal use of water and other complaints related to its service provision to its Area Offices through the above telephone lines ...

Easing customers' payments: We sent a clear signal to all our customers that we had become a flexible public utility so they could do business with us and we were prepared to listen. We had introduced convenient payment schedules and procedures for the sake of our customers. As pointed out in Chapter 3, we began by strengthening the instalment payment system to give a soft landing to heavily indebted customers to clear their bills. Because of negligence and poor debt management practices on the part of NWSC, most customers had accumulated huge arrears, which they could not pay easily. For those customers whose debts were genuine but who could not settle them at once, the Corporation accepted instalment payments to ease their burden. Most importantly, the customer was allowed to continue being supplied as negotiations were concluded to avoid loss of revenue and inconveniences to the customer.

This approach attracted considerable applause from customers. For example, a customer in Tororo Area, to show appreciation for such a payment arrangement, wrote to the Tororo management:

> This is to thank you for the good job of letting me pay the bill by half. This has made me very happy [because] at this period of time, we are supposed to pay [school] fees. I hope to pay the balance after this fees problem (letter from Uhuru Rebbeca, Agururu BI Village, to the Tororo Area Manager, 13 January 2005).

In addition, for the convenience of customers who could not pay at our offices, we instituted a system that allowed them to settle their water bills directly through banks. For customers with mobile phones, arrangements were made for them to access their account balances through the SMS facility and pay to the nearest cash office or bank ATM. The Corporation also introduced a direct debit system.

Strengthening branch/Area operations: In all our internal reform efforts, efficient service delivery was always the cornerstone of our improvement programmes. Previously, the over-centralisation of functions at the headquarters was a major constraint to efficiency at the Area level. We therefore found it necessary to decentralise most functions to the Areas to make service delivery to our customers more efficient. This approach gave the Areas the opportunity to be more creative and to design strategies that were suitable for their own local environment. On several occasions, we informed the public about our

Public Partnership Reconstructed | 311

decentralisation activities and expected benefits. For example, in an article that appeared in *The Monitor* in February 2005, entitled 'NWSC Decentralises', it was reported:

> Following complaints of bureaucracy by customers, National Water and Sewerage Corporation has decentralised its water services. Effective this year all water connections, payment of bills, and other related services will be done in various branches within the reach of customers. Customers will no longer have to throng the NWSC head offices. They will instead get services from the nearest water office, putting an end to the era of persistent delays in getting services from the Corporation.

To strengthen decentralisation further, I encouraged field managers to work with speed and defy any tradition of bureaucracy that slowed operations in the Corporation. For example, I pinpointed incidents where customers rung me to get water meters when responsible managers were there. It was crazy!

Furthermore, the billing function was also decentralised. Previously, customers' databases for most Areas were kept at NWSC headquarters. To open up this bureaucratic bottleneck, I insisted that computerisation would help us to decentralise the billing function to the service centres close to the customers. Fortunately, my managers understood this strategy and have implemented it extensively in all NWSC towns. Since then, billing, revenue collection, response to customer enquires and the speed at which new connections and reconnections are handled have improved rapidly.

Reviewing our prices: Our customers were very critical about our tariff. A number of questions were being asked about the rationale for pricing, especially connection and reconnection fees, minimum charges, etc. As already mentioned in the previous chapter, we responded to our customers' needs by reducing the connection fee from an average of UGX 150,000 to UGX 50,000 and also allowed customers to pay in instalments.

The tariff adjustment also incorporated rebates on reconnection fees and the abolition of a minimum charge. The high reconnection fee acted as a hindrance to disconnected customers to reconnect. For instance, a customer disconnected with a bill of UGX 10,000 was expected to pay a total of UGX 55,000 inclusive of VAT before reconnection. In most cases, such customers would collude with NWSC staff to steal water, or would look for alternative sources of water. This did not make any business sense. Consequently, we reduced the reconnection fee from UGX 45,000 to UGX 12,000, and encouraged owners of suppressed accounts to come back to supply. At the same time, according to the tariff, it was mandatory that all customers be charged for six cubic metres as minimum consumption regardless of actual water consumed. This caused many complaints from the public, and most of the low-consumption customers actually abandoned our water. In order to arrest and then reverse this trend, we abolished this minimum charge. This encouraged low-consumption customers to come back to supply. We made these tariff changes known to our customers and other stakeholders through the print and electronic media. In so doing, we kept our customers informed, and they supported us in our change management programmes.

We also surmounted the connection fee imperfection, as already mentioned in Chapter 12, through a new policy that entailed providing materials and labour for all customers within a distance of 50 metres from the NWSC service main. This policy was welcomed enthusiastically by the customers, stakeholders and the general public. To many of them, this was a great relief. However, the NWSC was the major winner because we experienced a great surge in customers' requests for new water connections. This has in turn increased our billing remarkably.

Measures pay off: moving in the right direction

Although water service delivery is a sensitive, even a controversial activity, in which it is difficult to satisfy all categories of customers at all times, the feedback that I received was exciting. It was slowly but surely becoming apparent that our measures had paid off and our customers were beginning to appreciate our services, giving their approval to what we were doing. The feedback indicated that there was a truly astounding shift in customer perceptions in favour of NWSC as a result of our performance improvement programmes. Through the monitoring mechanisms, we also established that customers had regained confidence in NWSC.

Even when there were differences of opinion, which in any case are inevitable, our staff now handled customers in a polite and civilised manner. *Due to our initiatives, customers no longer suffered in silence.* They openly aired their complaints about inept services and demanded prompt answers to their questions, which kept NWSC staff on their toes. Customers started to report staff misconduct as well as illegal connections by their neighbours.

Whenever I visited NWSC Areas upcountry or various branches in Kampala, I talked to customers from all walks of life – in schools, hospitals, hotels and restaurants, and even on the streets, to find out how they felt about the quality, adequacy and efficiency of our services. On formal NWSC occasions, I sought the views, opinions and suggestions of invited customer guests about the performance of the Corporation and what else needed to be done to continue improving the delivery of services. In most cases, our customers' responses were utilised to make NWSC *a more efficient and cost-effective public utility.*

Public Partnership Reconstructed | 313

It was usually gratifying to receive appreciation from the people we serve and I indeed looked forward to receiving positive individual ratings from customers in regard to our services. Much as the feedback we were getting from the direct interviews showed approval of performance by most customers, I was keen to read individual customer testimonies made on their own initiative. Every morning, I looked at the letters page of the print media to keep abreast of what our customers were saying. It was exciting and motivating for me to read of a generally high rating and approval by some customers of our performance. In much of the correspondence and in various fora, customers came out openly on their own not only to commend the Corporation for good performance but also to propose ways of improving service delivery. I have attempted to capture below a few reactions, among many, from the customers' own confessions regarding our services.

In April 2004, I received a letter from the Parish Priest of Holy Trinity Church, Kamwokya Catholic Parish, which read in part:

> our gratitude to the entire organisation for the good services of your staff... I was impressed by your staff's urgent response, cooperation, extra time and also a good working relationship they portrayed. In particular, I was much touched by the working skills of Francis Omondi, a plumber attached to the group based on the 6th Street, Industrial Area.

Mr Omondi had helped fix the water of this satisfied customer, whose supply had been cut off by road construction activity. For his labours, Omondi earned himself one-minute praise from his immediate supervisor.

Similarly, in a letter titled 'I'm grateful to NWSC, Kitintale', another customer wrote, on 3 July 2006:

> I'm grateful that we have such good staff working in my Area, Port-Bell. I'm yet to find such good people in other departments. I have been to some offices where, as soon as they see you coming, they quickly pick up a phone and dial someone so as to let you wait while they talk and laugh as you are waiting! I had a very good experience with the Kitintale National Water Office, which has made me pay water bills fast whenever they arrive. I very much thank the Manageress, whoever she is.

On 23 June 2004, the following write-up appeared in the Citizens' Alert column of *The Monitor*:

> The National Water and Sewerage Corporation, Jinja Branch, has received accolades from a satisfied customer. The gentleman, who identified himself as Mwanje, says you have only to pay for your

connection in the morning and by evening or next day you will be the happy recipient of piped water. He also says the body has the courtesy to send its customers reminders and warning notices before they are disconnected for non-payment.

In a letter dated 21 June 2005 to our Mbarara Area manager, the University Secretary of Mbarara University wrote:

> The University appreciates the manner in which you deliver your services. The patience and tolerance you exhibit deserve particular mention. Please keep this spirit up.

In a letter dated 9 February 2005, the town clerk, Kampala City Council, expressed appreciation for the prompt response by our water plant staff when he wrote:

> I hereby on behalf of the Mayor and the entire executive of Council thank the Manager Bugolobi Disposal Unit, Mr Kitgum, and the unblocking gang which is now seriously undertaking all the spots which have been flooding around the city... There is always immediate response whenever you are contacted. Please let the spirit continue. Thank you all.

In separate notes to the Greater Entebbe Water partnership, customers wrote the following:

> You have done the needful and you are living up to your motto – Water is Life. Please avoid unnecessary disconnection. I wish you got more funding to carry out network expansion in the villages (Male Henry, in a letter to Area Manager, Entebbe, 22 March 2005).

> Since March 2003, I have enjoyed the cordial services of my water partners. They have always treated me well. Keep the spirit behind 'Water is Life (Sabiti Charles, in a letter to Area Manager, Entebbe, 2 May 2004).

In other separate notes to the Kampala Water partnership, customers wrote the following:

> I take this opportunity to thank you for your services and the tremendous customer care that you have continued to give us. This commitment to your work has been reflected in the timely delivery of bills and kind reminders to settle the bills so as to have continued uninterrupted smooth services. You have a wonderful team! (From Joseph Masaalo to Branch Manager, Ntinda, dated 25 August 2006).

Public Partnership Reconstructed | 315

Samson Mpindi and Sarah Tebusweeke of Kinawataka, Kampala have something positive: This is to thank the management and staff of National Water and Sewerage Corporation for their tireless efforts to keep their customers satisfied. In particular we would like to appreciate the rapid response tactics you have adopted to minimise leakages and repairing of burst water and sewerage pipes. As if this is not enough, you have now given the public a hotline (977) to reach you in case of problems in your Area of Jurisdiction. Thank you very much. I pray that other companies ... take a leaf after you not only in having hotlines but actually following up on reported cases for immediate rectification (Citizens' Alert, *Daily Monitor*, 16 August 2005).

The Uganda Austria Foundation (UAF) also commended NWSC for the high level of efficiency. In a letter to the Corporation, UAF's chief coordinator, Francis Odida, said that earlier in the year, a tractor working on a new road had accidentally cut a water pipe on Gayaza Road, near Kampala, but NWSC had responded promptly. 'When contacted, your field staff quickly responded that night and reconnected the pipe before it inconvenienced your clients' *(The New Vision,* 15 July 2005*)*.

In an analysis of our call centre operations in Kampala, a customer wrote:

I wish ... was like Kampala Water Call Centre. When one calls 977, it is the sweetest welcome one can ever get. On Saturday afternoon, I had a problem with my water and called the helpline. First, Amelia welcomed me very well and took my details. After a few minutes, another lady, Susan, called to update me, after solving my problem. George called to find out whether my problem was over. Thank-you so much Kampala Water, Keep it up. This is truly the meaning of Customer Care. If only every service provider was like Kampala Water, Uganda would be a great country. (John Bosco Tamale, *The New Vision*, Tuesday, 12 July 2005.)

In Jinja, customers came out openly to compare service delivery between NWSC and other companies , when he wrote to the Citizens' Alert column of the *Daily Monitor*:

Make Customer King': Edith, Sarah, Gorretti & Grace Nyanga of Jinja, Uganda have a good piece of advice to [other companies]... We simply appeal to the officers concerned in ... [other companies] to go and visit the water supply offices and learn from them. These two organisations are different in all ways. With water supply, "The Customer is the King", but with ...[other companies], it seems to them the customer is an enemy. I advise the managers of ...[other companies] to learn from their friends of water supply (*Daily Monitor*, 5 August 2005).

Similar expressions of customer satisfaction have been received from all NWSC Areas across the country, indicating clearly that our change management programmes have had a positive impact on customer satisfaction. Customers used words such as *'appreciate', 'proud of', 'happy with', 'efficient', 'best company'*, reflecting a positive shift of their perceptions about and attitudes towards the Corporation.

The positive shift in customer responsiveness to NWSC services has also been manifest in the way some of the customers have volunteered to assist the Corporation in the fulfilment of its responsibilities. For example, in February 2004 one of the customers promised to be 'a thorn in the flesh' of a landlord whose water account was 'seriously overdue'. This was embarrassing and inconveniencing to the tenants. Owing to pressure from this customer, the landlord has been promptly and regularly settling the water bills since then.

Customers argue against complacency: demand for more

Of course, not all customers responded positively to the improvements brought about by our successive change management programmes. The letters of a minority of customers addressed directly to the Corporation and in the print media reflected residual dissatisfaction with the Corporation's delivery of water and sewerage services. Some of the customers continued to complain about, among other things, disruptions of water supply, high water bills, water leaks and bursts, sewage overflows and broken manholes, and failure by the Corporation to settle outstanding debts, and unsealed trenches that cut off roads. Some of the customers have even claimed to be 'frustrated' or, worse still, 'disgusted' with NWSC services.

The following were some of the messages sent to the print media to show customer demands for better services:

> We have spent a very long time without water in Kirinya, Byeyogere. We reported the problem to the National Water and Sewerage Corporation, but nothing has been done. Local leader should come to our rescue. *The New Vision*, Thursday, 9 August 2007.

Another customer in an expression of disgust said

> I am very disappointed by the way NWSC has punished Nsambya and Kabalagala residents. Water supply has been cut off for a week now and no explanation has been provided by the water body. We all know

that water is life and basically, there is no life in the Nsambya-Kabalagala zone! *Saturday Monitor,* 3 November 2007.

In a similar tone, a customer in Lira complained about NWSC services as follows:

> I paid UGX 205,000 on October 15, 2007 at NWSC in Lira for water connection for my sister. Unfortunately, to date, the sister has not been offered the service. The chronic excuse from management is that they do have materials for the connection in stores. *Daily Monitor,* Monday, 17 June 2008.

Criticisms like these are healthy because they motivate us to make more improvements and avoid resting on our laurels. Another good illustration of customer assertiveness was the case of a hotel manager in Mbarara, who, in May 2005, was hosting an important function attended by the president of Uganda. He accused the Corporation of gross inefficiency and negligence and even 'sabotage' because of water shortages during the function. Although the shortages were due to the Hotel's insufficient reservoir storage capacity rather than our fault, we still apologised to the customer for the inconvenience and offered to help install a water reservoir with an appropriate capacity for the hotel on all occasions. Our policy is to avoid arguing with the customer even when the complaint is not justified. Right or wrong, the customers' complaints must be addressed in a manner that does not offend their sensibilities. Fortunately, this policy has been rewarding and accounts for the tremendous improvement in the rating and enhanced image of the Corporation among customers in all its operational areas.

In order to address the residual dissatisfaction issue, I held a discussion with the management and we agreed to use more proactive approaches to capture residual complaints. We have established well-trained research assistants to sit in the front desk offices and track customers entering and leaving customer service centres. These independent research assistants interrogate the customers, in a cordial manner, with the intention of capturing their residual grievances and checking if the NWSC staff have assisted them. The research assistants are detailed to discuss their findings with the operating unit managers and then compile daily reports incorporating the agreed way forward with the managers.

The unit managers are expected to take action in a timely manner and give feedback to their superiors at the Head Office. We do not hold managers responsible for the intensity of complaints but rather the action velocity and reduction in complaints of the same nature. This approach has worked very well, especially in Kampala, where complaints are common owing to the huge scope of operations. The demand for prompt feedback has strengthened the customer orientation of managers and staff.

Our own independent assessment

Much as the complimentary testimonies of individual customers were very encouraging to my team and I, the scope was not wide enough to reassure us that customer perceptions about NWSC had shifted radically in the positive direction. I discussed my reservations with the managers concerned and we agreed that to back up these testimonies, we needed to carry out an independent assessment of customer perceptions about our service delivery. We resolved to gauge customer satisfaction with regard to the delivery of water and sewerage services through periodic customer surveys. The main objectives of these surveys included determining overall customer satisfaction with NWSC services and identifying customer-perceived areas of strength that needed to be enhanced and consolidated, as well as areas of weakness that needed to be addressed to improve services on a continuous and sustainable basis. In addition, the surveys would assist in developing a tool to measure, over time, how increased customer satisfaction impacts on repeat customer sales.

Using the questionnaire approach, a number of parameters were selected and investigated to test the shifting degrees of customer satisfaction with regard to water and sewerage services. These parameters included response time to bills, billing accuracy, payment convenience, staff-customer efficiency and rapport, water quality standards, water quantity adequacy and regularity, service coverage and response time to customers' complaints.

The customer survey findings in Tables 14.1 and 14.2 show the average percentages (proportion of sampled customers saying services are either good or very good) of customer perceptions in Entebbe, Mbarara, Kasese and Mbale Areas with regard to five of the parameters – response time to leaks, billing accuracy, bills payment system, regularity of water supply and response time to customer complaints. These findings are extracted from the customer survey evaluation reports of July–December 2004 and January–June 2005; representing the financial year 2004/2005. The survey results represent customer perceptions as at the month in question. The surveys were carried out on a monthly basis and the periods of analysis in this section are chosen to track progress in customer satisfaction levels.

From Table 14.1 above, the findings of the customer survey in July 2004 clearly show very high levels of customer satisfaction, ranging from 82 to 100%. There were minor variations from Area to Area, but in general the mood of the customers was buoyant and this was cause for satisfaction to my staff and management. The findings of the customer survey, which were reported in

Table 14.1: Customer Perception Analysis – July and December 2004

Parameter	Entebbe			Mbarara			Kasese			Mbale		
	July	December	% improvement	July	December	% improvement	July	December	% improvement	July	December	% improvement
Response time to leaks	82	93	11	77	98	21	96	100	4	94	94	0
Billing accuracy	84	100	16	87	100	13	95	100	5	96	100	4
Bills payment system	93	97	4	86	100	14	93	100	7	93	98	5
Regularity of water supply	88	93	5	88	100	12	93	100	7	93	93	0
Response time to complaints	83	100	17	82	100	18	83	98	15	97	100	3

Note: Figures represent those customers who said that services were either good or very good as a percentage of the total responses obtained.

Table 14.2: Customer Perception Analysis – January and June 2005

Parameter	Entebbe			Mbarara			Kasese			Mbale		
	January	June	% improvement	January	June	% improvement	January	June	% improvement	January	June	% improvement
Response time to leaks	67	83	16	100	91	(9)	92	99	7	100	100	0
Billing accuracy	74	100	26	100	98	(2)	89	95	6	90	100	10
Bills payment system	70	93	23	100	96	(4)	73	86	13	95	100	5
Regularity of water supply	85	83	(2)	100	93	(7)	75	90	15	90	90	0
Response time to complaints	85	100	15	93	90	(3)	74	95	21	90	95	5

Note: Figures represent those customers who said that services were either good or very good as a percentage (%) of the total responses obtained.

December 2004, were even more pleasing and reassuring. In Entebbe, Mbarara, Kasese and Mbale, there was an upswing for each of the five parameters that were investigated.

The biggest improvements in customer perceptions were registered in Mbarara, ranging from 12% in the regularity of water supply to 21% in the case of responses to water leaks. For Entebbe, the improvements in perceptions ranged from 4% in convenience of bill payment system to 17% in responses to customer complaints. The improvements for Kasese and Mbale were less dramatic than those for Mbarara and Entebbe, mainly because the levels of satisfaction were already remarkably high anyway.

However, the findings of the customer survey, reported in January 2005 (see Table 14.2), were rather disappointing compared to those of December 2004. For example, the customer perceptions for Entebbe were way down, ranging from 67% in responses to water leaks to 85% in responses to customer complaints. The customer perceptions in all parameters for Kasese and Mbale had fallen from the high levels of December 2004. Only Mbarara maintained high rates of satisfaction of 93 to 100%. The fall in customer perceptions in January 2005 was presumably due to the slow momentum typical of all business cycles at the beginning of the year. However, as expected, there were significant rebounds in customer perceptions in the customer survey whose findings were reported in June 2005, ranging from 5% in responses to complaints (Mbale) to 26% in billing accuracy (Entebbe). In Mbabara, there were relatively minor perception declines of 2 to 9%, but even in that Area the sense of satisfaction remained remarkably high.

To us, the face value of the figures was important because it gave a good indication of the level of satisfaction. Small declines and gains could have been due to sampling and other questionnaire administration problems, but the fact that the satisfaction levels remained above 80% (our benchmark) was a commendable achievement.

Honoured at home and abroad

The success associated with NWSC's change management programmes was not only recognised and applauded by the customers in all NWSC Areas but also by the public, both at home and abroad. The first public recognition of positive progress in NWSC service delivery came when the *Sunday Vision* of 29 December 2002 declared me to be one of the best managers in the country; whose management had set a good example for other managers of public enterprises to emulate. The paper noted that I had provided good leadership, which had turned the Corporation around through concerted improvements in water production, distribution, revenue collection and customer care. In 2002, the Corporation not only made a profit of UGX 9 billion but was also the first public company to submit its annual audited accounts to the Auditor General in accordance with the statutory requirements and at the stipulated time. This recognition no doubt symbolised the positive wind of change that was sweeping through the Corporation and the growing public appreciation of enhanced performance in its service delivery.

In addition, the Second National Integrity Report of March 2003 ranked NWSC the best performing organisation among government departments, ministries, local government councils and public enterprises. This was the most significant recognition of the radical transformation of the Corporation as a result of its change management strategy. This report was a result of a survey commissioned by the Inspector General of Government in July 2002, undertaken by K2-Consult, a Danish Consulting Firm, to generate information about household and institutional perceptions and experiences regarding, among other things, service delivery and the integrity of public institutions. In its assignment, K2-Consult carried out 12,190 household interviews in 55 of the 56 districts. The other source of information was interviews from a sample comprising 480 and 618 public and private sector organisations, respectively, in eight major towns and 16 sub-counties across the country. From the NWSC perspective, the findings of this countrywide national integrity survey were welcome and reassuring though, of course, not surprising.

In 2004, the NWSC once more hit the newspaper headlines when it won the 2003 Employer of the Year Corporate Social Responsibility and Business Award of the Federation of Uganda Employers – the first ever to be won by a public utility in Uganda. The Corporation was recognised for its exemplary service delivery and excellence. According to the Federation of Uganda Employers, the NWSC had 'disproved the hypothesis that the government is always a bad businessman'. The NWSC's success story had demonstrated that there was no reason why a well-managed public company with a team of dedicated managers of high integrity, commitment and professionalism could not succeed as well as a private one. The Federation went on to point out that the NWSC record had shown that 'sound management is not a monopoly of the private sector, nor expatriates. Good intentions, ethics, honesty, accountability and integrity on the part of a highly skilled and motivated workforce are essential prerequisites to good governance and commercial viability.' In its congratulatory message to me, the Federation of Uganda Employers stressed: 'There is no doubt that with your leadership in human resource management and business practices in this country, a lot of employers will have a lot to share and learn from your organisation.' This was very encouraging for me and gave me more energy to re-engineer performance improvement strategies. In 2005, we again won the

Employer of the Year Award 2004 of the Federation of Uganda Employers in the category of Productivity and Performance Management. This was another demonstration that the NWSC was continuing to be recognised by outsiders, a clear vote of confidence by the public in our reforms.

During the prestigious 8th East Africa's Most Respected Companies Survey of 2007, which is sponsored by the Nation Media Group and an international audit firm, Pricewaterhouse Coopers, NWSC emerged the Best Managed Public Sector Company in the Region. A total of 353 chief executives took part in the survey. In making their decisions, the judges considered quality of service, focused expansion strategies, strong leadership and management, innovative services and brand presence. Speaking at the award ceremony, Mr Charles Muchene, the survey director and country leader of Pricewaterhouse Coopers, Kenya, said that 'Respect is a value that cannot be bought or demanded; it has to be earned. Once earned, it must be jealously guarded, as it is fragile. The highest accolade that a business can get is to truly earn the respect of other business leaders.' In addition, at the Uganda International Trade Fair held from 4 to 10 October 2005, the Corporation was voted the best exhibitor for the services sector: a manifestation of and credit to our strategic customer care initiatives.

The success of NWSC has been internationally recognised and honoured. In July 2004, the International World Marketing Organisation awarded the NWSC the prestigious Gold Star Medal in recognition of its outstanding leadership and excellent performance. The marketing organisation selection committee comprises renowned businessmen and women, diplomats, education, marketing and communications experts, and government officials. Of the 21 finalists for the award, four were from Africa and only one – the NWSC – was from East Africa. The NWSC was the first-ever company to win the Gold Star award for management merit in East Africa.

Emulating our success

Our success has been a good example which other organisations have been eager to learn. Indeed, over time, the NWSC's experiences and successes in service delivery have attracted the attention of many people and organisations in the country. They are keen to learn how successful interventions can succeed in turning an ailing government parastatal into a thriving, profit-making organisation. We have received invitations from a number of companies and government departments to make presentations about our change management programmes. They have been interested in learning about the critical success factors, management style and approaches to our change management programmes. Consequently, our staff have conducted induction courses and seminars for a number of company employees.

In August 2002, the NWSC and the Uganda Electricity Distribution Company agreed to form a partnership to share experiences and ideas in order to provide efficient services to the diverse customers of the two organisations. Similarly, in December 2002, the mayor of Entebbe Municipality invited me to talk about the stretch-out management concept. Since then the mayor has adopted the stretch-out management concept for his municipality.

In February 2002, the Institute of Uganda Professional Engineers, Uganda, invited me to give a talk on 'Improving Performance through Internal Reforms by the Public Sector – A Case of National Water and Sewerage Corporation, Uganda'. I was amazed to see how interested the engineers were in learning how to balance commercial and engineering services management. These three examples show how the success of one public enterprise can have multiplier effects that are beneficial to public corporate governance in Uganda and the developing world at large.

Reports about our improved performance in service delivery have not been confined to Uganda. The NWSC has become a point of reference in many discussion centres as a utility promising to be a 'beacon of hope' in Africa. The Corporation has been commended for well-maintained facilities and 'management to international standards'. Not surprisingly, therefore, water utility managers from Kenya, Tanzania, Malawi, South Africa and other African countries have visited Uganda to learn from the NWSC change management experience and to seek management advice and technical assistance from the Corporation. For example, in April 2005, the top managers and board members of Nairobi City Water and Sewerage Company and Tanga Urban Water and Sewerage Authority visited NWSC to inspect its facilities, benchmark its methods of work and operations and to explore ways and means of cooperation in order to improve the delivery of water and sewerage services in Kenya and Tanzania. Indeed, since 2005, NWSC has provided external services to utilities in Kenya, Tanzania, Rwanda, Mozambique and Zambia, among others. Many positive compliments were given by visiting water entourages to the NWSC. For example, after making a benchmarking visit to the NWSC in June 2005, officials from Umgeni Water, South Africa, expressed their appreciation to NWSC in an e-mail:

> 'Thank you for such royal treatment when we visited your company. We are short of words to express our gratitude and appreciation of the professionalism that everyone at your company displayed. The team was overwhelmed by the proceedings for the 2 days and never stop spreading the good word about your company even today. We will certainly feedback our experience and lessons learnt to the rest of our organisation and our peers in the water sector in South Africa...

The right to know; the duty to inform

Our successes inevitably brought the Corporation into the limelight. We could therefore not avoid being followed closely by the press as we moved on with our successes.

When I took over the management of NWSC in 1998, I was media-shy, to say the least. In my previous jobs, I had had limited contact and interaction with the media, which was justified because I feared and mistrusted the world of journalism. For me, the distinguishing features of the media in Uganda, and elsewhere in the world for that matter, were *sensationalism, intrusiveness* and even *malice*. It is no exaggeration to say that the media are, by the nature of their profession, more inclined to blame, to find fault, than to praise. Journalists prefer to report shocking, eye-catching stories that sell – nothing makes for good news like bad news – than to recognise achievements, however outstanding or groundbreaking these may be. I was conscious that some of the unscrupulous, self-serving journalists had gone as far as publishing or broadcasting deliberate distortions and falsehoods that had dragged the names of many people through the mud and ruined the careers of distinguished public servants. In extreme cases, such journalists resort to blackmail and extortion. In any case, given the mushrooming of a variety of print and electronic media in the 1990s, it was difficult to choose which media to deal with or how to deal with them. For these reasons, my initial instinct was to tread carefully in my interactions with the media; to keep them at arms' length and to avoid contact with them as much as possible.

However, Miriam Kadaga, the NWSC principal public relations officer, had other ideas. On the day I reported for duty, her first briefing was to convince me that, given the nature and sensitivity of NWSC services, we had to cultivate and sustain contact with the electronic and print media, whether we liked it or not. For better or for worse, it was in our own corporate interest to keep in touch with the world of journalism, if only because it was the most practical channels of communication to and from the public. Ms Kadaga presented a sound and compelling case in favour of an immediate, continuous and sustained interface between the Corporation and the media world in all NWSC operational areas. She argued that at worst the media were necessary evils through which to communicate with the public to clarify issues and to combat any anti-NWSC distortions, falsehoods and propaganda. At best, the media could be a useful tool to inform the public, articulate our policies and programmes and publicise our achievements. It was, therefore, inevitable and imperative for the Corporation to work closely with the print and electronic media on a regular basis to present our case, and the sooner we did this the better.

The principal publicist suggested a number of ways of using the media to our benefit, especially as a medium for communication with our customers and the public at large. These communication messages could be conveyed by *press briefings, announcements, advertisements, press conferences* and *interviews*. They also included talk shows and phone-in programmes on radio and television, as well as invitations of journalists to our functions. In order to exploit these channels of communication, she asserted, it was necessary to cultivate and sustain media interest in our activities and programmes and to keep our doors open to journalists who took an interest in our work, and to provide whatever information they required about NWSC operations.

The purpose of this open-door policy was to build media confidence in the Corporation through transparency and accountability. The thrust of our policy was to assure the media and the public that the NWSC had nothing to hide. The public had the right to know; the Corporation was duty bound to inform them, and the media were welcome to access and use information to enlighten the public about NWSC corporate activities, programmes and statutory responsibilities. It was against this background that Ms Kadaga invited journalists to my inaugural address to members of the Board and staff at the NWSC training centre in Kampala.

Initially, the media was extremely suspicious of the objectives and intentions of the happenings at NWSC. As such, the Corporation received not-very-supportive coverage in the press. I almost regretted accepting the advice of the principal publicist to involve the media in our work against my initial instincts. One of the few consolations in the ensuing barrage of criticism was the report that I would not tolerate thieves and liars, which appeared in *The Monitor* of 19 November 1998.

The negative reactions against the 100 days Programme, mentioned in Chapter 2, were just the beginning of my media troubles. It seemed that our image in the media had to get worse before it got better. There were rumours that I had been appointed on political grounds rather than merit and that I would start witch-hunting and getting rid of workers who did not belong to my own ethnic community. I had to assure the workers, the public and the media that no worker would be victimised on grounds of race, gender or ethnicity. As reported in *The Monitor*'s article during the review of the 100 days Programme, I took pains to clarify that

> Whether you were from the moon or wherever, once you do your work diligently no one will touch you. However, even if you are a Muhairwe, but you fail in your work, I will reprimand you. No one should worry. Ethnicity or political beliefs will not influence our work methods in any way. We are going to be guided by professional merit.

However, despite my assurances, media scepticism and even cynicism continued unabated. My only hope was that once the change management programmes began to impact positively on NWSC's service delivery and performance, media perceptions and attitudes would also change for the better. In the meantime, both sides had to live and let live.

Despite negative reporting and criticism from the media, I kept the doors open. This open-door information policy gradually began to yield dividends once the media started noticing that our change management programmes and improvements were moving the Corporation in the right direction. Soon enough, as our change management programmes began to be felt on the ground, the print and electronic media warmed up to us and changed their tone regarding our service delivery, reflecting a dramatic shift in public opinion. We began to gain favourable media coverage from mid-1999 when complimentary stories started appearing in newspaper headlines.

Briefly, the newspaper stories focused on our growing profitability, commercial viability and effective pro-customer management. By the end of 2000, the papers were reporting on how the government had given a 'big thank you' to NWSC for its concerted change management efforts to transform the Corporation into a viable commercial entity. By the beginning of 2001, the media were praising the Corporation for demonstrating that it was possible for a public utility 'to make money out of water' and that there was no longer any justification for privatisation. Indeed the *Sunrise* newspaper of 17 November 2000 pointed out how NWSC had shown other parastatals how to escape being sold off. The paper made reference to an evaluation workshop in Lira Area where all managers and staff had gathered to reflect on performance achievements and constraints.

What began as a trickle of approval in May 1999, soon after the conclusion of the 100 days Programme, turned into a flood of favourable and enthusiastic coverage and commentaries! The favourite facts that supported the NWSC success story in 2002, for example, were the billings had shot up by UGX 8 billion and operating profits were up by 43%. By the end of 2002, one reporter, writing in *The Monitor* of 4 December 2002, was candid enough to concede that 'until recently the argument that a government parastatal could be profitably run would never have held any credibility in Uganda'.

Since introduction of the change management programmes the NWSC's public relations office has kept its doors open to the media. I have also kept my door open to the print and electronic media for interviews, because I believe that the public, especially the customers, have the right to know what is happening in the Corporation, and we are duty bound to keep them informed. In this context, the media has become an invaluable channel of communication to and from the public.

Of course the media, too, has gained from the growing rapport with the Corporation. We are a source of news and information for the media houses. For the public, including NWSC customers, water and sewerage services are always newsworthy since 'water is life' and, accordingly, its availability or the lack of it is of great interest to, and has far-reaching consequences for, the public.

Compliance with the Master's Voice

Getting information from the public through the media and customer surveys, as well as verbal and written contacts with customers and the general public, is one thing. Putting that information to good use is quite another. It is common for public servants to get information from clients and simply put it on the shelves or in filing cabinets to gather dust without follow-up action. At NWSC, we have tried our best to avoid this pitfall. To start with, our public relations office keeps a file of press cuttings, compiles and analyses each piece of information regarding NWSC that appears in the media. I personally keep my ear to the ground and my eyes wide open in order to get feedback from the customers and the public about the pluses and minuses of our service delivery. I insist that all the managers and members of staff in all NWSC Areas should do the same, for it is their duty to do so. By listening to the Masters' Voices, we are able to gauge shifts in customer perceptions and public opinion, and to respond accordingly.

Once we get the information, from whatever source, the challenge is to analyse and evaluate what the public, especially the customers, say about NWSC performance and service delivery. We try to study the nature and content of the public complaints, comments and suggestions. What do they say about our response to water leaks and bursts, and sewage overflows? What are the customer views about billing and billing accuracy? What do they think about the pros and cons of our customer care and the conduct of our field and front office staff? What do they think about the continuity of water supply? What suggestions do they make about the improvement of water production and distribution? To what extent is the public informed about what we do and how to access our services?

The next stage is to address these issues raised through various channels of communication. Usually, follow-up action depends on the source and nature of the information. The information gathered from various sources is passed on to the relevant departments for appropriate action. NWSC managers and members of staff respond promptly to verbal and written communications and answer queries to the best of their abilities with a sense of urgency and professionalism. The public relations office responds promptly to clarify issues, refute groundless accusations and provide accurate information and appropriate feedback to the public. Information with policy implications is discussed at management meetings and even at Board level. In this regard, information from various sources not only serves as a barometer of customer perceptions and public opinion, but also feeds into the development of a corporate customer care policy and into the construction of the customer care aspects of the change management programmes.

In conclusion

Since 1998, NWSC relations with customers, the public and the media have improved dramatically. Starting from the low ebb of mutual animosity, public perceptions, attitudes and opinion, it gradually warmed to the Corporation, culminating in mutual respect, cooperation and harmony. This favourable change of fortunes in our relationship with the media is partly due to the NWSC success story and open-door information policy, but it is also partly due to media publicity of this success, which has contributed immensely towards informing and sensitising the public. Although we were originally reluctant to interface closely with the media, the publicity and visibility we have achieved over the years have worked wonders for the Corporation, and we intend to keep it that way for the good of the Corporation, its customers and the public at large.

Of course, it was not my leadership alone that did wonders. It is the resolute, concerted and relentless efforts of the entire NWSC family that have transformed the Corporation into a household word. The East AfricaÇs Most Respected Companies Award in the Public Sector Category of 2007 was yet another proof, if any was needed, that our change management programmes were having the intended and desired effect, which merited celebration of its achievements and redoubling of our efforts to secure its future with or without privatisation. The fact that the NWSC is currently contemplating selling shares to the public bears testimony to the impressive progress in performance that has been made since 1998. Indeed, without the support of our customers, the press, teamwork and staff commitment, this progress would not have been possible.

Lessons learnt

- In public enterprises, or any other enterprise for that matter, *the public have the right to know* what is happening and the managers have the duty to inform them. Once they know that you are doing the right things, they will invariably become partners in your efforts to enhance performance.
- The media are *an indispensable tool* in shifting public opinion and perceptions. Adopt an *open-door information policy* to cultivate media support to articulate your programmes, publicise your achievements, clarify issues and correct any media distortions or misreporting. By so doing you will improve your image and render better services.
- Rather be guilty of overexposure rather than of underexposure. Nature abhors a vacuum. In managing change, inevitably, there will be losers who will be only too happy to shoot down the process. By celebrating new developments publicly through the press, you create a buffer of goodwill that will help when the inevitable bad press emerges.
- In service delivery, *know your customers* well, study their complexities and peculiarities, and go out of your way to satisfy their needs and expectations. Always remember: do not argue with your customers, even when they are in the wrong. Accept their complaints/criticisms as warning signals alerting you to the possibility that you are going down the steep slope to complacency and even backsliding.
- *Shifting customer perceptions and attitudes must begin inside your organisation.* Once you build a pro-customer culture among your staff and all your activities are driven by the quest for customer satisfaction, the public, especially the customers, will get on board as partners to enhance performance and service delivery.
- *Always ensure that customers have unrestricted access to you, listen to them, solicit their opinions/advice and show them that you care.* Customers are not only 'the reason we exist' but they are also our 'eyes and ears on the ground' and as such their partnership is instrumental in organisational success.

Chapter Fifteen

Behind the Scenes

The Invisible Partners

The Corporation workers as an organised force of Uganda Public Employees Union have been and are constantly calling meetings with the Corporation management in an effort to solve the problems that affect them, as has always been the case. And that in all meetings held, there have not been any deadlocks on any issues so far discussed.

Robert Wanzusi Mutukhu
General secretary, Uganda Public Employees' Union
15 March 1999

In books like this one, which provide insights into overcoming the challenges of managing public utilities, authors tend to concentrate on the big, visible actors – investors, shareholders, policy makers, government, donors, the media, managers and, last but not least, customers. Also, authors tend to focus on the big picture – the policies, programmes, management styles and strategies and the macro corporate change management programmes like those that have been initiated and implemented in NWSC, from the 100 days to IDAMCs. In my view, this only tells part of the story. The emphasis on the visible actors more often than not ignores or minimises the instrumental and indispensable shop-floor staff – the foot soldiers behind the scenes – who toil day and night, day in and day out, to deliver and sustain utility services around the clock and throughout the year. These actors behind the scenes are comparable to the vital body organs such as the heart, liver, kidneys and lungs that provide lifelong sustenance to the external, visible body. This chapter is devoted to the invisible partners behind the scenes who have played critical roles in the transformation of NWSC.

The invisible partners described in this chapter consist of internal partners – *the employees that make the Corporation tick by selflessly and efficiently managing the broad spectrum of the water supply and sewerage service chain.* The service chain spans from the point of water abstraction to the point of delivering water to the users. These internal partners include the *cleaners, plumbers, water booster attendants, laboratory and plant maintenance technicians, storekeepers and their assistants, plant engineers and distribution overseers and the managers at all levels, both in the Areas and at the Head Office.* Such people, technically referred to as 'internal customers' in NWSC,

© 2009 William T. Muhairwe. *Making Public Enterprises Work*, by William T. Muhairwe. ISBN: 9781843393245. Published by IWA Publishing, London, UK.

work to produce and distribute water to customers, repair water pipe leaks and bursts, deliver bills, procure and store inputs and maintain waterworks, as well as the physical facilities and offices. The managers in the Corporation, in the Areas and at the Head Office alike not only provide day-to-day guidance, supervision and leadership for the running of the Corporation, but also have been among the lead champions of the change management programmes that have progressively transformed the NWSC into one of the best managed public utilities in Africa. Without the dedication, hard work, teamwork and cooperation of the internal actors, our achievements would not have been possible and the whole NWSC water and sewerage services would surely grind to a halt. This book would therefore be incomplete without mentioning the significant role the rank and file of the indefatigable NWSC staff have played in transforming the Corporation.

Hard times: understanding who to work with

Understanding the traditions, beliefs, practices and behaviours of workers is one of the first points of entry for successful design and eventual implementation of change management. It is important that one knows what workers value and what it takes for them to be committed to the job. The following sections describe my initial analysis of the workers, who I was destined to work with to turn around NWSC.

Deep-rooted beliefs and practices: Before joining NWSC, my earlier work experience as the chief executive of the East African Steel Corporation and subsequently as the deputy chief executive of the Uganda Investment Authority had offered me adequate opportunity to assess the actions and practices of the staff in the public and private sectors in Uganda. Overall, I had learnt that the historical environment of our country had played a big part in shaping the practices of the staff. The 1990s could best be described as a decade of hard times. During that decade, Uganda had a low and depressed wage economy in which most employees found it hard to make ends meet, forcing them to resort to all sorts of mechanisms and contrivances, such as moonlighting and informal sector activities, in order to survive. Uganda's economy can also be described as an 'economy of affection or altruism' in which the *breadwinners have to carry social responsibilities beyond their immediate families to cater for extended families and relatives.* In this country, and indeed in other developing countries, working people are obliged to contribute to the costs of weddings, graduations, funeral ceremonies, etc. over and above the day-to-day struggle of meeting the costs of food, clothing, shelter, fees, as well as saving for a rainy day. This struggle for survival invariably impacted on the performance, concentration and productivity of employees at their workplaces. Such tendencies were worse in public enterprises like NWSC.

From my experience at the Uganda Investment Authority, I had learnt that the looming privatisation of public enterprises such as NWSC haunted their employees. The pillars of privatisation were not simply divestiture and the transfer of assets to private investors, some local but mostly foreign, but also downsizing, restructuring and rationalisation. For ordinary employees, what privatisation meant in practice was the loss of jobs, the source of income and the means of livelihood, and the premature termination of careers. *A person who has no job to support his or her family is believed to have no pride or dignity and the argument that privatisation improves economic performance is no consolation to such an individual.*

In extreme cases, joblessness has been a slippery slope to the bottle, broken families and self-destruction. It was therefore not surprising that in the 1990s the very mention of the word 'privatisation' sent shivers down the spine of every worker in public enterprises in Uganda, causing anxiety, fear, uncertainty and foreboding. I knew that NWSC workers were no exception. They had seen how privatisation had been carried out in organisations such as Uganda Posts & Telecoms Corporation, Uganda Commercial Bank, Uganda Grain Millers, Nytil and Pepsi Cola. To workers, privatisation in these organisations meant *anxiety, fear, loss of jobs and even, to a large extent, loss of terminal benefits.*

I had also heard on the grapevine that for many workers my appointment as the new NWSC chief executive did not seem to augur well for the future of industrial relations in the Corporation. As an economist with no professional engineering background and no management experience in the water business, I did not appear to be the right manager for the Corporation. Therefore, there were widespread fears that the new chief executive would not be able to getalong with the water engineers and technicians whose backgrounds and experiences were entirely different from his own. Although these rumours were soon put to rest, they were potentially detrimental to amicable industrial relations in the Corporation.

The working relations and staff behaviour I found on the ground: When I took over NWSC, the workforce consisted of management, line staff and unionised workers. The unionised staff were under the umbrella of the Uganda Public Employees Union (UPEU), NWSC branch. A chairman and a branch secretary headed the executive of the Union branch. At the time, the working relations between the workers and the NWSC management were unpredictable. Staff behaviour and attitudes were depressing, to say the least. Overall, there was a civil servant mentality, characterised by the 'I-don't-care' attitude, prevailing in the Corporation. Workers contrived to dupe the Corporation that they were working hard from 8.00 a.m. to 5.00 p.m. by the 'coat behind the chair' ploy as if such a trick was proof of efficiency and productivity. Work was programmed around the tea break and lunchtime. In summary, benevolence and paternalism were the central characteristics of work relations.

Given this sort of staff behaviour, I was not surprised that the performance of the Corporation was blinking red light. Such a culture could not enhance productivity. Some of the managers have retrospectively acknowledged this situation in their memories. For example, Charles Ekure, then Lira Area manager, recalls

> During that time the Corporation was making losses, it was overstaffed and general staff morale was low, hence poor performance.

The organisational behaviour that prevailed in the Corporation according to some of the managers was a clear recipe for poor performance. This was manifested in the manner in which things were being done. There was, for example, a lot of bureaucracy in decision making. Managerial decisions that could be made in minutes took days or even weeks. The most affected areas were procurement, write-off of bad debts and staff recruitment. Even where the staff could make quicker contact decisions, communication was through memos. *There was enormous difficulty in securing approvals regarding critical operational decisions and activities.* In the end, work suffered and the customer was lost. There was intrigue, political cliques, bitterness and infighting in the organisation.

Surely, something had to be done

With the type of a worker described above, I had to look for a strong ignition. I knew that the only way to succeed was to send a clear signal and work fast with my management to align practices, beliefs and operating environments of workers who, as we all know, always stand at the centre of any performance initiatives.

Sending 'curtain raiser' signal to all the staff: The first opportunity to do this was during my inauguration. Indeed, on that occasion, I made it very clear to all the staff that it was time for each individual to justify his or her presence in the Corporation. I reminded my audience, in the presence of the Board, that I had information regarding those involved in negative practices such as collusion with customers to defraud the Corporation, mistreatment of customers, arrogance and laziness, and warned that this would not be tolerated during my tenure in the Corporation. I promised swift action to reprimand and even dismiss staff found to be involved in such acts. In an article entitled 'New NW&SC chief to sack pen thieves', *The Monitor* of 19 November 1998 captured the essence of my message when it reported

> The newly appointed managing director of NWSC, Dr William Tsimwa Muhairwe, has said the first 100 days of his management will turn around the Corporation. He promised to sweep the Corporation of liars and thieves.... "I will sack on sight, even pen thieves. If I look directly into your eyes and detect a lie that will be your end because one plus one will be equal to two not three"... Muhairwe told top managers on his first day in NW&SC at a function held in the Corporation's human resource centre, Bugolobi, on November 18.

During a meeting convened to formally introduce me to the Area managers, I repeated my warning that the poor performance would not be tolerated in the Corporation and called on all the Area managers to pull up their socks. In their recollections, most of my managers concur that my introductory statements were very harsh and 'hair-raising'. They were indeed a 'curtain raiser' signal! I intended it so, that we all start from a right footing.

For example, Andrew Sekayizzi, then the Masaka Area manager, recalls

> The new MD in his address emphasised his hatred for thieves, rumourmongers, liars, lazy characters and non-performers. He declared that he could see such people among us and cautioned that those still in slumber-land were living on borrowed days ... However, on the positive side, and in clear terms, he consoled us that he likes efficient, effective and hardworking people, he was in for reward for good performance.

Similarly, Joel Wandera, then the Jinja Area manager, got the nepotism message loud and clear. He recollects that

> The Managing Director did not have any relative as far as work was concerned; he did not care where someone came from, whether from Mars, the moon or earth so long as one produced results.

Joseph Kaamu, the Mbarara Area manager, recalls that

> His first message was he hated liars, hated people who embezzled public funds and did not like people who reported others for favours. He was for hard work, efficiency, setting goals, monitoring and achieving them, not only for the organisation but also for individuals.

Balancing staff welfare and Corporation's interests: After sending the first signal, the next thing was to get to real work. Staff practices, actions and relationships which had impacted negatively on the Corporation's performance had to change. Apart from other high operational costs, the expensive, large and inefficient labour force was a big hindrance to rationalisation and revitalisation for the much needed corporate improvement. The situation in NWSC inevitably necessitated painful remedies that would affect the welfare and livelihoods of its employees directly. I had been told that my predecessor, Eng. Onek, had also had similar feelings and observations and that is why he had spearheaded major restructuring activities in NWSC, involving the use of PSP in Kampala. Therefore, as the lead manager, I had no option but to work with NWSC management and the Board to undertake the necessary remedies. I knew that in the short run it would be very painful and perhaps disastrous for some staff. Ultimately, the staff had to endure a lot of hardships for the benefit of the Corporation and the entire 'water family'.

We had to take action to correct extant staff-related institutional imbalances and distortions at the risk of hurting individual workers and their families. Staff levels were unrealistically high and our duty was to scale it down. It was time to strike a balance between staff welfare expenditures and promoting the Corporation's interests. A convergence of objectives had to be sought and I was determined to ensure that this happened. I was aware that staff support was essential to the survival of the Corporation. Nonetheless, the continued and unchecked expenditure on such high staff costs was bound to cause more financial problems. I made it categorically clear to all NWSC senior managers that we had to do the right thing however unpalatable it might be in the short run. Hence, in this respect, the viability of the institution had to be our priority. As detailed in Chapter 2, the Board had challenged me to turn around the Corporation and so we had no option but to reduce staff costs. This called for new strategies in cost reduction on staff expenditures. The Board and management agreed to revisit some of the policies regarding employee benefits and terms of service, e.g. staff medical scheme and transport for senior officers.

Staff rightsizing: As a first specific step, we had to optimise our staff levels immediately. Our observation was that it did not make business sense to continue with a large workforce without looking at the productivity side of it. By then we were operating at an unrealistic staffing ratio of *36 staff per 1,000 connections* against an African best practice of less than *10 staff per 1,000 connections*. Of the challenges that we were facing, none was potentially more harmful to the Corporation's industrial relations than the planned downsizing of the workforce within a period of less than two years. Yet, it was generally agreed that a workforce of 1,850 was far too large, unproductive and unaffordable for the Corporation. The plan was to downsize the work force by 50% in less than two years. However, the task of terminating workers' employment was a sensitive matter – the incomes and livelihoods of hundreds of people and their families were at stake – that had to be handled with utmost care.

We all feared that downsizing the workforce would provoke stiff resistance from the unionised workers and could lead to industrial action that would disrupt the operations of the Corporation. The challenge that we were facing was to persuade the workers and their union that downsizing in the long run was in the best interests of all the stakeholders of the Corporation however painful it was in the short term. *Those who were destined to lose their jobs could not share the view that turning around the Corporation and ensuring job security and better remuneration were sound objectives, for once they were shown the exit, they had no hope of benefiting from the Corporation's financial health except perhaps as customers.* Though their fears were understandable, leaving the status quo intact would have led to the collapse of the Corporation and loss of jobs anyway. *In business as in life, it is better to lose some than lose all.*

We sensitised our staff on the demerits of a huge and unproductive workforce. I informed the Union that the threat of privatisation was already looming large over the Corporation, and that we had to reduce our staff levels whether we liked it or not. Initially, the Union was up in arms on this matter but, eventually, they realised that with privatisation, we would all be losers. As already pointed out in the previous chapters of this book, I explained to the workers that if we all cooperated and worked very hard together, there would be no need to have NWSC privatised since our hard work would deliver the same outputs anyway! I told them that this was the only sustainable way in which they would secure their jobs and yet *fulfil the privatisation objective of improved water sector performance*. Eventually, they realised that being antagonistic to the inevitable wind of change was bound to be counterproductive. Accordingly, the first JNC (Joint Negotiation Council) meeting held in 1998 on staff retirement was not centred on whether some staff should be retired, but on how they could retire as honourably and painlessly as possible. The other issue was how to lure staff to retire voluntarily. The mandatory retirement age for all the staff in the Corporation, except those on contract, was 55 years. To single out staff to be retired before that age necessitated management to prove that such staff were more of a liability to the Corporation than an asset. This was practically impossible since almost all the staff had very clean records on their personal files. Although the authenticity of this data record was suspect, there was no sure way of disputing it.

Somehow, we had to come up with a win-win solution. We involved the Board, management and the Union to devise an attractive early-retirement formula to entice as many workers as possible to retire voluntarily. The Human Resources Department visited several organisations where retrenchment had been effected to construct an acceptable formula. These organisations included the former Uganda Posts & Telecoms, Uganda Commercial Bank, Bank of Uganda, British American Tobacco Co Ltd and the Ministry of Public Service. The Union, on the other hand, carried out its own surveys. When the two parties met at the next JNC, they hammered out what has turned out to be the best collective bargaining agreement (CBA) formula on retrenchment/retirement.

The retirement formula was so attractive that more than 50% of our staff were eventually retired without creating any animosity or industrial unrest. The negotiation exercise centred on a number of issues, most of them related to performance-inhibiting factors, such as the irrational performance appraisal system, redundancy/excess staff and incapacity due to illness. Others included staff who could not offer effective service to the Corporation owing to domestic problems and perceived redundancy.

Surprisingly, as the implementation process progressed, we realised that the scheme had become too attractive and some of our good and capable employees had started applying for retirement. I was worried about losing such staff. *As a result, we agreed in the subsequent JNC that, henceforth, the management reserved the right to approve or reject any application, though not unreasonably.* The scheme involved additional facets, such as continuous downsizing. It allowed staff/workers who had worked diligently for years up to the mandatory retirement age of 55 years to benefit, using the same computation criteria. In short, the elements of 'fairness' and 'justice' were the major drivers of the scheme.

The smooth running and acceptance of the scheme was a great relief to the workers and management because it made it possible for the staff to be motivated, work harder and put in long years of service knowing well that at the age of 55 years, they would walk away with a package commensurate with their long years of hard work and dedication. All stakeholders, namely the workers (Union), management and the Board agreed that undisciplined staff should not benefit from this scheme and their exit should follow the established disciplinary procedures stipulated in the terms and conditions of service strictly.

Our main challenge, therefore, was no longer how to retire staff but how to pay off those who were willing to go. The numbers were overwhelming and we did not have ready funds to pay them off. We appealed to the government for assistance in this regard. Consequently, we agreed on the mechanisms, principles and procedures with the Ministry of Finance, Planning and Economic

Development, which eventually provided the requisite financial support. Using a combination of financial support from the government and our own internally generated funds, we were able to retire over 1,000 staff from the Corporation under this scheme. This created room for replacement of less capable staff with more energetic and motivated workers. With a more optimal staffing level, productivity greatly improved to seven *staff for 1,000 connections by 2008, comparable to the African best practice.*

Reviewing the medical scheme: Next in the row of strategies was the alignment of the medical scheme. Initially, the Corporation had a generous medical scheme, which was being abused. As a result, the establishment costs associated with this scheme were unreasonably high. We resolved that the 'arithmetic and gymnastics' of this scheme needed rethinking and careful scrutiny. Under the scheme, every member of the staff was entitled to free medical services for self and the entire family. There was no limit to the number of spouses and children who were entitled to free medical treatment at the expense of the Corporation. In some cases, even relatives and friends benefited from this generous medical scheme. This irrational generosity had to stop. However, this was yet another very sensitive matter affecting the welfare of the staff. We adopted a transparent approach to resolve the conundrum. The matter had to be resolved through a joint effort of the Board, management and the Union.

Fortunately, the Union was supportive of the initiatives to turn the Corporation around. Accordingly, a resolution of the JNC of 1999, which agreed to limit the beneficiaries of medical services to a spouse and a maximum of six children, sailed through without much opposition. This limitation understandably caused shock waves among the employees and their families. For example, in an article entitled 'NWSC workers complain', *The New Vision* of 10 May 1999 reported

> Unionised National Water and Sewerage Corporation workers have asked the management to change the policy, which entitles only six children per worker to medical care. They said most workers had more than six children. 'Workers felt the restriction leaves out many children', the Lira Union branch chairman, Mr Angelo Banya, told Managing Director William Muhairwe and Board Chairman Sam Okech in Lira last week.

Revisiting the transport policy: Transport was another cost centre contributing greatly to the high administrative costs of the Corporation. Before 2000, chief managers and the heads of the departments were entitled to chauffeur-driven vehicles, free vehicle maintenance, free fuel and free upcountry trips. These privileges and benefits were not only a source of resentment among the rank-and-file workers but they were too expensive for the Corporation to sustain in the prevailing financial conditions. The vehicles were used to carry managers' children, wives and relatives in addition to doing other non-Corporation-related business such as ferrying building materials and helping friends.

Besides, this transport policy was not in the long-term interests of the managers themselves because the vehicles were tied to the office and, therefore, managers who left the Corporation lost their means of transport. For these reasons, we monetised the benefits and introduced a car loan scheme for entitled officers.

Under this scheme, the managers could get car loans from commercial banks, guaranteed by the Corporation. The car loans were payable in instalments over a period not exceeding 24 months and deducted at source from the Corporation's payroll. The Corporation gave a consolidated transport allowance to managers with car loans to buy fuel and maintain their vehicles. We technically referred to this initiative as a wet lease scheme. However, like all new policies, the introduction of the car loan scheme and the monetisation of transport benefits were initially treated with suspicion and scepticism, if not with outright opposition. To start with, the car loans were not enough to buy new vehicles. Managers could only afford to buy secondhand or reconditioned vehicles, which were expensive to maintain. The managers felt that the car loans had made them poorer and that, in devising the car loan scheme, the Corporation had taken them for a ride.

Although the car loan scheme was initially a potential source of conflict in industrial relations, as the benefits became visible, it became acceptable and popular among the managers and even among other members of the staff in the Corporation. With time and patience, managers were able to sell their old vehicles and buy newer, more roadworthy ones. In addition, commercial banks became more confident and willing to advance bigger loans once they realised that all parties respected the scheme. By 2005, all top and middle-level managers and a big number of secretaries were proud owners of 'my car'. The policy, in the end, gave rise to the challenge of inadequate parking space at the headquarters. As a result, we resorted to hiring parking lots outside the premises to ease parking congestion for our staff members' cars.

To the rank-and-file workers, the new transport policy was a welcome idea. By providing private means of transport to managers, the policy released the Corporation's vehicles for field operations. The new policy also checked the misuse of vehicles by drivers and managers outside working hours, which is a common practice in the government and public enterprises in Uganda. The availability of the Corporation's vehicles for official duties contributed to the improvement of response time to leaks and bursts and generally enhanced overall operational performance in the Corporation. Of course, some senior managers continued to use the vehicles outside working hours, which, apart from undermining official policy, was resented by junior employees because it amounted to a double benefit. The old adage, 'Do as I say, but not as I do', still remains a residual problem in NWSC. But this is more of a managerial control headache than an institutionalised policy dilemma, as it used to be in the past. We reprimanded the few culprits who committed this offence now and then.

Challenges with the measures taken: staff protests

While implementing the above drastic reform measures, I was nervous about the eventual reaction of the staff and the Union. Indeed, as I pushed staff reforms and other cost-cutting measures hard across the board, sections of the staff started being very critical of any expenditure related to the MD's office. Some protested against the cost-cutting measures while others resorted to the media, making wild allegations and accusations about the office and person of the MD. A case in point was the report of an imminent strike by NWSC staff protesting about my allegedly lavish spending on a trip to Germany and buying an expensive vehicle and office furniture. On 15 March 1999, for example, *The Monitor* reported

> National Water & Sewerage Corporation (NWSC) workers are planning a massive protest against the alleged 'lavish spending' of their new boss, Dr William Muhairwe. The Monitor has learnt that the planned workers' protest was mooted shortly after the new MD demanded a brand-new Pajero Intercooler at a mind-boggling UGX 69 million.

According to the media, the new MD was given a posh Prado (Land Cruiser), which he had rejected, and then he demanded that the Board purchase him a new one at the whopping figure of UGX 69 million. Although this purchase was approved in the budget well before the recruitment of the new MD, workers could not wait to investigate and put the record straight. They further complained about my salary (about UGX 5.5 million: an equivalent of USD 3,500 at that time) which they claimed was one of the highest in Uganda. There were also accusations that I spent UGX 3 million on furniture for my office, although the decision to buy the furniture had been made before I was appointed.

These protests did not surprise me! I had anticipated it, given the history of the Corporation. I had been told that the general perception about the Corporation was that it would be very difficult for me to succeed. Nonetheless, this did not deter me from my crusade to press for more reform measures.

A timely lending hand: the Board and Union support

While I had anticipated such reactions from the staff, I was not quite sure how to handle any industrial unrest that might arise in reaction to change. The prospects of a 'civil war' between the workers and the management within the Corporation could not be ruled out. At worst, the unionised workers were likely to resist the change management programmes and to disrupt water production and distribution, as well as delivery of water and sewerage services, through recurrent industrial action. Such action would inevitably have hindered the progress of our change management initiatives. Even if the workers did not resort to industrial action, including strikes and work stoppages, they were likely to be despondent and sulky, attitudes and behaviour that would have slowed down the recovery of NWSC. Given the apparently contradictory interests of workers and management, the stage looked set for confrontation. Fortunately, neither side chose to follow such a course. Instead, both the workers, through their union, and management opted for consultation and negotiation to resolve the industrial relations challenges that the Corporation was facing. This, in the long run, proved critical to the recovery and transformation of the NWSC.

Although some workers were divided in their opinions about my pluses and minuses as the NWSC chief executive, union leaders showed goodwill by giving me the benefit of doubt to the first of our change management initiatives – the 100 days Programme. To my pleasant surprise and relief, for example, the general secretary of the Uganda Public Employees' Union at that time, Mr Robert Wanzusi Mutukhu, went out of his way to dispel the rumours regarding any form of impending industrial action. In a lengthy press release, published in all print media, the Union secretary responded to the press reports that had appeared in the print media and made it categorically clear that, as an organised force, NWSC workers were too sophisticated to take industrial action on what the chief executive was entitled to. The letter went on to point out that the alleged reports to the press appeared to have been written by some disgruntled and dissatisfied individual (s) who might have lost some benefit as a result of the cost-reduction measures recently launched by the new leadership.

Consequently, the secretary general of the Union stated that the workers' union wanted to reassure the entire public that the article did not only intend to damage the person of the new MD, Dr William Muhairwe, but was also misleading and did not represent the attitudes of the entire workforce of NWSC. Furthermore, he asserted that NWSC had a very committed workforce with established grievance-handling procedures that could not go on a strike over the MD's entitlements. He reiterated that as workers of the Corporation, they would continue to serve the public with all dedication and support the new MD in his efforts to improve service delivery and the image of the Corporation.

By coming to my defence, Mr Mutukhu sent out a clear message that NWSC workers were serious partners. From his press release it was apparent that the disgruntlement within the Corporation did not reflect the general feeling of most workers or the official position of the Union.

Apart from the Union, the Board also issued a supportive press release published in *The New Vision* of Friday, 9 April 1999. The statement made clarifications on allegations that the MD had pressurised the Board to purchase a new vehicle, stating that the vehicle had been budgeted for before the MD was appointed. The Board also shed light on the purchase of furniture and clarified that the MD's furniture was part of the bigger planned purchase for other Corporation offices, which was also duly budgeted for prior to the MD's appointment. In addition, the trip to UK and Germany, which had become a centre of accusations, was explained as a necessary undertaking that was initiated and approved by the minister of Water to benchmark best utility management practices abroad. In addition, the trip was to be used as an opportunity to follow up on stalled donor-funded projects. Finally, the Board shed more light on allegations of corrupt tendencies (e.g. forgiving corrupt staff) and the transfer of staff who had overstayed in particular work places in the Corporation; all of which had also appeared in the press. It was clarified that staff transfers were normal and in accordance with approved terms and conditions of service. On corruption tendencies, the Board made it clear that as far as forgiving corrupt officials was concerned, the government had fully fledged machinery for handling criminals of all sorts. Therefore, the MD could not acquit any person from or condemn any person for any criminal offences as he did not have the machinery to do so. The Board, therefore, concluded that the management could only exercise grievance-handling procedures in accordance with the terms and conditions of service.

In a nutshell, the Board appealed to all the stakeholders either individually or collectively to assist management in the creation of an enabling environment in which the Corporation would provide the public with the required water and sewerage services. They advised that any member of the public or any employee of the Corporation was free to approach the Corporation directly and his/her problems or suggestions for improvement would be dealt with expeditiously. They further stressed to the public that problems could not be solved through anonymous letters and inaccurate reporting.

The campaign for change

These voices of support from the Board and the Union lifted my spirits and raised hopes for NWSC. Far from obstructing and slowing down the pace of the Corporation's change management programmes, *the Board and workers turned out to be a positive asset that eventually played a vanguard role in the transformation of the Corporation.* Although many workers initially saw the 100 days Programme as a big joke and adopted a 'wait-and-see' attitude, as soon as it began to yield positive results, they embraced the Programmes and played a leading role in ensuring ultimate success. The Union leadership was particularly helpful in persuading the unionised workers to abandon their 'business as usual' attitude and fully participate in the successive change management initiatives that ensued. The union leaders teamed up with the management to mobilise and sensitise the workers on the importance of discipline, hard work, setting goals and meeting targets. They emphasised the need for transparency and integrity in order to ensure and sustain the success of the Corporation to the benefit of all the stakeholders.

It is not surprising that relations between the management and the workers gradually and steadily improved. Once trust was developed between the management and the Union, the new Union Executive (2000–2001), headed by Mr Peter Werikhe as the branch secretary, embraced the change management programmes without reservation. On 6 April 2000, for example, the Corporation signed a fresh 'Recognition and Procedural Agreement' with UPEU. One of the provisions of this stated thus

> Both parties pledge their readiness to safeguard the interests of the Corporation, particularly promotion of a positive public image through participation in providing and facilitating effective delivery of good quality services, efficiency at work and improvement of productivity.

The Union developed new and proactive methods of work. They decided, henceforth, to make sure that all their 'brothers' and 'sisters', as they call them, understood the positive messages of the NWSC leadership in order to encourage togetherness and change ownership. They visited different NWSC Areas, sometimes accompanied by the Corporation's human resources manager, to exchange ideas with the workers and to explore ways and means of improving industrial relations. Furthermore, they organised workshops to discuss business ethics and principles, workers' rights and responsibilities and the progressive relationship between workers and management.

In these meetings, the Union officials always underscored the fact that the divisions between 'us' and 'them' had to be abandoned. The drive was to find a strong congruence between the workers' and managers' goals and objectives in the interest of NWSC. The background to this was premised on the fact that both sides were interested in increased productivity and enhanced corporate performance, which are the only ways to ensure job security and satisfaction, and to improve workers' remuneration. The union leaders and the NWSC management had no choice but to be partners in the development of NWSC. With the advent of IDAMCs, they even became 'shareholders' in the Corporation's undertakings, risks and incentive plans, as detailed in Chapter 9. During our performance improvement celebrations, workers' representatives, as partners in development, were always given a platform to address the gatherings. Their speeches were not confined to the welfare of workers, but often covered strategic issues aimed at increased productivity and customer satisfaction. Whenever workers heard those speeches, they took them as manifestations of goodwill and of management's commitment to involve them in NWSC change management initiatives. They saw themselves as part and parcel of the bigger picture of enhancing performance strategy.

This good partnership between workers and management could not have been possible without the crucial role played by key union leaders and Board members. Notably, all the shop stewards, led by Messers Laro Wod Ofwono, Peter Werikhe and Robert Matukhu, played cardinal roles in convincing the rest of the staff to work hard in order to increase efficiency and subsequently improve their welfare. On the other hand, all the Board members headed by the chairman Mr Sam Okec spent sleepless nights defending and protecting the new management efforts. *Specifically, Drs Henry Aryamanya and Abdullahi Shire, Eng Francis Openyeto, Mr Patrick Kahangire, Dr Kyabaggu, Messrs Lawrence Bategeka and David Sebabi and Mrs Hawa Nsubuga, all members of the Board, were always in touch with workers to mitigate their grievances.* This approach ensured that the harmony between workers and management grew unimpeded.

Doing the inevitable

We followed a number of staff-oriented approaches to mobilise buy-in of our programmes. In doing this, we used our experiences and results of the situation analysis to put in place things which we felt mattered most. In the following sub-sections, I highlight some of the approaches we used with a lot of success.

Enhancing worker involvement: The relationship between the NWSC workforce and management was anchored in the principle of mutual respect and reciprocity. 'You scratch my back, I scratch yours' had to be the operating principle in NWSC industrial relations. From the outset, we, as management, were conscious of the fact that it was important to cultivate and sustain the support of workers for the change management programmes to succeed. The only practical and realistic way of doing so was to involve and consult the workers at every stage of each initiative. This was done in order to tap their ideas and enhance their sense of ownership through participation. This was essential, if the initiatives were to be sustained. *It is a serious management mistake to ignore the ideas of workers.*

In any business, there is no monopoly of ideas. Ordinary workers have long and varied hands-on experience that should be tapped to inform policy and to initiate radical management changes of the type that has been unfolding at NWSC. It was, therefore, imperative for management and the Union to develop institutional mechanisms and channels through which workers' participation and contributions could be harnessed to facilitate the recovery of NWSC and to transform it into a commercially viable public utility to the mutual advantage of all stakeholders.

The conception of the 100 days Programme was the first test of worker involvement. I convinced management that we could not make any meaningful headway without sufficient buy-in from workers. *Fortunately, everybody agreed that the benefit of change ownership was a critical ingredient in this first attempt at short-term enhancement programmes. We had no choice but to move with the workers!* As has been described already in Chapter 2 of this book, after explanations and sensitisation of workers regarding the 100 days Programme, I noticed that NWSC workers had slowly but surely begun to warm up to it. They attended the workshops to work out the targets and strategies to achieve those targets. At the end of the 100 days, the fact that the Corporation posted a surplus was a great encouragement. Because of their involvement and support, what looked like an impossible task suddenly appeared achievable. We all noticed that with the new re-engineering approaches, we could, together, move the Corporation to greater heights.

The intensity of worker involvement was more pronounced during the SEREP. This time, workers and staff actively took part, not only in the setting of targets, but in their implementation as well. Workers divided themselves into 'cells', and the 'cells' were entrusted with different tasks such as revenue collection, billing, arrears collection and cost reduction. The mindset of the workers began to change dramatically.

If an Area did not win a trophy or cash award at the evaluation, its workers felt very much let down and disappointed. This challenged them to work harder and to make sure that they scooped something next time. Because these awards were publicly given out at a function held at the venue of the overall winning Area, the workers began to rally behind their Area managers and participate actively in the implementation process. In fact, this was the time when the real talents of Area managers, with respect to participatory management and action, began to show.

The implementation of Area Performance Contracts (APCs I) and Support Services Contracts (SSCs I) from October 2000 to August 2002 demonstrated that the workers were more committed to change than ever before. Because of this goodwill, it became easy for the management to work with the staff to incorporate rewards and penalties in the performance drive machinery during APCs and SSCs. If the staff worked hard, they would receive more money and if they worked less, they would lose part of their pay. The monthly incentive payment meant additional income to the workers and, therefore, increased their household income.

The stamina of workers became more apparent at the preparation stage of APCs and SSCs. Throughout this very daunting task, the staff worked hard for long hours, sometimes into the wee hours of the morning, to come up with comprehensive and tailor-made business plans. Amazingly, not a single staff member downed his/her tools or resigned as a result. If anything, each worker took the preparation of the business plans as a 'personal challenge'. This was a deciding moment for staff transformation from the 'I-don't-care' attitude to 'hard work with commitment'.

Empowerment of workers: No programme was as pro-worker as the Stretch-out Programme! When I joined NWSC, many workers were highly qualified but lacked empowerment and facilitation. There was too much control at the centre. This meant that there was dependency more than empowerment and accountability. We deflated the dependency syndrome and made each worker accountable for his/her actions. I was fully convinced that once a worker is challenged to be accountable, the chances are that he/she will endeavour to make correct decisions. That is why we decided to empower the Areas through operational autonomy. The change management programmes were 'home-grown' and participatory, and left no room for dissent. Workers appreciated the new emphasis on Area self-management without 'interference' from the headquarters.

The elimination of bureaucracy and the inculcation of *tenets of simplicity and worker involvement were the main motivators of staff morale, commitment and productivity.* The 'hump-crushing' Stretch-out Programme, which blurred hierarchical identities through, for instance, the wearing of NWSC T-shirts by both managers and plumbers on Fridays created a sense of equality and industrial relations harmony and participation in the Corporation.

Encouraging worker creativity: Throughout the process of our change management programmes, it was thrilling for me to see how workers came forward, on their own, to contribute superlative ideas aimed at continuous improvement. Staff demonstrated a strong sense of ownership, and sometimes I was overwhelmed by their contribution to the success of the programmes. The IDAMCs experience, for me, provided the most important insights. For example, workers took the lead in coining the names of their partnerships; names such as Greater Entebbe Partnership and others mentioned in Chapter 9 were workers' innovations without guidance from the management. That is why each Area was very proud of its new identity. Areas also crafted their own work schedules and came up with their own slogans. For example, the slogan for Kampala Water was 'Every Drop Counts'. All this created a strong sense of autonomy and ownership, thereby contributing greatly to the performance improvement efforts of the Corporation.

Keeping industrial relations doors open: The most important institutional mechanism for governing industrial relations in NWSC was the JNC, which consisted of the Corporation's Board and management, on one side, and Union representatives, on the other. The purpose of this bilateral council was to negotiate the workers' terms and conditions of service, including their remuneration packages and welfare and retirement benefits. Over the years, this Council played an important role in promoting industrial relations harmony and cooperation within the Corporation, and in facilitating the implementation of successive change management programmes. It is usually through this Council that the Corporation and the workers' union representatives negotiate and review various aspects of industrial relations and sign binding collective bargaining agreements.

To ensure transparency, fairness and natural justice, other institutional mechanisms have been the *disciplinary committees* at the Area and headquarters levels, on which the workers are represented and whose task is to curb misconduct and wrongdoing. These institutional mechanisms have gone a long way in creating industrial relations harmony and this has, in turn, underpinned our change management achievements to date.

Since 1998, the management and the Union leadership reached a number of agreements. These agreements contributed greatly to the smooth and effective implementation of change management programmes in the NWSC. The most interesting agreement was the 'Recognition and Procedural Agreement' of 2000. The objective of the agreement was to promote the principle of mutual recognition and acceptability based on agreed practical procedures and regulations for the benefit of the Corporation and its workers. *Under this agreement both sides pledged 'their readiness to safeguard the interests of the Corporation, particularly the promotion of a positive public image through participation in providing and facilitating effective delivery of good quality services, efficiency at work and improvement of productivity'*. Furthermore, the two parties committed themselves 'to attaining continuous improvement in workers' earnings and terms of service, including the general safeguard and promotion of social aspects of workers'.

In practical terms, the Union recognised the right and duty of the management to run the Corporation in the best interests of all the stakeholders. On its part, the Corporation recognised the Union as the sole negotiating body on behalf of all the unionised workers, with the right to organise and represent them in matters regarding their terms and conditions of service. This agreement formed the basis for industrial relations in the Corporation and for full worker participation at every stage of all its successive change management programmes.

Enhancing staff capacities and qualities: No organisation can register continuous efficiency gains without continuous skills development. Thanks to the digital revolution, the quest for new skills and new knowledge has been gaining momentum around the world. *In this rapidly changing world, those with the mastery of requisite technological skills, especially in information communication technologies, will forge ahead and enhance their performance, productivity and competitiveness, and those without will be relegated to irrelevance and oblivion, if not extinction.* I, therefore, knew very well that the survival of NWSC's performance gains depended on deliberate investments in its human resources. Those who do not invest in their people will do so at their own peril. Fortunately, in Uganda there has been an upsurge in the 'back-to-school mentality' in order to quench the thirst for knowledge and new technologies. In the 1970s and 1980s, the typical member of the Ugandan elite was proud to be 'an importer/exporter'. *Nowadays the typical educated Ugandan is likely to be a consultant, management expert or computer specialist.* In the popular imagination, 'the importer/exporter' has been relegated to the dustbin of history.

Right from 1998 we recognised that staff training and development in NWSC were not a luxury but an absolute necessity if the Corporation was to transform into a viable and sustainable public utility. Accordingly, one of the tasks of the Corporation's human resources manager was to identify suitable courses at home and abroad that would be beneficial to the Corporation and its workers at all levels. The next task was to carry out training needs assessment across the Corporation to identify the training requirements of the institution at all levels, in all Areas, and in consultation with the workers' union. Since 1998, training needs assessments have become annual exercises aimed at identifying the training gaps and matching training with the staff development requirements of the Corporation.

In addition, other development opportunities covered corporate governance, strategic planning and management, public and private sector partnership, marketing, water and environmental management, purchases and supplies, public relations and customer care, taxation, and fire prevention and fighting. This training improved and refined tremendously the skills and professional competence of the NWSC staff and managers at all levels and boosted their performance and productivity. All in all, hundreds of staff and managers benefited from these courses since 2000. *In short, organisational behaviour and corporate culture at NWSC have been revolutionised greatly.*

Focusing on skills performance: When I joined NWSC, I found a strong cadre of professional staff with a lot of potential, capacities and capabilities. My assessment indicated that the staff had potential for performing well at corporate planning, business planning, institutional development, monitoring and evaluation, project management, operations and maintenance, water treatment and quality management, financial modelling and management, stakeholder mapping, etc.

All the requisite water management skills were potentially available, although such skills had not been fully utilised to enhance water and sewerage services delivery. I discovered that the missing link lay in bringing together these skills and mobilising them into a team 'tube'. *I knew that even the best footballers could never win a match if they do not play as a team.* I also discovered that the staff had massed theoretical skills that needed to be converted into practical reality. *In my assessment, it was time to move from talking to doing.*

I resolved that the Corporation could not have 'food' while the customer went 'hungry'. The rich staff capacities had to be utilised. We had to harvest this productivity. We needed to have a boundary-less organisation in order to build strong synergies. I inculcated the need to make *performance our destination, instead of qualifications.* Through my work at East African Steel Corporation and Uganda Investment Authority, I had learnt good mobilisation skills for such 'troops'. Much as I still knew little about water management, the assessment of these gaps convinced me that I was on the right track. I had the task of mobilising the 'troops' to fire the 'bullets' to win a battle that had been raging for a long time. This is what the principle of participatory management entails anyway! I told my staff and management that every *skill was a strong building block in our performance drive.* During my first months at the workplace I, therefore, played the role of a mobiliser as my mastery of the water business evolved.

It was a wonderful experience for me. My contribution hinged on initiating action and insisting on avoidance of counterproductive initiatives. I was, however, careful not to discourage my staff from thinking 'big'. I urged them to look for what works best and provided timely guidance. I followed up what was happening on the ground to make sure that work targets were achieved in a timely manner, giving one-minute praises where staff excelled and one-minute reprimands where they reneged. My work was like that of a farmer who provides manure and waters the flowers so they blossom. This approach paid off. Staff and management had the feeling that they were part and parcel of the delivery process of our subsequent efficiency gains.

Other staff concerns

In a big and complex organisation like the NWSC, it is difficult, if not impossible, to satisfy all the workers at all times. It is, therefore, not surprising that despite the existence of excellent industrial relations in the Corporation, there have been instances of staff grumbling and complaining about some of the policies of the Corporation. Apart from issues related to the new medical scheme, transport policy, new managing director's entitlements and foreign trip described earlier in this chapter, staff had other concerns that had to be tackled resolutely.

The role of international private operators in the delivery of water and sewerage services in Kampala posed great challenges to both the NWSC and its workers. First of all, the workers were not familiar with the concept and practice of private sector participation and they were not sure how the international private operators would affect their terms and conditions of service. Under the management contracts, the workers continued to be employed by NWSC. For example, under the KRIP contract, staff were not seconded to Gauff; yet they had to work for, be supervised by, and report to Gauff. *In this dichotomous situation, the lines of responsibility and accountability regarding industrial relations were bound to be confused, blurred and, ultimately, prone to generate conflict.* In addition, under KRIP, workers were confused by the introduction of a different working culture. The expatriates, mostly Germans, had a strong disciplinary approach, which was a shock to the workers who were used to being handled with kid gloves. However, when OSUL expatriates, who were mostly of French nationality, came in 2002–2004, they were seen as simple, democratic and *laissez-faire* oriented. This change of working culture caused confusion among staff and I used to get constant reports about the trend of events in Kampala. I had to tell workers that change was a fact of life and they needed to adapt accordingly.

The question of remuneration was another intractable source of conflict between the NWSC and its workforce. For workers, remuneration packages are never, and can never be, enough. Human needs, especially in a country like

Uganda where peer pressure is very high and the economy of affection is very strong, are infinite and no employer, however generous, can ever afford to satisfy workers' needs. Nevertheless, in NWSC we have progressed a long way to improving the salaries, performance bonuses, allowances, gratuities and other benefits of the NWSC workforce at all levels to reflect our enhanced corporate performance, productivity and profitability, and in accordance with the overall remuneration structure of public enterprises. Over the years, we have progressively increased the salaries and incentives of our employees. In 2000, salaries went up by 10% across the board. While this appeared to calm the nerves of workers for a while, by 2003 they were becoming restless once again. In January 2003, the workers demanded a salary increment of 28%. Not only was this on the high side but it could not be granted without the approval of the Ministry of Finance. Their main argument, which I found sound, was that the Corporation had registered high productivity gains, without commensurate sharing with the drivers – the workers. This, at face value, appeared utterly exploitative and management was duty bound to do something about it.

For the next two years, there were protracted negotiations and consultations between NWSC management and the workers' union, on the one hand, and between NWSC management and the officials of the Privatisation and Utility Reform Units of the Ministry of Finance, on the other hand, regarding the issue of salary increments. By the beginning of 2005, the workers' union was losing patience and threatening that the workers would lay down their tools unless the NWSC Board and management met their salary demands. On 10 March 2005, the Union representatives stormed out of a meeting of the JNC, accusing the NWSC Board of subterfuge and procrastination and threatening to call a strike. Fortunately, the strike was averted because on 22 March 2005 the two sides reached an agreement in which the unionised workers and management were awarded salary increments of 20% and 5%, respectively. This was the only occasion that NWSC workers threatened to go on strike since 1998, which is testimony to the industrial peace that has prevailed in the Corporation, accounting for its success and change management achievements.

As part of the strategy to streamline governance in field operations, I introduced clandestine mechanisms of catching those involved in illicit activities in the field. These included giving selected customers marked money notes in the field so that if our field staff demanded cash before carrying out work they were supposed to do, the money could be identified as exhibits later. Initially, the union leaders were deceived by the culprits and thought that I had intentions to fire staff for no good reason. I had to explain that this mechanism was meant to rid the Corporation of bad elements that were causing customer dissatisfaction. In the end, we all agreed that as long as this exercise was well explained to the staff beforehand, those caught could not rely on Union leadership's support to defend them. This approach helped us to reduce illicit field activities.

In their own words

Nothing illustrates the instrumental role of NWSC members of staff, from the rank and file to top management, better than their personal testimonies since 1998. I sent out a structured questionnaire to a representative sample of workers and managers across the Corporation .while writing this book. Many of the target respondents had been deeply involved in the design, implementation, monitoring and evaluation of successive change management programmes. I therefore sought to find out what, in retrospect, their views about the successive change management programmes were and what roles they had played in the radical transformation of the Corporation.

I received over 30 written responses from a cross-section of officers of varied experiences and lengths of service ranging from 3 to 33 years, both from the Head Office and in the Area partnerships. I analysed all the responses and have selected a sample of those who were deeply involved in the change management programmes. All the authors of the testimonies presented in this section were already working at senior level when the change management programmes commenced, and they had gained varied experience across the spectrum of NWSC operations.

David Isingoma

I head the Department of Corporate Planning. I joined the NWSC in 1997. Being an economist in a predominantly engineering-based institution, my initial challenge was to prove my relevance. The initiation of the change management programmes gave me ample opportunity to prove my worth as an economist and planner – and also to play a leading role in the turnaround of the Corporation.

From the outset, I perceived these programmes, especially the 100 Days Programme, as ambitious. Given the magnitude of the challenges that the Corporation was facing in 1998, I believed that turning it around within 100 days was like daydreaming! Therefore, the fact that the programmes have turned dreams into realities is an astonishing achievement in the history of public utilities in Uganda. Under the overall guidance of the managing director, and in consultation with senior colleagues and external experts, we believe we have contributed to making the NWSC fit for the purpose for which it was established.

I have attended courses and workshops in USA, Germany, Mozambique and Tanzania on a wide range of subjects. I have therefore gained management and planning skills, refined my analytical skills and learnt how to deal with people with divergent and often competing interests.

I have also become convinced that staff participation, which is blended with a sense of ownership, in addition to individual and collective effort, is the key to the success of any organisation. In this regard, the NWSC has been moving in the right direction and the challenge is to sustain the momentum in the years to come.

Silver Mugisha

I head the Research, Monitoring and Evaluation Department, and have hence closely followed the ups and downs of all successive change management programmes in the Corporation. I joined the Corporation in 1994 as an assistant resident engineer on the Kampala Network Rehabilitation Project. Since then I have served in various capacities.

From the outset of the 100 Days Programme, it was apparent to me that the lacklustre performance of the Corporation would soon be history. These initiatives were a refreshing wind of change and I have no doubt in my mind that they account for the transformation of the Corporation from looking 'terminally sick' into a success story without precedent or parallel in the corporate history of Uganda.

Having specialised in water utility management at Master's degree level, I have played a pivotal role in championing most of the change management programmes since 1998. I was at the forefront of the design, planning and implementation, as well as monitoring and evaluation of SEREPs, APCs, Stretch-out, one-minute management and IDAMCs. Under the leadership and guidance of the managing director, and the collaboration of other champions, I spearheaded the packaging and selling of the Programme to all the Corporation's employees at all levels with considerable success.

The NWSC, in collaboration with its development partners, has over the years facilitated my postgraduate training and sponsored my attendance of workshops and conferences at home and abroad, which have greatly enhanced my change management and institutional development capacities.

To start with, thanks to my employer, the Netherlands government sponsored my Master's degree studies in 1998/2000 in sector utility management. Then, in 2002 I attended a World Bank-sponsored course in regulation and performance at the Public Utility Research Centre (PURC) of the University of Florida, USA. During the same year, the GTZ sponsored me to attend a public–private sector partnership course in Berlin, Germany. To crown it all, the German Scholarship Programme (DAAD) sponsored me to do a PhD in performance management and regulation at Makerere University in collaboration with PURC, which I completed successfully in 2005.

My leading role in successive change management programmes and the skills and knowledge that I have gained through postgraduate studies, and other courses, have transformed me from a purely technically oriented engineer to an all-encompassing manager with the professional capacity, competence and motivation to handle complex challenges with confidence and effectiveness.

My deployment on external services assignments has enhanced my exposure and capabilities in the regional and international arenas. I am now well placed and well equipped to discuss international issues with other water utility management experts around the world. As a satisfied manager with a clearly laid out career path and supportive leadership, I have no doubt that I will continue contributing to the success of NWSC.

Since 1998, I have personally witnessed a radical shift in organisational behaviour in the NWSC, especially in the area of industrial relations and worker participation. By inculcating the spirit of staff commitment, performance, participation, customer care and sense of ownership, the NWSC has become a unique public utility with rare qualities in Africa. Although the Corporation remains a public utility, it behaves like a private business enterprise driven by profitability and financial viability. By shifting emphasis to 'the customer is the reason we exist' paradigm and focusing on financial viability and worker partnership without compromising its strong tradition of engineering acumen, the NWSC has become a fortress of outstanding performance and achievements that have made it famous throughout Africa.

Charles Odonga

I am currently the general manager, Kampala Water Partnership. I joined the NWSC as an assistant engineer in 1982. Over the past 23 years, I have worked in various capacities before assuming my present position. I have progressively gained varied professional and managerial experience. I am one of those who can speak with authority on the radical transformation of the Corporation since 1998.

As part of top management, I was from the outset deeply involved in successive change management programmes to improve NWSC's performance, productivity and service delivery. As chief engineer, Operations, and then director of Engineering Services, I chaired the task forces spearheading the planning, implementation, monitoring and evaluation of all successive change management programmes. I was therefore one of the key champions of these programmes, whose task was to conceptualise and internalise the essence of the change management initiatives, to articulate them and then to sell them to all NWSC stakeholders.

The success of the NWSC can be attributed to not only a supportive Board and the inspiration of the managing director, but also the teamwork, commitment sacrifices of the workers to meet the demands and expectations of the Corporation.

My career development through training also taught me the virtues of self-reliance, patience and endurance, the pitfalls of patronage or 'baby sitting' and the importance of separating the personal from the employer's interests.

Thanks to the change management programmes, the NWSC has become a unique public utility in Uganda. *What distinguishes the Corporation from other utilities is its receptivity to new ideas and practices in management and its boldness to try them out.* Other distinguishing features include the emphasis on customer orientation, the inculcation of a sense of ownership of the Corporation among the members of the staff, the centrality of workers in all the Corporation's activities and the behaviour of the Corporation as a private sector business enterprise. These unique virtues have transformed the Corporation into one of the best employers in the country, with a contented and motivated work force, improved service delivery and satisfied customers. By transforming organisational behaviour, work attitudes and ethics, the NWSC has become a renowned success story and a joy to work for.

Johnson Amayo

I joined the NWSC as a sanitary engineer in 1990. I was posted to Masaka as Area engineer. Two years later, I was transferred to Kampala to serve as counterpart sewerage engineer in a European Union-funded Sewerage Rehabilitation Project. Between December 1992 and December 2001, I worked as a plant engineer, Gaba Works, head of the Water Distribution section in Kampala, Area engineer, Kampala, Area manager, Entebbe and chief engineer (Operations) at the Head Office. In December 2001, I was promoted to my current position of manager, Operations, with additional responsibility for static plant maintenance.

I first became involved in the NWSC change management programmes when I was the Entebbe Area manager. During the 100 days Programme, together with the Area task force teams, we performed our duties to the best of our abilities to the extent that Entebbe Area emerged winner in March 1999.

At the end of the 100 Days Programme, Entebbe won another trophy and cash prize for being the best performer in revenue collection countrywide. The success of the 100 Days Programme transformed me into a champion of the subsequent change management programmes, from SEREP to IDAMCs. Indeed, once I got used to change, accepted and believed in change, learnt how to anticipate and adopt to change, I became one of the key drivers of the programmes that have transformed the NWSC into an astonishing success story.

My main contribution to the progressive improvement of NWSC's service delivery and performance has been in the spheres of programme planning, implementation, monitoring and evaluation. My passion for business growth through exemplary performance has driven me to do whatever is within my reach and capacity to ensure the success of successive change management programmes in all Areas. I have even gone the extra mile to enhance my financial, technical and operational capabilities to sustain the commercial viability of the Corporation.

Like many of my colleagues, I have benefited tremendously from the NWSC capacity-building and staff development programmes, which have been funded by the Corporation and its development partners, notably the World Bank, the Netherlands government and Water Utility Partnership. Apart from four short courses in water engineering and management in UK, USA and South Africa, I was sponsored for an MSc degree in sanitary engineering in the Netherlands. The Corporation has financially and morally facilitated me to participate in water business-related seminars, workshops and conferences, both inside and outside the country. These training programmes have enhanced my professional confidence and competence. I am now conversant with the issues and trends in water service delivery and management around the world.

That the NWSC employees at all levels have played an indispensable role in the success of the change management programmes is indisputable. What is needed is to continue nurturing and cultivating the employees' confidence in the future of the Corporation through training, better remuneration packages and refined group and individual incentives mechanisms.

* * * *

David Isingoma is a quiet, simple but resolute fellow without whom all our corporate plans, strategies and information management would have grounded to halt. Mr Charles Odonga has since retired to private practice and consultancy. On the other hand, Silver Mugisha (now Eng. Dr after attaining his PhD) has since been promoted to head a newly created division of Institutional Development and External Services (IDES). In the same vein, Eng. Johnson Amayo, one of the strong supporters of change management programmes, is good at convincing and telling the truth to the staff. He has since been promoted to head the new division of planning and capital development (P&CD).

All the foregoing sample testimonies, and many others that have not been included here, tell the same story. Workers have played vital roles in the success of the change management programmes. They are the key operators behind the scenes who are seldom seen or heard but who would bring the Corporation to its knees if they chose to down their tools. Accordingly, they should be trained, nurtured, empowered, motivated and rewarded adequately to do their work. They should be given a voice and platform to participate fully in all corporate decision-making processes.

A contented and participatory workforce is invariably a productive one. *The challenge that all the managers face, whether in the public or private sectors, is to make their employees feel that they belong to and own their sources of livelihoods and that the destiny of their employers is inseparable from their own.* In the NWSC, we have tried our best to cultivate confidence, commitment and job satisfaction among our workers and to continuously upgrade their skills, and this worker-centred approach to industrial relations has proved rewarding in terms of individual and group performance, productivity and corporate loyalty.

Encouraging managers and staff initiatives

Nothing gave me more hope that NWSC will scale greater heights than staff and managerial commitment. During the stretch-out, for example, I introduced what came to be known as 'dream' targets and challenged staff to give them a try. I made sure that I moved around with the Board members to listen to varied expressions of commitment from managers and staff, at the time of signing stretch-out memoranda of understanding. In a way, we were testing the waters with these 'mad' ideas. During the meet-the-staff tours, we stressed the importance of stretching out for performance and incentives. We were after 'superhuman' effort and we were not ashamed of this 'dream'! The positive feelings and perceptions of the staff and management that resulted from our human resources management and motivations were cause for satisfaction. Interestingly enough, the operating teams quickly got on the bandwagon. They were no longer afraid of me whom they had perceived as the 'hangman'! This transformation of the work culture and worker attitudes was the secret of our change management success. *When NWSC workers and managers had become 'shareholders', they were able to move mountains.*

Lessons learnt

- Involve your entire workforce at all levels in every stage of decision making and implementation of every change management initiative in order to optimise the utilisation of their vast energies, the potential for enhancing performance and catalysing organisational change.
- Embrace new technological and management strategies through concerted and continuous skills development of the workforce in order to survive and flourish in the age of information and communication technologies.
- Harmonious industrial relations are the secret to enhanced performance productivity and profitability in any commercially oriented enterprise. The workforce will accept painful changes, including rightsizing, restructuring and retrenchment for the sake of restoring the health of their organisation and revitalising its dynamism and sustainability provided they own those changes. So be transparent and straightforward with internal customers and win their support for the right decisions, which may entail short-term loss of jobs and benefits.
- Once you have empowered your staff through training and staff development, give them space to make decisions and take risks, keeping your overall guidance discreet and intervening only when it is absolutely necessary to do so.
- Always instil a sense of trust and confidence in your staff, and keep your doors open to your employees at all times, listen to their ideas to get feedback, which may inform subsequent change management initiatives.
- Focus on the skills you have at hand. Use them, value them, nurture them, mobilise and motivate them. They are the most important assets you have, not the ones you are expecting to recruit.
- Listen, listen and listen to your workers/staff and do not take them for granted.
- Learn to use a situational management approach since an organisation is made up of people with different personalities.

From their business to our business: staff involvement and empowerment during the change management programmes created a sense of belonging and ownership of the business. NWSC staff led by the corporate secretary, Mr David Kakuba (fourth left) and the union chairman, Mr Larwod Ofwono (first right), receiving the Employer of the Year Award for 2005.

NWSC staff joining the MD (right) in the celebration of the achievement of being voted East Africa's Most Respected Company in 2007.

Chapter Sixteen

Friends in Need

Suppliers and Service Providers

We provide legal and advisory services to other public utilities and we can unequivocally state that the NWSC is one of the most professionally run public utilities we have the pleasure dealing with. The staff are responsive, very dedicated and very professional in handling their affairs, and this change has largely occurred in the last 7 years and explains why NWSC stands out among all other public utilities in the country.

Katende, Ssempebwa & Associates Advocates

We have been doing business with several public utilities in Uganda since 1991, but we have not yet come across an organisation like NWSC where supplies and payment procedures are handled efficiently and professionally. We have had many bad experiences with other Corporations, for example when it comes to deliveries the stores departments refuse to accept our materials either because of lack of space or their original copies have not yet reached the stores, which means our loaded lorries come back. We also had problems with payments for it takes six months and, in some cases vouchers go missing from accounts departments. We have not faced such problems with NWSC. There was only one incident where deliveries were lost in the store but true copies were presented and verified and later the matter was resolved.

Gentex Enterprises Ltd

In Chapters 12 to 15, I described the roles of the donors, government, the media and the workers in the transformation of the NWSC from a Corporation that looked 'terminally sick' into the most commercially viable and thriving public utility in the country. But the success story of the recovery of NWSC would be incomplete without mentioning the equally important role of its friends in need – the suppliers and service providers. A large water and sewerage service organisation such as NWSC invariably requires a lot of material and service inputs from external suppliers and service providers to do its work efficiently and effectively, to the satisfaction of all its stakeholders. In the case of NWSC, the material inputs include chemicals, pipes and fittings, equipment, spare parts and accessories, vehicles, computer hard and software, office furniture, filter

© 2009 William T. Muhairwe. *Making Public Enterprises Work*, by William T. Muhairwe.
ISBN: 9781843393245. Published by IWA Publishing, London, UK.

media (sand) and stationery and office supplies. Also, the Corporation requires legal, technical, transport, media, cleaning, fumigation, clearing, accounting and auditing, customs and computer services in order to fulfil its commercial and social missions.

In performing its day-to-day statutory duties, the NWSC requires reliable internal and external suppliers and service providers to supply quality goods and services on time, at competitive prices and on reasonable conditions, to render technical and professional advice and to offer after sales service and training when requested. The Corporation has well over 200 local suppliers and international service providers, who have made it possible to keep its wheels rolling. Since Uganda is an agricultural and landlocked country, most of the key material inputs imported from abroad are shipped to Mombasa or Dar es Salaam and then transported from the coast by road or rail to Kampala. This means that the goods must be pre-inspected before shipping, insured and cleared both at the coast and at the customs in Kampala. This calls for legal, clearing and forwarding services, as well as placing orders at least six months in advance before deliveries. Thus, without efficient and reliable international service providers, working closely with our procurement and legal departments, the provision of our water and sewerage services would surely grind to a halt.

This chapter details how friends in need, namely, the suppliers and service providers, have contributed to the recovery and success of NWSC in a number of ways. They have done so not only in terms of credit facilities, discounts, professional advice, training, and after-sales service but also in the provision of new knowledge in the water industry. Their expertise and advice have been particularly useful, especially with regard to new technologies, new processes in water production and distribution, billing and revenue collection, accounting and computerisation and, last but not the least, customer care.

Need for credit facilities and discounts

No public utility such as NWSC, or any large enterprise for that matter, whether public or private, can afford to do business without credit facilities. It is patently obvious that credit facilities are the lifeline of all business operations. Credit facilities are essential to bridge the gap between earnings and spending, to ease cash flow constraints, to manage corporate planning and budgeting and to comply with standard public procurement and payments procedures and practices. But to build and sustain a free-flowing credit system requires mutual confidence between suppliers and customers, clearly defined rules and procedures, procurement specifications, quality control assurance mechanisms, timely orders and deliveries, expeditious payments and general mutual compliance with contractual obligations. Failure to meet any of these imperatives clogs the credit system with disastrous consequences for day-to-day business operations and the smooth delivery of services. That is why a corporation such as NWSC must keep backup credit facility channels open so that they can be utilised if and when needed to sustain the delivery of services.

In addition, any public enterprise such as NWSC, which buys large quantities of production inputs on a continuous and predictable basis, must use its bargaining power and take advantage of bulk purchases to minimise production costs by negotiating substantial discounts. Naturally, suppliers love to do business with long-standing customers whose annual orders are predictable and whose share of total sales is substantial. This puts a customer in a strong bargaining position and it may make commercial sense for the supplier to trade off the profit margin for turnover. Therefore, it is a common business practice for suppliers to give attractive discounts of up to 20% or even higher for bulk purchases. It is also imperative for bulk purchasers to bypass intermediaries and buy directly in bulk, from source – from the manufacturer of the required input – in order to minimise costs. Of course the buyers themselves must keep their side of the bargain. They must give clear procurement instructions and specifications, place their orders in time, take goods on charge in a professional manner and process payments expeditiously.

Shortfalls in the procurement system

Prior to the onset of the change management initiatives, the Corporation's credit facility system was clogged to the limit and was not functioning properly. The procurement process was overcentralised and was suffering from excessive bureaucratic procedures and practices. Our regular suppliers were up in arms because of lack of payment. Some of them had withdrawn credit facilities or reduced the credit period to 14 days and were insisting on advance payments, which were contrary to public procurement procedures, or they were insisting on letters of credit. Those who were still giving credit were overpricing their goods and services in order to cater for delayed payments. Processing payments was a nightmare. Payment documents such as purchase orders, invoices and delivery notes often disappeared into thin air and copies had to be resubmitted. Suppliers had to chase their payments through the maze of the stores and accounts departments. Rumours and allegations of kickbacks, 'chai', 'air supply' and collusion between some suppliers and stores and accounts officers were universal. In practice, all this meant that shortages of supplies were frequent, under-invoicing and overpricing were common practices, and some of the goods supplied were of inferior quality. As a result, the Corporation was not getting value for money and the water production and distribution functions were erratic as a result of lack of inputs. Accordingly, one of my first challenges was to restore the suppliers and service providers' confidence and to transform them into reliable business partners.

One-by-one corrective measures

Consider all the shortfalls and procurement system predicaments, I knew that we could not continue unless our friends in need were brought on board. We had to make streamlining of the logistics supply system one of our priority actions. We had to build the confidence of these friends to conduct business with the Corporation without fear and suspicion of bad faith and reneging on obligations. In the following sub-sections, I outline some of the key measures that we undertook to strengthen the logistics system.

Reorganisation of the Procurement and Stores departments: At that time, procurement and stores functions rested in not only different departments but were also physically located far from each other. There was virtually no communication between these two departments. I was not impressed about the location of the procurement and stores offices. I had been informed that good accounting principles demanded separation of the two functions. Much as this was not a problem to me, I did not see the logic in the case of NWSC because the two departments are functionally closely related, yet their locations were miles apart. This physical separation created many operational setbacks.

It made it practically impossible for the procurement manager and staff to verify stock levels in a speedy manner before clearing procurement requests. Conversely, it was equally hard for the store manager to establish procurements in process before initiating procurement requests. In short, the physical distance between the two departments created a bureaucracy that negatively affected the efficiency of the whole procurement function. Documents would take weeks, if not months, to be moved from and to each of the two departments. While this was an obvious problem, I was perturbed that nobody else seemed to care about it. To me, these two functions were not only related but they were also, in fact, inseparable. In my view, to alleviate these problems, the best thing was not only to house the two functions in one office but also to bring them under the umbrella of one big section or department.

In one of our weekly top management meetings, I announced my intention to merge the procurement and stores departments. The reaction of my top managers was negative. I was told that I was risking putting 'goats and leopards' together. My proposal was strongly resisted, but as a compromise we agreed to locate the two departments in the same place while keeping them functionally separate. This would at least solve the communication problem! I moved the procurement department to 6th Street in Industrial Area to join the stores department as agreed. The existing stores office was repartitioned to cater for all stores and procurement personnel. *In the process, the leopards did not devour the goats.* On the contrary, with time the two sections worked harmoniously.

Documents that used to take one to two weeks to move from one section to the other were now taking hours and minutes. Communication was now by word of mouth and not telephone calls. The main direct beneficiary was the supplier, who found it a lot easier to check on payments, and he now had a one-stop centre for answering questions. This made the suppliers happy and their lives easier in serving NWSC.

Later I was told that these two sections used to arrange parties together, meet and make merry. We came to realise that actually, with good management, *a leopard can live with a goat very peacefully.* This was proved at our stores and procurement offices. Because of the positive lessons learned from the proximity of the two sections, we resolved, as management, to merge the two sections into a Department of Logistics under one head. In order to maintain the tenets of good governance, we maintained clear and separate lines of procuring and storing and only created a convergence head of department level, ensuring strong synergies and coordination. The merger worked well.

Reviewing management approval threshold: At the time of my appointment, the management had the authority to approve contracts only up to a value threshold of UGX 10 million, which was far below the average value of most of our procurements. This policy was worse for Areas whose value threshold was even lower and yet they were the centres of operations. As a result, most procurements were greatly hampered by the need to arrange Board meetings to seek approval for transaction values above the management threshold. Most operations at Area level were constrained by such delays and, worse still, the price validity of goods and services quoted by suppliers sometimes expired before procurement could be concluded. In such situations, the processes would have to begin afresh, our operations suffered greatly and the Corporation lost time. NWSC Area managers were always complaining about delayed procurements. Once again, I could not just let sleeping dogs lie. I could not stand this irrational bureaucracy. With the support of my top management, I requested the Board to revise these management thresholds. In response the Board increased the threshold to UGX 18 million and that of the Areas to UGX 10 million. This policy shift greatly helped to speed up the procurement process and, in turn, enhanced our logistics supply efficiency. Our strategy was that, with time, this threshold ought to be increased as the quest for more autonomy gained momentum.

Streamlining procurement and accounting processes: Our another priority was to review the procurement processes and information channels through which management could establish number of committed purchase orders, deliver purchase orders pending payments, invoice purchases outstanding for payments, etc. I was not happy with the length of the procurement cycle at the time. The procurement guidelines needed to be rewritten. This called for a significant overhaul of the whole procurement and accounting information systems and processes. We had to get rid of unnecessary bureaucracies in approval procedures to reduce procurement lead time. All this was carried out with the objective of making our procurement and accounting functions efficient and responsive in providing an effective support system to the centres of activity – the Areas. Furthermore, the accounting system was customised so that outstanding payments to suppliers, committed purchase orders pending delivery, etc., became more visible. In this respect, this strategy was designed to address the overwhelming number of complaints from suppliers regarding delayed payments. At first, we thought this was understandable given the Corporation's fragile cash flow position. But as we improved our cash flow through the successive change management programmes, we continued to receive numerous complaints regarding delayed payments to suppliers. I did not take this lightly and warned the finance and procurement staff to change their ways or else face instant dismissal. Most of the delays were due to lax processing of vouchers for the suppliers. Through our one-minute goal strategy and daily task planners, we tasked the procurement and finance staff to improve their act or face the consequences! I insisted that Departments hold weekly meetings to plan and make timely payments to suppliers. Indeed, with time, most of our suppliers became happy with our system.

Decentralising procurements: As I mentioned earlier in this chapter, at the time of my appointment, the procurement process was overcentralised and was burdened by bureaucratic procedures and practices. Areas carried out minor procurements of non-core items such as sugar and tea. All the major goods and services were procured at the Procurement Department at the headquarters. Areas had no control over prices, quality of goods, delivery cycles, etc. They were at the mercy of the procurement manager and his team. I was not impressed by this counterproductive arrangement and had to correct it! We therefore adopted a protracted approach aimed at empowering our Areas to carry out their own procurement functions. In line with the PPDA, we appointed delegated contracts committees at the Areas to oversee the decentralised procurements. This decentralisation process was most effective during the era of APCs and IDAMCs. Through the implementation of these programmes, we were able, to a large extent, to streamline the procurement function and since then suppliers have been free to deal directly with the Areas. This has speeded up procurement, and allowed Area managers to buy supplies or hire services locally.

Adherence to the Procurement Act: During the early stages of my role as NWSC chief executive, the procurement law was reviewed and strengthened through the Public Procurement and Disposal Act (PPDA). This Act introduced new and adjusted a number of previous procurement procedural guidelines. Our procurement transactions had to follow suit. We enforced compliance with the PPDA, which incorporated many standard guidelines on initiating procurements, sourcing quotations, selecting and awarding tenders, communicating with bidders, etc. The suppliers, in turn, welcomed the strengthening of procurement principles such as open and competitive bidding that conform to international best practices. These principles have proved beneficial, both to the Corporation and to the service providers. I had no doubt that these principles were actually good for the Corporation and the individuals involved in the procurement transactions.

Because of the introduction of PPDA rules and procedures, the conduct of the procurement function in NWSC had to be revised and modified accordingly. We made sure that we complied with the new *modus operandii*, to the benefit of all our service providers. It was no wonder that the PPDA, through their routine audits, named NWSC as one of those few government institutions which was compliant with the PPDA Act and suggested that many international organisations could benchmark with us in the area of procurement. Our procurement staff were, time and again, invited to give lectures and conduct seminars alongside PPDA staff on best practice in procurement.

Clearly, PPDA rules and procedures shielded all parties involved from the charges of *favouritism, kickbacks and other corrupt practices*. Furthermore, through competitive bidding the burden of quality assurance, fair prices and invoicing and compliance with bidding specifications has shifted from the Corporation to the suppliers. Under the new procurement system, the supplier is sure of getting a better deal, knows the rules and regulations, knows what to expect and is confident of being the winner provided he/she has the best positioning in the market. The open, competitive bidding system has relaxed the procurement atmosphere, to the benefit and satisfaction of all stakeholders. This has reduced the scope for rumour mongering, suspicion and mistrust, which used to be characteristic of the procurement processes and practices in the Corporation.

The turnaround: confidence restored

After the systematic implementation of the structural changes in the procurement process, the suppliers and service providers gradually gained confidence about doing business with the NWSC. Most of our suppliers and service providers have dropped conditions such as advance payments, cash on delivery and bank guaranteed letters of credit. They have instead reverted to credit facilities of up to 30 days with effect from the delivery date. This revival of a predictable credit facility system has greatly helped the Corporation to plan its procurement, to assess the supply needs and budget accordingly and to effect payments on time. With a good and rolling payments schedule, the suppliers and service providers know when they expect to be paid so that they too can plan ahead to meet their production and supply obligations.

Given the new pro-supplier and pro-service provider culture that had been cultivated in the Corporation, many suppliers and service providers have offered to enter into long-term credit agreements with the Corporation in order to guarantee the continuity of their market. However, on our part, we have been reluctant to enter into such arrangements for several reasons. Firstly, the existing public procurement regulations do not allow for such arrangements even if we wanted to enter into them. Secondly, it is not in the interest of the Corporation to accumulate huge debts for these would, sooner or later, constrain our forward cash flow planning and spending priorities. Thirdly, long-term credit agreements would tie our hands and limit our freedom to consider the products or services of new entrants into the market. For this reason, I have impressed it upon the NWSC management to keep all our options open when it comes to the procurement of goods and services. Moreover, with our improved financial performance since 1998, we can afford to meet our credit obligations without undermining our overall cash flow position.

Although open and competitive bidding is the usual method of procurement in the Corporation, I insisted that we must retain the flexibility to resort to the direct procurement method if and when necessary. I emphasised that open tendering is not a suitable method for the purchase of vital inputs such as water meters, specialised laboratory equipment and information communication technologies and software. However, the real world is always different. We are still being constrained by the PPDA rules that insist on open tendering. Therefore, in such situations, we have retained the services of our 'fire fighters' – suppliers and service providers – who are always on standby to supply goods and services on credit and on short notice to enable us to cope with emergencies, and to sustain water and sewerage service delivery without interruption.

'Fire-fighter' suppliers have enabled us to avoid unnecessary stockpiling, thus tying up scarce resources in stock and warehouse costs. By supplying goods and services promptly, on demand, and on credit, our 'fire-fighter' suppliers and service providers have not only helped us to *reduce procurement and warehouse costs but they have also helped us to cope with emergency situations, to the satisfaction of our customers.*

Apart from the provision of credit facilities, our suppliers have provided other procurement-related services to facilitate the work of the Corporation and improve corporate performance and service delivery. Some of our suppliers have given large discounts of up to 20% and, by doing so, have contributed to substantial reductions, not only in the costs of inputs but also in overall operational costs, to the ultimate benefit of the customer. The suppliers and service providers have participated fully in pre-shipment inspections in order to ensure the delivery of high-quality goods in accordance with technical specifications. Insurance service providers who have covered the goods against loss and damage en route to NWSC have facilitated this process. Suppliers have also provided free installation and after-sales services to the Corporation on request. The provision of these services on a continuous basis reflects the good working relations and mutual confidence that have been revived, cultivated and sustained by the NWSC management. That the suppliers and service providers have expeditiously provided goods and services to keep the wheels of the Corporation turning all these years has been no mean contribution to its overall success.

Advice, information and training

Suppliers and service providers do not simply supply goods and render services. They do more than that. In the case of NWSC, suppliers and service providers have rendered technical and professional advisory services, provided useful information and trained our staff, both in-house and elsewhere. Technical and professional advice has covered a multiplicity of legal, engineering, accounting and computing issues that impact on the work of the Corporation on a regular basis. Technical and professional service providers have helped reduce operational costs, improve service delivery and minimise or contain conflict. Suppliers have helped the Corporation to identify and procure new water production and distribution inputs and technologies, adapt new methods and processes of water production, service delivery and customer care. These technical and professional services have, over time, contributed to the improvement of the Corporation's performance, efficiency and productivity, and laid the foundation for its change management achievements.

Take the case of our external legal service providers, for example. The NWSC used to spend a lot of money in litigation involving the Corporation and other individual and corporate parties. A lot of money and time were spent on the preparation of legal cases and court appearances. Thanks to the professional advice and assistance of our external lawyers, Katende and Ssempebwa Advocates, MMAKS Advocates and Turyakira & Co Advocates, the litigation caseload involving the Corporation has greatly been diminished. All legal cases have been handled professionally and expeditiously at a reasonable cost to the Corporation. Other cases have been settled out of court through arbitration to avoid lengthy, costly and protracted litigation. By 2005, 90% of all the cases in the courts of law had been resolved, and there were only few cases filed against the Corporation. Most of the cases against the NWSC have revolved around breaches of contract, accidents, negligence, disputes, damage to property, trespass and wrongful dismissal. The fact that fewer cases are being filed against the Corporation is a testimony to the professional effectiveness of our lawyers. They have saved the Corporation a lot of money, time and sleepless nights.

As a corporate entity, the NWSC can be sued but it can also sue to enforce its rights. Over the years, the Corporation has sued some of its customers, such as the Departed Asians Custodian Board, a Chinese engineering company and a hides and skins factory, either to recover debts or, in the case of the last example, to stop a nuisance originating from its premises. With regard to the Custodian Board and the Chinese engineering company, the cases were settled out of court and both defendants settled their bills in accordance with their respective consent agreements. In the case of the third example, the defendant was warned in writing and the nuisance stopped forthwith. Through their legal advice and advocacy, the lawyers have facilitated the recovery of the Corporation. They have stood by us in hard times, endured any delays in the payment of professional fees and, above all, advised us to meet our contractual obligations and exercise due diligence in our work to avoid costly litigation.

Being legal practitioners as well as water consumers, lawyers are a vocal section of the community. They demand quality and efficient services. They shape public opinion through their participation in public debates and seminars. As members of parliament and members of commissions of inquiry, they are influential in moulding the content and course of legislation, and in public decision making. Therefore, the fact that many law firms have been knocking at our doors to render their professional services suggests they have as much confidence in the future of the Corporation as we have in their services and professional advice. Recently, in order to avoid putting all our eggs in only a few baskets, we retained a couple of other legal firms in order to instil the spirit of competition, to tap the diversity of legal skills and talents, and to maximise value for money. The continued partnership between our internal and external lawyers, and the Corporation's management is one of the ways of consolidating and enhancing the progress that has been registered since 1998.

Suppliers know best the new and cost-effective products in the water sector market. They know what is available and where and how it can be obtained. In the case of NWSC, suppliers and service providers have helped to source new water production inputs and technologies to provide logistics, transport and insurance. They have endeavoured to deliver goods and services on time, in recognition of the fact that time is money. Some of them have not hesitated to report laxity and corrupt tendencies and practices on the part of procurement, stores and accounts staff, which has enabled us to make appropriate corrective interventions before it is too late. Suppliers have also rendered technical assistance and advice in the installation and maintenance of plant and equipment, the handling of potentially dangerous chemicals, the use of new water production and sewage treatment, and the servicing and upgrading of computer hard- and software applications. By providing up-to-date useful information relating to the supply, installation and maintenance of water production inputs, the suppliers and service providers have gone beyond the call of duty to transform NWSC into a success story that is the envy of other water utilities in Africa.

Furthermore, the suppliers and service providers have contributed to the training of NWSC staff in handling and using water production inputs. For example, some suppliers, such as ARAAD, Jos Hansen and UTEC, have trained NWSC staff on how to handle repair equipment for chlorine gas drums. Similarly, Desbro Uganda organised sensitisation workshops on the proper handling methods of laying pipes. ARAAD, Siemens and UTEC marketing executives supplied CDs as reference materials containing product specifications, manufacturing processing, information on handling chemicals and how to install switch panels and electrical fuse components. The suppliers taught NWSC staff how to use the CDs, which are now in the Corporation's library. Computer service providers have helped install Scala, Custima and Lotus Notes accounting and billing programmes and provided hands-on training, both in-house and in Kenya and South Africa. All in all, over 62 NWSC staff members have been trained under supplier-sponsored training programmes. This training has enhanced capacity building in the Corporation, resulting in improved performance and productivity.

The black sheep

Despite commendable contributions by suppliers and service providers to the transformation of NWSC, there are always some black sheep in any group who want to spoil the good work and to tarnish the image of their colleagues. While NWSC suppliers are generally honest and reliable, a small minority have turned out to be unscrupulous, to the disappointment of the Corporation. For example, one supplier delivered 50-page writing pads instead of the 80-page ones that had been specified in the purchase order. Fortunately, the store manager was alert enough to detect the error before taking the pads on charge. He took a sample from the delivery and counted all the pages physically. The supplier, who was caught red-handed, was asked to make up for the shortfall, and the trick, whether deliberate or inadvertent, has never been repeated. Another supplier delivered 180-metre rolls of HDPE pipes instead of the 200-metre ones that had been specified in the purchase order. That supplier too was asked to make good the shortfall, and was cautioned.

More serious and worrying than the undersupply of goods has been the occasional deliberate collusion between stores staff and suppliers to defraud the Corporation. Some storekeepers have been tempted to 'cut deals' with unscrupulous suppliers and to take bribes in order to condone undersupply, over-invoicing or even outright 'supply of air'. One supplier was caught at the central stores delivering boxes containing stones instead of computers. We discovered that some of our staff had been involved in this attempt to 'supply air'. To avoid tedious litigation by arresting the culprits and initiating criminal charges, we took administrative action immediately. We terminated the services of the guilty staff and blacklisted the supplier. Since then that supplier has never done any business with our Corporation, and the colluding staff lost their jobs and terminal benefits.

In another incident, one of our suppliers delivered a consignment of 380,000 kg of aluminium sulphate. When the store manager, with the assistance of an auditor, counted the delivery bag by bag, he discovered that the consignment had a shortfall of 2,000 kg. The supplier was asked to make good the shortfall, but we could not establish whether this was deliberate and, if so, how long the cheating had been going on. We gave that particular supplier the benefit of the doubt but henceforth instructed the stores staff to plug the loopholes that might be exploited by the black sheep among the suppliers. Since then we have trained our staff not to take any deliveries for granted; they must thoroughly check every consignment, however bulky, and verify all the contents.

They must cross-check the deliveries against the purchase orders and delivery notes before taking the goods on charge. Once the goods have been delivered and accepted, the store manager and his subordinates assume responsibility for any shortfalls or losses.

All the above anomalies pointed to the need for a comprehensive re-alignment of staff behaviour and actions. These incidents were eye openers and we needed to be more resolute and vigilant in curbing such tendencies. We henceforth adopted a 'zero-tolerance' approach to such behaviour and made the policy known to all concerned. The procurement and stores managers talked about these incidents and the possible repercussions in every meeting with their staff. In the end, this made what we stood for more visible to all sections of the staff. This approach, together with significant re-engineering of transactional procedures, has yielded good results. *By tightening up the delivery rules and procedures, we have virtually eliminated incidents of irregularities in the procurement system and made it clear to all suppliers that they must clean up their acts before they can do business with NWSC.*

In accordance with our one-minute management principles and practices, we adopted the idea of the carrot and the stick – of praise and reprimands – in the procurement system. We praised the good and honest suppliers, paid their bills on time, honoured our contractual obligations, tried as much as possible to reduce bureaucracy, and instilled a sense of professionalism and urgency to eliminate any bottlenecks in dealing with procurement, stores and accounts staff of the Corporation. Time and again, I called on the stores staff to see themselves as secondary suppliers to the Areas and other departments at headquarters. *The stick against the wrongdoers in the procurement system includes immediate dismissal of the guilty staff as well as warning and blacklisting suppliers and removing them from the suppliers' list.* It also entails enlisting the intervention of the offices of the inspector general of Government and the auditor general and even the CID to investigate allegations of corrupt practices among the NWSC staff and service providers alike. The central purpose of these efforts is to build and sustain a clean procurement system that is beyond reproach.

Two tales of friends in need

During the course of writing this book, I sent out a questionnaire to a number of key NWSC suppliers and service providers. As was the case with the employees, the purpose of the questionnaire was to capture suppliers' views about their working relationship with the NWSC and to demonstrate how they have been indispensable partners in the recovery and transformation of the Corporation into a success story. From the responses received, I selected two – one from a service provider and the other from a supplier – edited versions of their testimonies, which are reproduced in this section for purposes of illustration.

The service provider's perspective

We have been providing legal and advisory services to NWSC since 1992 to date and we are, therefore, in a position to provide an informed opinion on the Corporation. During the time we have worked with NWSC, we have witnessed its dramatic transformation from an ordinary public utility provider to a world-class client-centred service provider, which is better run and managed than most firms in the country.

We basically deliver our services to one centre, that is NWSC headquarters, since their legal department, which is centralised, hands down most cases to us. In a number of instances, however, we have had to go to upcountry courts to handle matters on behalf of the Corporation. Every service delivery is noted and the Corporation goes a step ahead to thank us for the continual good services that we render to them. This is professionalism. The managers and other staff members with whom we deal on a regular basis, particularly those in top management and the legal department, are very competent and knowledgeable in the way they handle their work and they perform their duties exceptionally well. This top management is the main reason for the remarkable turnaround that NWSC has enjoyed in the last seven years and they do not suffer from the usual problems (e.g. bureaucratic procedures, I-don't-care attitude and customer non-responsiveness) faced by the managers of parastatals or semi-government bodies. It is very rare to find managers of this nature in the public sector. It is, therefore, no surprise that their example is being quoted around the world.

It is our considered opinion that we have contributed to the improved service delivery of the Corporation, firstly, by helping them to settle a number of claims that were brought against them in court and through out-of-court settlements and in defending the Corporation in numerous litigation claims. Litigation is usually protracted and expensive but we have, in many instances, succeeded in having the claims reduced, thus saving the Corporation additional costs and helping them save billions of shillings, and various Corporation properties, which money, we believe, has been used to better their service delivery. We have also handled various cases which have protected the name and image of the Corporation. We have rendered advice to the Corporation on a wide range of transactional matters, enabling them to streamline and provide their services efficiently and in a proper manner.

We provide legal and advisory services to other public utilities and can unequivocally state that NWSC is one of the most professionally run public utilities we have had the pleasure of dealing with. The staff are very responsive, very dedicated and very professional in handling their affairs and this change has largely occurred in the last seven years and explains why NWSC stands out among all other public utilities in the country. We can say that we have enjoyed being service providers to NWSC and we look forward to a future with this prosperous and successful Corporation.

Katende and Sempeebwa Associates Advocates, Kampala

The supplier's perspective

We have been NWSC's supplier of uPVC and HDPE pipes and fittings for several years and we have been supplying on a regular basis nearly every month. The pipes and fittings are used for the expansion and maintenance of water distribution networks. We are one of three manufacturers of these products. Our products were tested and prices were compared before we were chosen to supply to the Corporation, and over the years we have proved ourselves by providing good quality products and timely deliveries.

We think that the NWSC procurement procedures follow a good procurement policy. In our opinion, the NWSC procurement system, from ordering to delivery, is managed very professionally. The NWSC employees with whom we deal on a day-to-day basis are cooperative. So far we have not encountered any instances of corrupt practices or misconduct from any of the staff of NWSC.

We have been doing business with several public utilities in Uganda since 1991, but we have not yet come across an organisation like NWSC where supplies and payment procedures are handled efficiently and professionally. We have had many bad experiences with other corporations. For example, when it comes to deliveries the stores departments refuse to accept our materials either because of lack of space or because their original copies have not yet reached the stores. This means our loaded trucks come back and this leads into double work and waste of time. We also get problems with payments for it takes six months to be effected. In some cases vouchers go missing from accounts departments. We have not faced such problems with NWSC. There was only one incident where deliveries were lost in the store but certified copies were presented and verified and later the matter was resolved.

We have contributed to the success of NWSC by providing good quality products within the shortest possible time. We highly rate NWSC as our major business partner and we look forward to their continuous support. At same time, we hope and know for sure that the Corporation will continue to register improvements and hence offer more business opportunities to us. We are partners in Uganda's development and NWSC plays a meaningful role in the development and growth of communities touching the lives of all Ugandans, bringing with it the gift of life-sustaining water.

Gentex Enterprises Ltd

Looking back

The two testimonies, one from one of our service providers and the other from one of our suppliers, which have been presented at length, clearly show that doing business with 'friends in need' is based on the principle of mutual reciprocity. The suppliers and service providers have played their part in the recovery and transformation of NWSC. They have been providing high-quality goods and services, giving credit, discounts and advice, delivering goods and services on time, and providing useful technical, professional and market information to the Corporation. We have reciprocated by putting in place an open and transparent procurement system with clear and fair rules and regulations. *As a result, we have created a 'win-win' situation for both the Corporation and its suppliers and service providers.*

In Uganda, there has been a widespread but erroneous impression, in business and the public at large, that suppliers and service providers are the main beneficiaries of procurement. Of course, suppliers stand to gain from selling their goods and services, to make money and earn a living. After all, that is why they are in the business to begin with. But suppliers and service providers stand to benefit even more if their customers or clients are performing well. *When the businesses of customers or clients are doing well, there is increased demand for goods and services, boosting market opportunities for suppliers and service providers.* Therefore, it is in the best interest of suppliers and service providers in any business to do everything possible to enhance the performance, productivity and profitability of their customer or client organisations. That is why the overwhelming majority of our suppliers have been keen to see the transformation of NWSC into a commercially viable public utility that excels in service delivery.

However, the customers or clients also stand to gain a lot from good and reliable suppliers and service providers. Imagine a situation where the suppliers and service providers supplied poor quality goods or provided shoddy services; for example, they either did not comply with order specifications, delivered too little too late or were indifferent to what happened to their clients or customers. There is no doubt that the customers would suffer the costs of goods and services. At the same time, the costs of production would shoot up, performance standards and targets would not be met and productivity and service delivery would go into free fall. *For me, therefore, our suppliers and service providers have been indispensable partners and we recognise them as part and parcel of the NWSC success story.* We value their services, and appreciate their support and cooperation. We share their difficulties and challenges. We wish them the best of luck in their businesses. We have no doubt that they too mean well by us. We are partners in performance improvement and our successes are theirs too.

In conclusion

This chapter marks the end of NWSC story on change management programmes for the period 1998–2008. Let me take this opportunity to say thank you to all the stakeholders who played a role in supporting the performance improvement initiatives in one way or another. Special thanks go to the development partners, my partners in the government, the workers, the Board, esteemed customers, the public, suppliers and service providers. I am very sorry I could not mention all of you by name, institution or country. But I do appreciate and cherish the tremendous job you did to turn around NWSC. I am grateful to you all.

Lessons learnt

- Suppliers and service providers are indispensable to the success of a business. Be creditworthy, honour your word, spell out your procurement rules, procedures, specifications and expectations and you will earn their goodwill, confidence and credit.
- Suppliers and service providers are not simply sources of goods and professional services, respectively, but they are also sources of information on new products and market trends and they provide on-the-spot training and after-sales support. Cultivating their friendship and tapping their knowledge and expertise are sure ways of enhancing organisational performance and success.
- Diversify your suppliers and service providers base to benefit from competition, discounts and the provision of quality but cost-effective goods and services. In a liberalised market, the buyer has plenty of room to shop around for goods and services that give value for money and reduce production costs.
- Watch out for the black sheep among the suppliers and services, plug the procurement loopholes, and blacklist the wrongdoers when they are caught to pre-empt future losses.
- Keep the supply chains flowing by upholding the principles of reliability, simplicity, speed and predictability to eliminate or minimise supplier/service provider-related constraints and bottlenecks in production, marketing, distribution and customer care.

Part Five
YES WE DID

Chapter Seventeen

Can Public Enterprises Perform?

Final Thoughts

> Indeed, to cross the cost-recovery boundary and start making money and delivering services to shareholder and customer satisfaction, public enterprises must embrace new ideas and management techniques, change with change and break the chains of old habits, tradition, conventional wisdom, fashion and the rules and procedures of bureaucracy.
>
> Author's viewpoint

This book has addressed the change management initiatives and programmes, from the 100 days to IDAMCs and Checkers System, which transformed NWSC from where it was in 1998 into a success story that has become the pride of Uganda and the envy of sister water utilities in Africa. But what lessons can be drawn from this 'management revolution' in NWSC since 1998? What can the NWSC family, from the Board, through management, to the 'foot soldiers' of the Corporation, learn from what has been achieved so far? How can these experiences be used to inform and shape the future of the Corporation? What can the NWSC partners – the government, the donors, the customers, the suppliers, the service providers and the public at large – learn to inform the management of other public enterprises such as NWSC? What can sister utilities in other countries learn from the NWSC experience?

What can policy makers and the public, and even the private sector, learn from the Corporation's 'management revolution' to inform the ongoing debate on the pros and cons of the divestiture and privatisation of public enterprises in Uganda and elsewhere in the World? This chapter is devoted to answering these questions, and draws the conclusion that public enterprises can perform too, provided they operate in an enabling environment, set the right priorities, focus on the customers, seek home-grown solutions to challenges, invest in the workers and have focused, visionary leadership.

The privatisation debate

During the 1980s there was heated debate about the future of public enterprises and what ought to be the proper role of government in economic management. The consensus that emerged from this debate was that public enterprises were performing poorly – far below expectations. According to this line of argument,

© 2009 William T. Muhairwe. *Making Public Enterprises Work*, by William T. Muhairwe.
ISBN: 9781843393245. Published by IWA Publishing, London, UK.

it was not the business of governments to run or manage enterprises. The proper role of governments was macroeconomic management by taming inflation and regulating money supply, and by creating a business and legal environment capable of attracting and protecting local and foreign investment. These arguments, which were supported by substantial but untested empirical evidence, led to the conclusion that governments should disengage from the ownership and management of public enterprises in favour of private operators, who were supposedly better qualified and motivated to manage such enterprises.

Given the common, almost universal, paradigm that public enterprises, especially those in the developing world, 'could not perform', *the obvious answer to the public enterprise dilemma seemed to be divestiture and privatisation, preferably to foreign investors with loads of money and vast management expertise.* Short of wholesale privatisation of public enterprises, the only sound alternative was private sector participation by international operators with lots of experience and technical know-how. Therefore, in 1998, the only panacea to the financial and management challenges facing the NWSC appeared to be privatisation or at least private sector participation. *For this reason, the Corporation was on the block!*

The NWSC experience over the last 10 years shows that the problem with public enterprises is not the nature of ownership. Properly managed by strong, committed and inspiring leadership with a clear sense of purpose and vision, public enterprises can perform as well as, or even better than, private ones. *Experience has also shown that, like public enterprises, private companies can also fall prey to the vices of patronage, corruption, subsidy handouts and monopolist practices that shield them from transparency and open competition.* Look at the bailouts of private companies all over the world, both in developing and developed countries, under the overall framework of the so-called 'economic stimuli'. Some economists actually regard some of these approaches as 'economic spending packages'. It is like an extravaganza and the recent (2008–2009) credit crunch, globally, has posed enormous challenges to those who side with either private or public-led economies. The big question largely remains: which is the better way to go? According to *Business Daily* (Nairobi, Kenya) of Monday, 26 January 2009, James Saft argues for nationalisation of weak banks in Britain and United States. Many big private businesses in the US and Britain have continued asking for bailouts from governments, in form of economic subsidies to sustain their competitiveness in the market. What a contradiction for those who thought that privatisation is the only way to go! Yes, privatisation is a strong economic development tool but it must be applied at the right time and leave 'doors' open for any possible business re-engineering reform. The elements which will always remain central to any business, be it public or private, are, among others, autonomy, market/commercial orientation, customer focus and productive efficiency.

Indeed, in public enterprises, the nature of ownership is not the issue. *The challenge is to define and separate the roles of shareholders and management.* Shareholders should determine and set corporate goals and expectations and ultimately hold management to account. Management should be left to get on with the job of running the enterprise on a day-to-day basis as they see fit and to the best of their ability. In the case of NWSC, government wisely left the Board and management to transform the Corporation without undue interference. In any case, all over the world, even in advanced countries such as the Netherlands and Sweden, governments are beginning to rethink the wisdom and efficacy of entrusting key public utilities and services, such as water, electricity and railways, to private hands. In Uganda, the NWSC success story has made a modest contribution to reopening the debate about the case for and against the divestiture and privatisation of public enterprises, especially utilities.

Reforming public enterprises without changing ownership

The NWSC change management revolution has shown that there is no single textbook solution to the challenges facing public utilities. *Privatisation or private sector participation may be two of the possible roads to corporate transformation.* Besides, privatisation and private sector participation options are not always available, nor do they always work. First, very few foreigners have shown interest in investing in water utilities in Africa; and even those interested in doing so have not been able to get suitable managers willing to work on the continent. In our case, one of the international private operators in Kampala had to change chief executives three times in one year! Second, the two trials involving international operators in the management of Kampala water services, despite very many positive experiences we had with them, did not meet performance expectations. The management fees were too high, and not enough qualified personnel from abroad were flown in to run the operations. As a result, the operators failed to deliver the targeted value in terms of enhanced performance.

Our experience over the past 10 years has shown that it is possible to successfully reform public enterprises without *changing ownership*. Interventions such as the introduction of sound management principles and practices, receptivity to change and to new ideas, the overhaul of internal processes and systems, the transformation of corporate culture and the revitalisation of the human resource function can do wonders for public enterprises. By focusing on the customer, motivating staff through a combination of group and individual incentive mechanisms, and building a culture of mutual confidence, respect and reciprocity among all stakeholders, NWSC has become a viable, dynamic and sustainable institution. This was inconceivable in the early 1990s.

One of the justifications for privatisation is that competition and profitability drive private enterprise, and are indispensable promoters of management efficiency and cost-effectiveness. However, public enterprises can also benefit from an injection of the principles and practices of competition and quasi-private management. For example, right from the 100 days Programme, through SEREP, APCs, stretch-out, one-minute management to IDAMCs and the Checkers System, we inculcated the spirit of competition through best performer trophies and cash bonuses, group and individual cash incentives as well as one-minute praise and reprimands.

Over the years, this competitiveness has worked wonders and enhanced performance. Other public enterprises – any public institution for that matter – stand to gain by emulating this NWSC example of practising internal competition. Similarly, through the NWSC government performance contracts, APCs and IDAMCs, we introduced quasi-privatisation in which Area partnerships enjoy operational autonomy, draw up their business plans and budgets, set targets and priorities, procure their inputs and manage the infrastructure in accordance with the IDAMC provisions. In this regard, NWSC has, in effect, become a 'union' of quasi-private and autonomous business units, without changing the essence of public ownership.

Arguably, the debate about public ownership versus divestiture and privatisation of utilities has been misdirected. Privatisation is not sacrosanct. It is one policy option to be measured against other options. It should not be allowed to become an ideological orthodoxy set in stone. *Policy makers and public enterprise managers should always be flexible and pragmatic enough to go for what works, what delivers, rather than for the 'fashion' of the day.* The NWSC experience has shown that it is possible to kill two birds with one stone. It is possible to retain public enterprises and yet achieve performance improvements that can equal or even exceed those of private enterprises through well-conceived, well-designed and concerted management innovations like those that have transformed NWSC beyond recognition since 1998.

Proactive management

In Uganda, and to a large extent in Africa and the developing world as a whole, managers of public enterprises have tended to be slaves to tradition. More often than not managers brought up in the British tradition of shunning familiarity on the grounds that it breeds contempt have tended to be remote and removed from ordinary employees. They have adopted the top-down leadership style of dictating, directing, instructing and bossing subordinates around. We learnt to be receptive to ideas, to shop for them from books, the streets and private conversations – from wherever, for no one has a monopoly on ideas – to market these ideas to colleagues and staff and to adapt them to our multiple management challenges. We dispelled the orthodoxy that the pursuit of ideas and knowledge should be the exclusive preserve of academia. *We learnt to negotiate, consult and involve colleagues and staff in order to build confidence and commitment, undertake the relentless quest for excellence and accept the inevitability of change, recognising that the only constant factor in business, as in life, is change.*

Through reading the works of business gurus, we learnt the virtues of simplicity, speed and work in an environment free of boundaries, inhibitions or restrictions, in other words, an environment that is open and devoid of the boss element and bureaucratic straightjackets. This injection of new ideas and thought processes, which gave birth to a new culture and created a free and refreshing atmosphere in the NWSC world of work, has paid off handsomely. *The lesson to draw from receptivity to ideas is that public enterprises, or any other organisations, for that matter, had better look for and embrace new ideas and techniques, lest they are left behind in the race for success and excellence.*

Setting focused priorities

Turning a lacklustre enterprise around is bound to be a daunting, uphill task. There are no ready-made answers, no shortcuts to progress or success. What is important is setting and continuously focusing on organisational priorities to deliver the desired results. Our experience shows that it is possible to transform a 'terminally ill corporation' into a viable and dynamic one within less than a decade. Through daring, perseverance and commitment, supportive staff and a helping hand from other stakeholders, we took the bull by the horns, relentlessly pushing the drivers of change forward, and turning the Corporation around, to the astonishment of most stakeholders in the water industry. We did this by choosing the strategy of successive *short, sharp and focused programmes,* from the 100 days to IDAMCs, with specific performance objectives and targets to achieve within agreed time frames. We began our change management initiatives by setting SMART priorities and targets. We went on to enhance revenue collection to achieve financial viability, stretched to the stars through superhuman effort, boosted productivity through group and individual incentives and crowned our change management strategy by introducing the principles of delegation and decentralisation through APCs and IDAMCs.

The results of these change management initiatives were spectacular and astonishing beyond our wildest dreams. We made our staff more productive, stepped up revenue collection, reduced arrears, controlled NRW and response time to emergencies, brought more customers on board through new connections and, above all, enhanced customer satisfaction. I believe that other public enterprises can learn from our experience to achieve the impossible without plunging into the untested world of 'foreign' economic theories. After all, what we have achieved with our own resources is not a miracle. It is a product of hard work, commitment and determination to succeed, whatever the challenges encountered along the way. Other public utilities can match or even surpass what we have done, provided they look inside their own organisations for home-grown solutions, including the identification and nurturing of their own expertise and the optimisation of resource utilisation, and the resolve to anticipate change and to change with change.

Managing change in a public utility such as NWSC requires meticulous, systematic planning and preparation. It requires investing plenty of time in reading, consultation, dialogue, documentation and constant feedback across the Corporation. It requires the identification and empowerment of dedicated lead champions to spearhead the preparation, implementation, monitoring and evaluation of change management initiatives. In this connection, it is imperative to bring everybody on board, to involve them at every stage of the Programme, from conception to evaluation, and to make all stakeholders develop a sense of ownership of every change management initiative. At the evaluation stage, it is important, not only to record the achievements but also to pinpoint the constraints, bottlenecks and shortfalls, and to draw lessons for informing the next programme to sustain the change management momentum.

Putting the customer first

In initiating change management, it is important to focus on the purpose and mission of public utility service delivery. Always remember to put the customer first for, after all, 'the customer is the reason we exist'. Since customers are the ultimate employers, the performance and success of any Corporation, or any business enterprise, for that matter, depends on customer satisfaction. It also depends on the support it receives from customers, on their willingness to pay and to report incidents in the field, including the conduct of staff, on customer honesty and integrity and, above all, on their feedback through complaints submissions and customer surveys. It goes without saying that customers are the eyes and ears of the Corporation on the ground and what they think, say and do is indispensable to the success of corporate operations and service delivery.

From the very outset of our change management programmes, we went out of our way to focus on the customers, cultivate their support, listen and promptly attend to their complaints, seek their opinions and suggestions and involve them in our corporate functions and activities. It is gratifying to note that, over the years, our efforts to focus on customer satisfaction have paid off handsome dividends. We have expanded our customer base, reduced illegal connections, curbed defaulting, meter defilement and vandalism of the Corporation's properties, increased revenue collection and improved the process of water bills delivery and clearance. *Indeed, customer support has become one of the key secrets of our success.* I believe that other public enterprises as well as the private sector can learn a thing or two from our putting the customers first and *making them the heart of all our service delivery operations.*

'Do-it-yourself' approach

From our experience, public enterprises have often underperformed owing to lack of confidence and a dependency syndrome. Managers tend to have little faith in the capacities and abilities of their corporate human resources. They do not look for 'home-grown' solutions to their problems. They do not invest enough in their people through training to produce a critical mass of professionals to spearhead change management and corporate transformation. Public enterprise managers tend to look to foreign technical assistance and expertise to solve local problems. They appoint consultants to study why public enterprises do not perform as well as they should and to recommend appropriate solutions. These experts make 'flying working visits' to Africa, consult a couple of documents, talk to corporate and government officials, prepare and submit reports in standard textbook formats and return home smiling, with fat consultancy fees in their pockets.

At NWSC we rejected this dependency syndrome for several reasons. Foreign experts and foreign-inspired solutions are expensive and not all of them give value for money. Many foreign 'flying' consultants have limited and superficial knowledge about local corporate management realities and some of them tend to be insensitive or indifferent to local sensibilities. They invariably rely on information from local corporate officials, which they rebrand and repackage in standard consultancy report formats with no substantial added value. In our view, these high-sounding diagnoses and prescriptions, more often than not, tend to be a waste of time and money. Foreign experts do not have a monopoly on ideas and knowledge about how to enhance performance and increase productivity or how to manage enterprises more efficiently and effectively. *Good performance has no colour, has no diagnoses and prescriptions set in stone and has no single source of origination. In reality, there are many ways of skinning a cat, some of which can or ought to be home grown, if only because they are cost-effective compared to the costs of importing 'first- or business-class experts'.*

Although we appreciated and indeed continued to receive invaluable external assistance, we adopted the 'do-it-yourself' approach to the challenges facing

NWSC. It has worked well. We developed and implemented our change management programmes. They worked. We developed our own financial model, as well as business and corporate plans, and transformed our organisational behaviour without much outside involvement, except financial and moral support. We identified, inspired and empowered our change management lead champions to spearhead the management revolution in collaboration with the Area managers and partnerships. We involved all our staff, from the messenger and plumber to the top management in developing, implementing, monitoring and evaluating our change management initiatives. They delivered the results and much more. We did all this with limited, though useful, foreign expertise and technical assistance. *By so doing we dispelled the myth that no change can take place in public enterprises without external origination and stimulation.*

Investing in the workforce

Of course, maximising the human resource potential in a corporation like NWSC calls for substantial and sustained investment in people. It is imperative to launch staff development, training and skills development programmes as well as to sponsor and facilitate staff participation in local and international conferences, seminars and workshops. Remember that learning is a lifelong human endeavour. The saying that 'old dogs cannot be taught new tricks' is patently false. We need to expose people to best practices and standards in the management of businesses. We need to expose them to new ideas and innovations from various sources around the world. At NWSC, we have done our best to train our staff up to the PhD level. We have encouraged and sponsored them to attend local and international seminars, workshops and conferences. We have established a training centre in Kampala to provide continuous training and refresher courses to update and upgrade our staff through skills development. As a result, we have, over the years, developed a highly trained, motivated and committed professional team comparable to their best peers around the world, which is already offering expert external services to sister water utilities in Africa and elsewhere in the world. In short, at NWSC we have learnt the virtue of self-reliance. And we have married practical hands-on experience to a continuous search for knowledge and professional self-improvement.

In addition to providing opportunities for training and skills development, we have transformed the Corporation's organisational behaviour and work culture through the provision of career development and cash incentives. Since 1998, NWSC employees have learnt to be clean and tidy, to keep time, to maintain a clean working environment and to be receptive and responsive to customers. At NWSC, we have encouraged the staff to own the programmes and to be partners in business rather than mere worker spectators. Nowadays, the NWSC workforce has a stake in NWSC at the headquarters, division and Area partnership levels. All the employees in the Corporation know that its success is also their success. They realise that if the Corporation generates more revenue and becomes more profitable, their take-home pay and benefits will improve correspondingly, and this would obviously result in the betterment of their lives and those of their families. The overall shift in organisational behaviour has worked wonders for the Corporation. Staff morale and motivation are buoyant. *Productivity has continued to improve and customers have never had it as good in terms of service delivery and staff responsiveness as they obtained from change management initiatives.*

Stretching dreams and plans to the 'stars'

By revolutionising organisational behaviour at NWSC, we taught our people to 'think outside the box and think big', to aim for the stars and to work not only harder but also smarter. Until the initiation of Stretch-out we were content with setting modest SMART targets. In some cases, Area managers set the targets too low in order to register high achievement scores. We soon realised that this was not good enough. We therefore resolved to exhort our people to stretch themselves to the limit through superhuman effort in their work, to work beyond the call of duty and *to make the impossible possible.* As a result, our people transcended the confines of SMART targets by dreaming big; formulating individual, departmental, division visions and missions; and setting, achieving and even surpassing stretch targets. Stretching worked wonders for NWSC's performance, productivity and service delivery.

Indeed, to cross the cost-recovery boundary and start making money for shareholders and delivering effective services to customers, public enterprises must embrace new ideas and management techniques, *change with change and break the chains of old habits, tradition, conventional wisdom, orthodoxy and the rules and procedures of bureaucracy.*

Making the case for public enterprises

Managers of public enterprises seem to take their usefulness or their roles in the economy and society for granted. They seem to assume that they deserve to exist simply because they were established to perform preordained statutory mandates. Otherwise, why would they have been established in the first instance? Managers see their corporations as victims of government interference and meanness and undercapitalisation, of donor conditionalities and of technical and capacity constraints. This self-pity, mourning and 'blame-storming' can only lead to a dead end. Public enterprises have to earn a living in accordance with the rationale for which they were established in the first place. They have to justify their existence and continuity through their performance and achievements. They have to make their case to win over the shareholders, development partners and other stakeholders through *results, not words*.

If a public enterprise performs well, other stakeholders will notice and back it. Should public enterprise managers present credible diagnoses and prescriptions, other stakeholders – including government and donors – will listen or at least give such propositions the benefit of doubt. Since 1998, our change management efforts have demonstrated that even sceptics and doubting Thomases can be converted into friends, allies and partners by incremental performance improvements.

Take the case of the government, the sole shareholder of NWSC. By 1998, the government was bent on privatising the Corporation, preferably to 'international operators' with lots of resources and expertise. As a colleague once remarked, the privatisation of public utilities was supposed to resolve management and performance problems at a stroke. However, the government was stuck with NWSC because there were no serious international bidders. Public utilities like NWSC are much more than ordinary commercial enterprises, whose paramount purposes are to make money for the shareholders. Over and above commercial considerations, public utilities are vested with 'social-mission' objectives and few international operators are willing to carry such a burden. In any case, at the moment our water market is too small to attract foreign investment.

When our home-grown internal reforms strategy began to turn the Corporation around, the government gave us 'a big thank you' for a job well done. Not only did the government grant us operational autonomy through successive performance contracts (2000–2009), they also agreed to settle the arrears which had accumulated over the years, and henceforth to settle new water bills on time. As a result, the spectre of privatisation that had haunted the Corporation began to recede. This gave us time to consolidate our performance improvements through even more ambitious change management initiatives. Thanks to our incremental successes, by 2005 the government had become a supportive owner in implementing internal reforms, which goes to show that governments are *flexible enough to back the winning horse when they see one*. This has been a useful lesson to us as an organisation, and hopefully our experience will serve as a good example for other public enterprises in Uganda and, indeed, in Africa as a whole.

Similarly, at the beginning of our change management programmes, donors, especially the World Bank, had grave reservations about the chances of success of home-grown corporate reform initiatives. As we have seen in Chapter 2, the

100 days Programme was dismissed as an ingenious gimmick to hoodwink the public and to delay divestiture and privatisation. Fortunately, actions always speak louder than words. Therefore, as our change management initiatives began to transform the Corporation into a viable entity, donors realised that we were dead serious and warmed to us. In addition, instead of seeing conditionalities as a suffocating yoke around our neck, we saw them as challenges to overcome. Surely, conditionalities such as improving revenue collection, reducing NRW and arrears and increasing staff productivity were meant for our own good rather than for that of the donors.

We also took up the World Bank challenge to us to draw up a financial model, in addition to developing our budget frameworks and corporate business plans. We won the war of nerves and brought the donors on board our change management bandwagon. Today, the donors are among the most enthusiastic supporters of our change management initiatives, and they no longer insist on the panacea of privatisation. *If anything, they seem to be having second thoughts about the wisdom of privatisation to the exclusion of other corporate transformation possibilities.* If we had not made our case by delivering the results and meeting the conditionalities, the World Bank and other donors would have dropped us like a hot potato and the privatisation hammer would definitely have fallen.

In making the case for public enterprises, it is important to cultivate the support of the print and electronic media, for example by providing information, keeping an open-door information policy, publicising your activities and achievements and participating in phone-in programmes. Needless to say, the media are the eyes and ears of the public. The media raise public issues, shape the nature and direction of public debate, mould public opinion and can make or break the reputations of public enterprises. Since 1998, we have done our best to cultivate media support and to win their confidence. We have used the media as the channel of communication and feedback. Through the media we have publicised our activities, programmes and achievements, and have responded to public criticisms and complaints. *Thanks to the media, the public has become more sensitised to what we do and, by so doing, our corporate image has been boosted in the popular imagination.* The Corporation has become a favourite household name and an oft-quoted example of what other public enterprises and authorities should be doing to improve their public approval ratings. Managers who ignore the media do so at their own peril. They miss opportunities to make the case for their organisations.

A move to access financial markets

The successful implementation of change management initiatives in NWSC has strengthened the overall performance of the Corporation as demonstrated in

Chapter 11. No wonder, during a shadow credit rating workshop for selected water utilities in Africa, held in Dakar, Senegal on 24–25 November 2008 put NWSC under category AA for short-term financing and A for long-term financing. This ranking positioned NWSC among the best three water utilities in Africa. This means the Corporation is in a better position to access financial markets and eventually repay the loans. Already, the Corporation has been cleared by the Ministry of Finance and Economic Planning to issue a commercial bond for viable investments in its operational Areas, through a series of tranches based on priority investment activities. Many banks have also expressed a strong desire to partner with us to provide investment funds at negotiable rates.

The NWSC case, therefore, is a clear sign that working on performance problems is a sure way of preparing public enterprises for accessing financial markets. Many investors are interested in investing in businesses, whether public or private, that show consistent positive operating margin trends. At NWSC, we resolved to be in this bracket as part of our performance objectives throughout the whole stream of initiatives. It has paid off and we hope for even better results in future!

Looking back to forge ahead

Upon looking back, we realise that we have made tremendous progress over the last 10 years, by transforming NWSC into one of the most successful public enterprises in Uganda. The NWSC has indeed become a model public utility in the water industry and has begun to share its experience with sister utilities in Africa and other parts of the World through its External Services Department. We have rendered consultancy services not only to sister water utilities in Zambia, Tanzania and Zanzibar, Rwanda, Kenya and elsewhere in the world, but also to private enterprises in Uganda.

For example, in April 2005, we signed a contract with Alam Group of Companies in Kampala to 'provide a wide range of professional support in respect of development and implementation of organisation behaviour change management programme'. Similarly, in partnership with ARD Inc. of the USA, and with the financial support from US Agency for International Development (USAID), we won the assignment to provide water utility technical and management support services to two towns in Northern Uganda: Kitgum and Pader. In addition, NWSC External Services was contracted by Kampala City Council to help review and recommend strengthening measures in revenue generation activities of the City. The External Services Department has also been actively involved in providing various forms of capacity-building programmes to local private operators in Uganda. All these activities clearly point to a 'management revolution', which has been carried out through local initiatives and programmes, without prompting or guidance from the outside world. We have revolutionised the Corporation's organisational behaviour and transformed its work environment.

We have built a professional team of managers and staff with the capacity, skills, brains, knowledge and experience to match or even outdo their best peers around the world. We have won battles and scored victories. We have proved that it is possible to transform public enterprises using home-grown solutions. We have also proved that it is possible to apply private sector principles, techniques and motives to public enterprises through delegation, decentralisation and operational autonomy. The emphasis on 'home-grown' change management initiatives does not mean that our development partners have not made useful contributions to the transformation of NWSC since 1998. As Chapter 12 shows, donors have provided significant technical assistance in terms of money, inputs and expertise.

Without donor support to some of our service delivery activities, we would not have accomplished as much as we have done. *We know we cannot do everything on our own without a helping hand from outside development partners. They have been useful in the past and the present and no doubt will continue to be so in future.* However, our point is this: change must be conceived, born and nurtured from within. *The lesson to draw from this is that whatever they do, outsiders cannot be a substitute for local management initiatives.*

Of course we recognise that we have not yet completed the quest for excellence or corporate perfection. Our story marks the early stages of a long journey to the ultimate realisation of our corporate vision of *'being one of the leading water utilities in the world'* through our corporate mission of 'providing efficient and cost-effective water and sewerage services, applying innovative managerial solutions to the satisfaction of our customers in an environmentally friendly manner'. NWSC ought to continue to aim for perfection in its service delivery, and not be oblivious of the challenges, or even internal and external threats, to the sustainability of its performance in the years ahead. In any case, our experience has shown that every success is invariably pregnant with new challenges that require new solutions. After all, enhancing performance to improve service delivery is bound to be a process without end, for perfection is always elusive and shifting, even in the most efficient of business enterprises. However, a strong, solid foundation has been built and this has helped the staff to gain sufficient experience, confidence, capacity and momentum to forge ahead into a prosperous and sustainable future for NWSC. Nevertheless, the current and future managers of the Corporation must take care because there is always a possibility of sliding back if hard work is not sustained.

When we initiated the 100 days Programme in 1998, we knew that we were taking the first step of a long and tortuous journey full of surprises, uncertainties and unpredictable outcomes. The road on which we have been travelling has not been a smooth one. It is heavily potholed, with sharp corners and major roadblocks. It is even shared with 'reckless road users'. However, with steady hands on the steering wheel of the change management vehicle, with technical and professional backup from the entire NWSC family and with helping hands from the government, donors, suppliers and service providers, it has been possible to successfully navigate all the obstacles along the way. We have moved steadily forward instead of sitting back and sulking about 'who moved my cheese'. We believe that we have built the confidence, accumulated experience, strengthened our resolve to succeed and gained the momentum to sustain progress to our destination as defined by our vision of becoming one of the leading water utilities in the world.

The message is clear. We must always aim for greater heights. Working for 'continuous improvement' is the key for any progressive business, and water and sewerage is not an exception. Nothing is impossible and if you go for it, you will surely get it. I wish all the readers the fruitful reflections and application of relevant lessons. I believe that the NWSC case study, which is admittedly limited in the way it addresses what we have encountered during our lifetimes has something positive to say about our working endeavours.

Can Public Enterprises Perform?

Creating partnerships to share experiences: Dr William Muhairwe (right) exchanging a partnership agreement with Ms Nirasha Sampson and Mr Peter Bahr of Umgeni Water, South Africa, for the two sister utilities to share knowledge and experience.

NWSC Board chairman, Mr Sam Okec (right) exchanging a partnership agreement with the Board member of Nairobi City Water and Sewerage company, Mr Titus G. Ruhiu as MD-NWSC, Dr William Muhairwe (second right) and MD-NCWSC, Mr Francis Mugo look on.

Dr William Muhairwe (right) exchanging a partnership agreement with the vice chancellor of Makerere University, Prof. Livingstone Lubobi, for the two institutions to work together in strengthening research and capacity building.

Yes we did ... cross the line! NWSC turnaround story was a marathon and required a lot of courage and determination to succeed. Dr William Muhairwe (left) with NWSC staff crossing the finishing line during the annual MTN marathon.

Behind every successful story there are unsung heroes: family support was invaluable during the turnaround.

Right: Dr William Muhairwe (center) is joined by his two daughters, Lucie (left) and Lisa (right) to celebrate another achievement.

Left: my wife, Vicky, has been by my side throughout, from the first shocking encounters to the recognition of NWSC as East Africa's Most Respected Company in 2007.

Acknowledgements

No book of this substance and content can mature to this stage without the inspiration and support from friends and colleagues. The initiative to write this book came from many quarters whose words and actions, subconsciously, contributed to a tremendous thrust to the process of writing this book. I had the opportunity to discuss strategic and operational issues regarding the management of organisations, both public and private, and sometimes this process stimulated debates that have helped to shape the arguments in this book. These arguments did not only keep my own change management 'brain engine' alert but also, together with the contributions from my management team and staff, translated into real actions that have led to the historic turnaround of NWSC. I cannot thank everybody who played this important role. I will only single out a small representative sample as a demonstration of my utmost appreciation of the part played by a bigger set of 'heroes'.

I cannot forget those who provided me the opportunity to visit and/or discuss operational strategies concerning the running of our organisations and, in one way or another, encouraged me to record the NWSC turnaround story for peer discussion. I thank Archard Mutalemwa, Alex Kaaya, Mutakaekulwa Mutegeki, Dennis Mwanza, Bernard Chiwala, George Bene, Francis Mugo, Lawrence Mwangi, Diru Magomere, David Onyango, Mamadou Dia, Usher Sylvain, Mounir Zouggari, John Mirenge, Ambrose Akandonda, Richard Byarugaba, Juma Kisame, Paul Maare, Godfrey Tumusiime, Frank Ssebowa, Kitili Mbathi, Nick Mbuvi, Nicolas Okwir, Joseph Kitamirike, Allen Kagina, Abdelali Zerouali and many others for their insistence that I tell the NWSC story in full. Their continuous encouragement has given birth to this book.

I am equally grateful to organisations and a number of development partners that provided invaluable and positive thinking experts who later became friends and allies in the fight to provide improved services to our people. We are greatly indebted to organisations such as the World Bank, ADB, KfW, GTZ, DFID, Water Aid, USAID, UN-Habitat and WSP, who did not only provide such support experts but who also invited me to present and share the NWSC experience with the rest of the world at different fora. We greatly benefited from their financial assistance, technical guidance and moral support. Their experts walked with us throughout the change management process at NWSC. To this end, I would like to thank my great friend, Ato Brown, and Alain Locussol, my mentor. I would also like to thank my 'spiritual' brothers and sisters in the utility management business, who include Philip Marine, Atem S. Ramsundersingh, Jan Janssen, Fook Chuan, Solomon Alemu, Meike Van Ginneken, Suzanne

© 2009 William T. Muhairwe. *Making Public Enterprises Work*, by William T. Muhairwe. ISBN: 9781843393245. Published by IWA Publishing, London, UK.

Mauve, Joerg Dux, Herman Plum, Fridtjof Behnsen, Yogita Mumssen, Claude Jamati, Prof. Sanford Berg, Prof Karl Vairavamoothy, Annet Windmeiser and Prof. Nemanja Trifunovic, among others. Thank you for being there for NWSC. Let me hope that receiving a copy of this book will inspire you to visit us more and help us to move NWSC to greater heights.

I am also indebted to all my friends, dating back to our youthful days, who persistently pressured me to invest in writing a book that presents a coherent story of the NWSC turnaround. I must say, I was challenged and this moulded my determination to reward my colleagues' good intentions and desires. To this, I pay profound homage to Edward Rutayungwa, Aloysius Lubowa, Joseph Turyabahika, Frank Bagambisa, Bannet Bagombeka, Bannet Ndyanabangi, Daniel Bwete Kasaasa (RIP) and his wife Vania, Adrian Bangirana and Pontian Bakeine. Some people did not only constantly inquire when the book would be published but also inquired whether it would ever be published at all! Can you imagine that kind of friendly concern? These were none other than my brothers and cousins: Remmy Rwankote, Emmanuel Byanyima, Stella, Benjamin, Jimmy, Anthony, Mary, Lydia, Peace, Rita, Ruhweza-Mathias and Kyomugisha; to all of you, I say bravo and thank you for the valuable support, guidance and patience.

Nothing that was carried out and achieved in NWSC during the change management process would have been possible without a solid infrastructure base that my predecessor managing director, Eng. Hillary Onek, and his Board put in place. Eng. Hillary Onek is actually regarded as the 'father' of the rebirth of NWSC after a long period of political turmoil and dilapidation. The job he did to keep the water assets alive, especially during the early 1980s, can be equated to 'getting water from a stone'! Without his foundation, neither subsequent efficiency reforms nor this book would have been possible.

In the same vein, I pay glowing tribute to those who created a lustrous operating environment to write this book. I want to thank the previous Board (1998–2005) for giving me the chance to serve NWSC and, above all, for providing an enabling environment for me to lead a complex water business and use lessons and experiences that led to this book. In this regard, I am deeply beholden to Board chairman Sam Okec, Francis Openyto, Henry Aryamanya-Mugisha, Hawa Nsubuga, Dr Joseph Kyabaggu, Lawrence Bategeka, Patrick Kahangire, David Ssebabi, Z. Akol, Sarah Mangali (RIP), and Abdullahi Shire. I may not have given sufficient credit for your very valuable input to this book. But I know, and we all know that you made the turnaround initiatives at the NWSC possible – my role was that of an implementer. You laid the base for the vibrant, vigorous and viable NWSC that we have today.

I am also grateful to those who kept the ball rolling. The drafting process for this book went through a number of different Board terms at NWSC. Without continuity, the process would have been disrupted and, above all, the story would have been dented by poor performance trends. To this end, I want to pay special reverence to the new Board (2005–2009) which comprised chairman Ganyana Miiro, Elizabeth Madra, Yorakamu Katwiremu, Victor Kobel, Christine Nandyose, Florence Namayanja, Sottie Bomukama, David Ssebabi and our very able Corporation secretary David Mpango Kakuba. Thank you for maintaining the 'vibrant-vigorous-viable' path of NWSC, which has enabled me to tell a longer turnaround story. I am very much honoured and grateful to have been given another chance to serve and hence be able to tell the ten-year story.

Special gratitude also goes to those who provided an enabling political governance framework. Those who ensured that the framework was right all the time include Henry Kajura, Gerald Sendaula, Ruhakana Rugunda, Kahinda Otafiire, Baguma Isoke, Jeje Odongo, Maria Mutagamba, Jennifer Byakatonda, Peter Kasenene, Manzi Tumubweine, Matthew Rukikairre and Akiika Othieno. Their inexorable 'political guidance' rather than 'interference', about which parastatals usually complain, enabled us to move in the right direction. Of course, I cannot forget the great job and input from the permanent secretaries and the technical staff in the ministries in charge of finance and water over this period. Particularly, I would like to single out Emmanuel Tumusiime Mutebile, Bill Kabanda, Chris Kassami, Mary Muduli, Keith Muhakanizi, Emmanuel Nyirinkindi, Florence Luzira, Margaret Kobusingye and Patrick Ocailap for the technical facts that gained us political support.

I would also like to extend credit to a team that provided a lot of editorial and research support. In this respect, I must say, if it were not for the days and nights, my editor and late friend Dr Justus Mugaju put into it, the content and detail would have been different. He set it up in its current structure, and led information gathering and analysis of numerous questionnaires, Corporation programme documents and reports. He conducted intensive interviews and face-to-face discussions with me and colleagues at NWSC to clarify and concretise positions. Death robbed our country of a great, unparalleled mind. May his soul rest in eternal peace. Other editors who supported me include Alex Bangirana, Paul Busharizi, Tom Gawaya-Tegule and Prof. James Ntozi. To all of you and your teams, thank you for a job well done. However, I still take full responsibility of the content in this book.

I also want to recognise people who were part of the NWSC change management process and who provided invaluable support to the development of this book. I am wholeheartedly thankful to the editorial team at NWSC, headed by Dr Silver Mugisha, which supported the drafting process. They include Johnson Amayo, Mahmood Lutaaya, David Isingoma, George Okol, Dr Josses Mugabi and Dr Martin Kalibbala. Your input into the development of this book went beyond the call of duty since your contribution was voluntary and in addition to your official Corporation duties and responsibilities, which were already stretched by our change management programmes. Silver and your team, thank you very much for your brilliance, resilience, understanding and hard work in shaping this book. I would also like to take this chance to thank all my chief managers: Jackson Opwonya, Evelyn Otim, Prossy Aketch, Alex Gisagara, Charles Odonga and the Union chiefs, Brothers Peter Werikhe and Laro Wod' Ofwono (RIP) for their constant support. I am also grateful to the rest of the Corporation staff, especially those who were interviewed by our editors and all those who sent me their useful recollections, codenamed 'memories of the turnaround'. Your recollections, comments and observations added humour and enriched the entire book.

To those who provided a good and supportive office environment, I give special deference to your contribution. I know that when the 'boss' is troubled or confused, it is normally the assistants who suffer most. In this respect I would like to thank all my assistants, previous and current (Hope, Irene, Monica and Jessica), my previous and current logistics officers Livingstone Musisi, Geoffrey Odeke, James Mpora, and my security assistant Rashid Konoha; thank you for being there for me in very many ways during those hard times.

'Not alone' indeed! My entire family suffered during the writing of this book. Even as I write this last acknowledgement, the children, Lucie, Lisa, Lynn, Laureen and Monica are wondering whether their father is actually around. They have been asking me, especially when we were on holiday, why my book is an endless endeavour! They always wondered when the book would finally come to an end so that I could become part of the family! Thank you very much for your patience with my absence. I would also like to thank my parents: Cecilia, Tereza and Nazarius for the family and parental support. My dear nieces and nephew: Grace, Daisy and Augustine, I know I should have been more available. I understand how much you have missed my guidance and support, especially on school-visiting Sundays. Thank you for your understanding.

Let me end with my dear wife Vicky. Being married to a very busy person is bad enough. Being married to a manager of an organisation that faced enormous challenges, who then chooses to engage in the arduous task of writing a book, requires the patience of an angel – one prepared to wait on you to eternity. Vicky did not only wait but also read through almost all the drafts of the manuscript and gave me guidance and input that contributed deeply to shaping this book. With every passing day, I understand more fully just how lucky I am to have Vicky as my wife, a companion and a friend who understands and cares. Even after 15 years of marriage, she remains my most helpful critic, my deepest and most enduring support. Thank you, Vicky.

Acronyms

ACR	Authorised Contract Representative
ADB	African Development Bank
ADCs	Area Disciplinary Committees
AMCOW	African Ministers' Council on Water
APC II	Area Performance Contract II
APC III	Area Performance Contract III
APCs	Area Performance Contracts
ATM	Automatic teller machine
BMZ	Bundesministerium Für Wirtschaftliche Zusammenarbeit *(German Federal Ministry for Economic Development Cooperation)*
BPCs	Branch Performance Contracts
CBA	Collective Bargaining Agreement
CD	Compact disk
CID	Criminal Investigation Department
DAAD	Deutscher Akademischer Austauschdienst *(German Academic Exchange Service)*
DAWASA	Dar es Salaam Water and Sewerage Authority
DAWASCO	Dar es Salaam Water and Sewerage Company
DFID	Department for International Development
DWD	Directorate of Water Development
EoI	Expression of interest
EU	European Union
FDI	Foreign direct investment
GEP	Greater Entebbe Partnership
GoU	Government of Uganda
GTZ	Deutsche Gesellschaft für Technische Zusammenarbeit GmbH *(German society for technical cooperation)*
HDPE	High density polyethylene
IDES	Institutional Development and External Service
IDAMCs	Internally Delegated Area Management Contracts
IGG	Inspector General of Government
IHE	International Institute for Infrastructural, Hydraulic and Environmental Engineering
ISO	International Organisation for Standardisation
IT	Information technology

© 2009 William T. Muhairwe. *Making Public Enterprises Work*, by William T. Muhairwe. ISBN: 9781843393245. Published by IWA Publishing, London, UK.

JBG	Jadgbombergeschwader
JNC	Joint Negotiation Committee
KfW	Kreditanstalt Für Wiederaufbau *(German Development Bank)*
KRIP	Kampala Revenue Improvement Programme
LCs	Local Councils
LPOs	Local Private Operators
LVNWSB	Lake Victoria North Water and Sewerage Board
LWSC	Lusaka Water and Sewerage Company
MDGs	Millennium Development Goals
MoU	Memorandum of understanding
MD	Managing director
MP	Member of Parliament
MSc	Master of Science *(degree)*
MWLE	Ministry of Water, Lands and Environment
NCWSC	Nairobi City Water and Sewerage Company
NEMA	National Environment Management Authority
NRA	National Resistance Army
NRC	National Resistance Council
NRM	National Resistance Movement
NRW	Non-revenue water
NWSC	National Water and Sewerage Corporation
NWSC-Z	Nkana Water and Sewerage Company – Zambia
ONEP	Office National de l'eau Potable
OSUL	ONDEO Services Uganda Limited
P&CD	Planning and Capital Development
PC I	Performance Contract I
PC II	Performance Contract II
PC III	Performance Contract III
PCRC	Performance Contract Review Committee
PEAP	Poverty Eradication Action Plan
PERD	Public enterprise restructuring and divestiture
PhD	Philosophiae Doctor *(doctor of philosophy)*
PPDA	Public Procurement and Disposal of Public Assets
PRSC	Poverty reduction support credit
PS	permanent secretary
PS/ST	permanent secretary/secretary to the Treasury
PSP	Private sector participation
PURC	Public Utility Research Center
RIP	Rest in peace
SEREP	Service and Revenue Enhancement Programme

Acronyms	
SMART	Specific, measurable, achievable, realistic and time-bound
SMS	Short Messaging Service
SOEs	State-owned enterprises
SSC	Support Services Contract
SWOT	Strengths, weaknesses, opportunities, and threats
TA	Technical assistance
UAF	Uganda Austria Foundation
UEB	Uganda Electricity Board
UEDCL	Uganda Electricity Distribution Company Limited
UGX	Uganda Shilling
UIA	Uganda Investment Authority
UK	United Kingdom
UMA	Uganda Manufacturers' Association
UNBS	Uganda National Bureau of Standards
UN-Habitat	United Nations Human Settlement
UPTC	Uganda Posts and Telecommunications Corporation
UPC	Uganda Peoples' Congress
UPEU	Uganda Public Employees' Union
uPVC	unPlasticised Poly Vinyl Chloride
URA	Uganda Revenue Authority
URU	Utility Reform Unit
US	United States
USD	United States Dollar
USAID	United States Agency for International Development
UTEC	United Technical Engineering Company
VAT	Value added tax
WAN	Wide area network
WHO	World Health Organisation
WSP	Water and Sanitation Programme – World Bank

Bibliography

Berg, S.V. and Muhairwe, T.W. (2008). Promoting High Performance in SOEs: Lessons from Africa. Department of Economics, PURC Working Paper. Place of Publication. University of Florida.

Berg, S.V. and Mugisha, S. (2008). Pro-poor Water Service Strategies in Developing Countries: Promoting Justice in Uganda's Urban Project. *Water Policy* (forthcoming).

Blanchard, K., Carew, D. and Carew, E.P. (2000). One Minute Manager Builds High Performing Teams. Place of Publication: HarperCollins Business.

Blanchard, K., Zigarmi, P and Zigarmi, D. (2001). Leadership and the One Minute Manager. Place of Publication: HarperCollins Business.

Blanchard, K. and Spencer, J. (2007). The One Minute Manager. Place of Publication: HarperCollins Business.

Global Credit Rating Company (2008). Uganda Water Utility Diagnostic Report. Sandton, South Africa.

Kayaga, K. (2008). Soft Systems Methodology for Performance Measurement in the Uganda Water Sector. *Water Policy*, 10: 273–284.

Mugisha, S. (2001). Possible Avenues of introducing competition in Water and Sanitation Sector in Uganda. UIPE May quarterly Magazine.

Mugisha, S. (2005). Effects of Managerial Incentives Intensity and Performance Monitoring Modes on Water Utility Efficiency and Productivity Growth: Findings National Water and Sewerage Corporation, Uganda. Ph.D. Thesis, Makerere University.

Mugisha, S. (2007). Interpretational Hitches in Using Partial Performance Indicators for Water Infrastructure Regulators: Empirical Evidence from Uganda. Working Paper, NWSC – Uganda, Kampala.

Mugisha, S. (2007). Turning Around Struggling State-owned Enterprises in Developing Countries: The Case of NWSC – Uganda. In: Going Public: Southern Solutions to the Global Water Crisis, 15–25. London: World Development Movement.

Mugisha, S. (2007). Performance Assessment and Monitoring of Water Infrastructure: An Empirical Case Study of Benchmarking in Uganda. *Water Policy*, 9 (5): 475–91.

Mugisha, S. (2008). Creative Approaches to Problem Solving in Water Utility Reforms: Application of Lateral Thinking Techniques. *Leadership and Management in Engineering Journal* (in press).

Mugisha, S. (2008). Development and Regulatory Challenges in Water Services to the Urban Poor: Examples from Uganda and Tanzania. Proceedings of the International Urban Water Conference, Heverlee, Belgium. Water and Urban Development Paradigm, Towards an Integration of Engineering, Design and Management Approaches. Place of Publication: CRC Press, Taylor and Francis Group, pp. 535–542.

Mugisha, S. (2008). Effects of Incentive Applications on Technical Efficiencies: Empirical Evidence from Ugandan Water Utilities. *Utilities Policy*, 15(4): 225–233.

Mugisha, S. (2008). Infrastructure Optimization and Performance Monitoring: Empirical Findings from Water Sector in Uganda. *African Journal of Business Management*, 2 (1): 13–25.

Mugisha, S. and Berg, S.V. (2008). State-owned Enterprises: NWSC's Turnaround in Uganda. *African Development Review*, 20 (2): 305–334.

© 2009 William T. Muhairwe. *Making Public Enterprises Work*, by William T. Muhairwe. ISBN: 9781843393245. Published by IWA Publishing, London, UK.

Mugisha, S., Berg, S.V. and Katashaya, G.N. (2004). Success for Uganda from a Series of Short-term Initiatives. *Water,* 21: 50–52.

Mugisha, S., Berg, S.V. and Skilling, S.V. (2004). Practical Lessons for Performance Monitoring in Low-Income Countries: The Case of National Water and Sewerage Corporation, Uganda. *Water,* 21: 54–56.

Mugisha, S., Brown, A. and Kiwanuka, S. (2005). Water Reforms in the Three East African Capital Cities. World Bank Working Paper, presented at World Bank Water Week, in March 2005.

Mugisha, S., Berg, S.V. and Muhairwe, W.T. (2007). Using Internal Incentive Contracts to Improve Water Utility Performance: The Case of Uganda's NWSC. *Water Policy,* 9 (3): 271–284.

Mugisha, S., Phillipe, M., Muhairwe, W.T and Mugabi, J. (2007). Transforming Public Water Utilities through Private Sector–like Management Principles: The National Water and Sewerage Corporation, Uganda Experience. World Bank Water Sector Board Discussion Paper. Washington D.C.: The World Bank, (under review).

Mugisha, S., Berg, S.V. and Janssens, J. (2008). Assessing Governance of Urban Water Service Provision: Practical Issues in East and Central African Countries. Working Paper, NWSC – Uganda.

Muhairwe, W.T. (1982). Overhead Costs Analysis Methods in Business. A Research Paper Presented at the Institute of Industrial Business Economics and Accounting, Germany.

Muhairwe, W.T. (1983). Strategic Product Planning – With Special Reference to Business Branch Analysis. A Pilot Study. University of Munich, Germany.

Muhairwe, W.T. (1984). Strategic Product Planning: An Innovative Concept for Planning and Realisation of New Products with the help of Venture Management Strategies. Master's Thesis, University of Munich, Germany.

Muhairwe, W.T. (1985). The Structure of the new Style Joint Venture and the Management Problems Involved. A Pilot Study Paper, University of Munich, Germany.

Muhairwe, W.T. (1987). The Planning and Realisation of Industrial Projects. A Paper presented at the Seminar of Uganda Scientists and Entrepreneurs, Organised by Ugandan Embassy, Cologne, Germany.

Muhairwe, W.T. (1988). New Style Joint Ventures: Ansatze zu einem Internationalen Innovationsmanagement. Ph.D. Thesis, Wartaweil 25a, 8036 Herrsching, Germany.

Muhairwe, W.T. (2006). Cost Recovery Mechanisms: The Success of the NWSC – Uganda and its Relevancy for Other African Countries. World Bank Working Paper, Washington, D.C.

Mutikanga, H. and Mugisha, S. (2005). A Phased Approach to Efficiency Improvement in Low-Income Countries: The Case of NWSC Fort Portal Town in Uganda. *Water Science and Technology: Water Supply,* 5 (3–4): 1606–1749.

National Water and Sewerage Corporation (1999). 100 Days Programme. www.nwsc.co.ug

National Water and Sewerage Corporation (2000). Area Performance Contracts. www.nwsc.co.ug

National Water and Sewerage Corporation (2005). Checkers Programme. www.nwsc.co.ug

National Water and Sewerage Corporation (2000). Government of Uganda Performance Contracts. www.nwsc.co.ug

National Water and Sewerage Corporation (2003). Government of Uganda Performance Contracts II. www.nwsc.co.ug

National Water and Sewerage Corporation (2006). Government of Uganda Performance Contracts III. www.nwsc.co.ug

National Water and Sewerage Corporation (2004). Internally Delegated Area Management Contracts. www.nwsc.co.ug

National Water and Sewerage Corporation (2006). Internally Delegated Area Management ContractsII. www.nwsc.co.ug

National Water and Sewerage Corporation (2009). Internally Delegated Area Management ContractsIII. www.nwsc.co.ug

National Water and Sewerage Corporation (2003). One-Minute Management Programme. www.nwsc.co.ug

National Water and Sewerage Corporation (1999). Service and Revenue Enhancement Programme. www.nwsc.co.ug

National Water and Sewerage Corporation (2002). Stretch-out Programme. www.nwsc.co.ug